W9-AKS-436

Renewing the Left

Renewing the Left

Politics, Imagination, and the New York Intellectuals

Harvey M. Teres

New York Oxford
OXFORD UNIVERSITY PRESS
1996

Oxford University Press

Oxford New York
Athens Auckland Bangkok Bombay
Calcutta Cape Town Dar es Salaam Delhi
Florence Hong Kong Istanbul Karachi
Kuala Lumpur Madras Madrid Melbourne
Mexico City Nairobi Paris Singapore
Taipei Tokyo Toronto

and associated companies in
Berlin Ibadan

Published by Oxford University Press, Inc.,
198 Madison Avenue, New York, New York 10016

Oxford is a registered trademark of Oxford University Press

Library of Congress Cataloging-in-Publication Data
Teres, Harvey M., 1950–
Renewing the left : politics, imagination, and the New York
intellectuals / Harvey M. Teres.
p. cm. ISBN 0-19-507802-0
1. American literature–20th centuery–History and criticism–
Theory, etc. 2. Radicalism in literature. 3. Politics and
literature–New York (N.Y.)–History–20th century.
4. Literature and society–New York (N.Y.)–History–20th century.
5. Authors, American–New York (N.Y.)–Political and social views.
6. Authors, American–20th century–Political and social views.
7. Radicalism–New York (N.Y.)–History–20th century.
8. Criticism–New York (N.Y.)–History–20th century.
9. New York (N.Y.)–Intellectual life–20th century. 10. Imagination. I. Title.
PS228.R34T47 1996
810.9′358–dc20 95-17522

1 3 5 7 9 8 6 4 2

Printed in the United States of America
on acid-free paper

For Susan and Caitlin

Acknowledgments

Portions of chapter 2 have been published in *American Literature* (vol. 64, no. 1 [March 1992]). The bulk of chapter 6 has been published in *The Wallace Stevens Journal* (vol. 13, no. 2 [Fall 1989]). The "Tess Slesinger" section of chapter 8 has been published in *Jewish American Women Writers: A Bio-Bibliography and Critical Sourcebook*, ed. Ann R. Shapiro (Westport, Conn.: Greenwood Press, 1994), and is reprinted by permission of Greenwood Press. A portion of chapter 10 draws on a review of Cary Nelson's *Repression and Recovery* for *Modern Philology* (vol. 89, no. 1 [August 1991]) and is reprinted by permission of the University of Chicago Press, copyright © 1992 by The University of Chicago Press; all rights reserved. Other portions of chapter 10 include reviews of Barbara Foley's *Radical Representations* and Walter Kalaidjian's *American Culture Between the Wars*, originally published in *Modernism/Modernity* (vol. 2, no. 1 [January 1995]) and are reprinted by permission of the Johns Hopkins University Press. Chapter 11 will be published as the entry on Lionel Trilling in *The Cambridge History of Literary Criticism* forthcoming by Cambridge University Press.

It has been a long road, and my journey would not have been completed without the help of so many along the way. I want to thank Kathleen Manwaring for her timely and enthusiastic assistance with material from the extraordinary George Arendts collection at Syracuse University's Bird Library. My thanks to staff members at the Newberry Library and Harvard's Houghton Library for their eager assistance and permission to quote material from, respectively, the Floyd Dell Papers and the Leon Trotsky Archive. For their help and for allowing me to consult the Dwight Macdonald Papers and the Wallace Stevens Papers I thank the staff members at Yale's Sterling Memorial Library and the Huntington Library, respectively. I wish to thank William Phillips, who a long time ago gave several hours of his time to a young academic with a long list of questions I'm sure he had heard many times before, but which he answered with humor and grace. To Walter Jackson, Michael Walzer, and Cornel West: your words have made a difference, and I thank you. Many colleagues have taken valuable time out of their schedules to read portions of this book, and, whatever its quality, it is very much better as a result. My sincere gratitude to Susan Albertine, Brian Bremen, Maria DiBattista, Eduardo Cadava, Susan Edmunds, Al Filreis, Claire Fowler, Joseph Frank, Danny Goldberg, Jennifer Hochschild, Sam Hynes, Michael Jimenez, Victoria Kahn, Mary Karr, Richard Kroll, Walt Litz, David Quint, Arnold Rampersad, Andrew Ross, Sanford Schwartz, Janet Sharistanian, Elaine Showalter, Robert von Hallberg, and Alan Wald. My appreciation to Princeton University for its generous support of this project, through the Surdna Foundation and other research grants, which made it possible to hire the several outstanding research assistants who contributed mightily to this project: Gwen Burgner, Maria Doyle, Richard Kaye, and Tammy Serwer. To Dean Robert Jensen at Syracuse University, my thanks for some last-minute financial assistance. I thank my indexer, Laurie Winship, whose eye is the equal of Tony Gwynn's. And my gratitude to my editors, Elizabeth Maguire, Paul Schlotthauer, and Brian MacDonald, for their patient support and hard, meticulous work.

To ACLS and NEH, in the words of e. e. cummings: no thanks.

Finally, to my wife and daughter: my inadequate but heartfelt appreciation for all you've given.

Contents

Renewing the Left

You and I have been breasting hills; we have been climbing upward; there has been progress and we can see it day by day looking back along blood-filled paths. But as you go through the valleys and over the foothills, so long as you are climbing, the direction, —north, south, east, or west, —is of less importance. But when gradually the vista widens and you begin to see the world at your feet and the far horizon, then it is time to know more precisely whither you are going and what you really want.

W. E. B. DuBois, *"Criteria of Negro Art"*

How easily the blown banners change to wings . . .
Things dark on the horizons of perception,
Become accompaniments of fortune, but
Of the fortune of the spirit, beyond the eye,
Not of its sphere, and yet not far beyond

Wallace Stevens, *"To an Old Philosopher in Rome"*

Introduction

I

This book had its beginning in the early 1970s when I began working in the factories of Chicago's south and west sides. It was there I received my education in American culture and the political left, and it was there that I learned of their many discrepancies.

With the breakup of Students for a Democratic Society in 1969, members of the Revolutionary Youth Movement II (RYM II—RYM I became the Weathermen) moved off campus into working-class communities to do "point of production" and neighborhood organizing. Increasingly guided by "Marxism-Leninism–Mao Zedong–Thought," small collectives numbering from ten to thirty began forming in urban industrial areas across the nation with the ultimate intention of forming a new, "genuine" Communist Party, which would replace the "revisionist" Communist Party aligned with Moscow. Fed up with the drug culture and frustrated by the anti-intellectualism and isolation of the student movement, I became part of this political tendency. In 1971 I joined a group of middle-class white students like myself who were earnestly attempting to master the profundities of Mao's "On Practice" and "On Contradiction," and a handful of other canonical texts by Marx, Lenin, and Stalin. Fueled by a boundless reserve of idealism and energy, about twenty of us one day loaded our meager possessions into a caravan of VWs and made the unlikely journey from Ithaca, New York, to the mean streets of Chicago.

Upon our arrival we were assured by a former national secretary of Students for a Democratic Society that we would have Chicago "sewn up" within five years,

Mayor Richard J. Daley and his political machine notwithstanding. Six years later, having worked as a spot welder, punch press operator, and millwright apprentice; having attended hundreds of meetings; and having helped organize several mildly successful campaigns on behalf of union democracy, health and safety, increased wages and benefits, and the rights of the unemployed, I left the organization to which I was affiliated and ultimately returned to school to pursue an academic career.

Measured by degree of commitment, hard work, persistence, idealism, selflessness, or organizational skill, there was much that was admirable about my and my comrades' enterprise. We had crossed social boundaries that few with our backgrounds were willing to cross in order to fight the good fight, and in doing so we brought some hope, righted some wrongs, and perhaps changed some lives. But measured by other standards—namely effective political strategy, intellectual rigor and integrity, independence of mind, creativity and self expression, spiritual sustenance, or moral vision—the movement of young radicals who embraced doctrinaire Marxism during the 1970s was a dismal failure. Viewed from the perspective of twentieth-century leftism, this neo-Stalinist revival, for all its sacrifices and modest accomplishments, in the end lived up to only one of Marx's precepts: history repeats itself as farce.

Saddled with a political line that called for the violent overthrow of American capitalism and the establishment of a dictatorship of the proletariat, even the most savvy political organizer was soon forced to confront the limits of his or her appeal. But more than an egregiously sectarian political line was responsible for the failure of our effort: the entire enterprise had the odor of orthodoxy about it, not only in its long-range political goals, but in terms of its whole culture. Solidarity was not based upon free-thinking individuals who were encouraged to analyze issues with all of the resources society made available to them; rather, unity was forged in a political culture that discouraged intellectual inquiry and internal dissent and of course forbade public dissent. Reading outside the canonical texts was grounds for suspicion, and this included reading novels or poetry, for which a committed revolutionary purportedly had no time. Political discourse had to be based on the "science" of Marxism-Leninism, and the dominant method used in solving political problems was that of exegesis. Democratic practices such as the secret ballot and parliamentary rules were condemned as bourgeois. Leadership, which constituted a rigid hierarchy, was not democratically elected, and rank-and-file members had no right of recall. Personal commitments, leisure time, and vacations were wholly contingent upon the needs of the organization. Moral or spiritual values and vocabularies were strongly discouraged.

Today the American left consists of hundreds of organizations that operate independently but sometimes unite for joint actions and campaigns both at the local and national level. These groups include minority, women's, student, worker, environmental, neighborhood, lesbian, gay, academic, single-issue, political, and traditional cadre-style groups. Although diverse and often effective in terms of pushing for specific reforms, the contemporary left is small and fragmented. Nonetheless, with smaller size and visibility since the 1960s it has managed to weather the conservative tide, if only to defend old victories against new threats. Forced into a defensive posture by both the conservative offensive and the liberal retreat, it has won precious few victories during the past two decades.

But the weakness of the current left should not be attributed solely to the conservatism of the Reagan-Bush-Gingrich years: any frank diagnosis of what ails the left should take into account certain perennial internal problems that remain untreated. Nearly all of these problems, in my view, have their sources in the American left's susceptibility to sectarian and even authoritarian trends of the kind I myself experienced. These trends have manifested themselves as a historic ambivalence toward the very liberal values that define American ideals and that fuel many movements for change abroad. This is, in short, to say that the American left has never consistently incorporated the symbols, vocabulary, and practices of classical liberalism, by which I mean a set of attitudes (as opposed to doctrines) concerning the scope of government, the efficacy of the individual, the freedom of conscience, the free exchange of ideas, the pragmatic frame of mind, and the reasonable limits of liberalism itself. This has been the case, I contend, certainly since the influence of Bolshevism during the 1920s put advocates of democratic socialism and liberal values on the defensive. It was also the case, although less egregiously so, during the nineteenth century, when utopian schemes to remake societies and selves clashed with the hesitations, doubts, desires, talents, and imperfections of individuals who wished to maintain a primary allegiance to what Emerson famously called "the infinitude of the private man."

If the current American left is no longer as rigid as it once was—in the recent past evidenced by, among other things, its general and open (though not always enthusiastic) support for the breakup of communism in the former Soviet bloc—it is a basic truth that the American left's experience with sectarianism has been a defining and lingering one. Although doctrinaire Marxism as an official ideology and set of practices is confined to the margins of today's left, the fact that the left continues to be beset by persistent dogmatism, reductive materialism, and ambivalence toward liberal values links the present to the past.

Although he frames these matters slightly differently than I, Richard Flacks has examined the chronic condition to which I refer in *Making History: The American Left and the American Mind* (1988). In this thoughtful and remarkably frank study, Flacks points out that democratic-centralist or not, all major American leftist organizations have been undermined by a tendency to encourage forms of ideological conformity at the expense of the direct experience of individuals. Thus, for example, although organizations such as the Socialist Party, the Industrial Workers of the World (IWW), the Student Nonviolent Coordinating Committee (SNCC), and SDS were not vanguard organizations, each of these groups expected conformity to established principles, the violation of which was reason for expulsion. The Socialist Party forbade members from cooperating with any other party during election campaigns; the IWW did not permit any of its locals to formalize its cooperation with a capitalist employer by signing a contract; SNCC eventually refused to cooperate with advocates of the more moderate tactics of Martin Luther King; and SDS, despite its early commitment to organization pluralism, was in the end torn apart by factional strife.

The left's ambivalence toward liberal values has also manifested itself in the persistent belief that the left will succeed only when it has built an all-encompassing national organization with enough power and authority to unify oppressed groups and wrest power from the capitalist class. As Flacks demonstrates, it has long been believed that such an organization would offer the left a single coherent strategic perspective laying out the only "correct path" for the revolution to follow. Not surprisingly, Flacks is pleased by the pluralism of the current left, which has remained distinctly decentralized for several decades now and has also eschewed a monolithic strategy.

But the question remains as to whether decentralization and the rejection of party-line politics has solved all of the problems of ideological rigidity on the left. My view is that they have not. These changes have rid the left of the apparatchiks, and they have helped restore a degree of badly needed civility to a movement that by the mid-1970s was fractured, exhausted, and full of bitterness. But often it is difficult to distinguish civility from conformity, ideological peace from intellectual laziness, the absence of recrimination from the absence of compelling debate, both internal and public. Flacks similarly puts the issue confronting the contemporary left as follows:

> The problem is to create sufficient structure to facilitate coordination; sharing of resources and information; mutual clarification of vision, strategy, and program; the maintenance of collective memory and identity—while avoiding the encapsulation, rigidity, intellectual deceit and distortion, interpersonal abuse and personal alienation that have been the plagues of the organized left throughout its history.[1]

Thus far, the left has not solved these organizational problems, despite the fact that the new left's healthy distaste for centralism has filtered down into the contemporary left. Thus organizations strenuously object to the whole idea of adopting a "political line," but this has meant that they avoid formulating long-range goals altogether. Most groups go to great lengths to avoid the kind of rigid hierarchy I have alluded to. It is common to see revolving leadership, co-leaders, short-term leaders, weak leaders, or no leaders at all. Whenever feasible the membership makes the decision, and every effort is made to encourage a unanimous outcome. Many groups, especially student groups, will extend debate indefinitely in order to develop a strong consensus. A bare majority vote is nearly always unacceptable, and there is often an unwillingness to tackle difficult issues in order to avoid schisms, differences, or even discomfort. All of this adds up to a successful struggle against Leninist and Stalinism centralism, but it has meant organizational forms and politics that lack the structure needed to encourage constructive debate, intellectual rigor, the creative solutions to problems, and active, accountable leadership.

As a solution to some of these problems, Flacks looks to Rudolf Bahro's notion of the party as a "collective intellectual":

> What is needed . . . is that the party . . . must be organized as the *collective intellectual*, which mediates the reflection of the whole society and its consciousness of all problems of social development, and which anticipates in itself something of the

> human progress for which it is working. . . . [I]ts achievement is subject to precisely the same criteria and has also the same general conditions as the work of a group of scientists. . . .
>
> [The organization] must therefore be open to all those having the need to go beyond the pursuit of their immediate interests, having recognized that the barriers to their self-realization bear a social character. . . .
>
> Now that there are millions and millions of intellectualized people . . . [who] are themselves hungry for a more comprehensive social communication, it must already be possible to *work out through discussion* the necessary compromise of interests and carry it through primarily with the "gentle power of reason."[2]

With their stress upon free and open discourse, full participation, compromise, and the importance of grounding future plans in present realities, these passages contain much that deserves our assent. But quite beyond the residual scientism, there is a disturbing note here as well: the diminution of private life implied by Bahro's very traditional claim that only the organization will allow individuals to "go beyond" their personal commitments and truly realize themselves. However much it may be true that socialization can serve to enhance the quality of an individual's life, it is equally true that crude socialization can damage that quality of life. Moreover, Bahro fails to see that socialization in the name of common interest, but not suffused by the personal interests and commitments of individuals, is doomed to failure. As is so often the case with leftists, Bahro is more acutely aware of the dangers of individualism than of collectivism, even though the history of twentieth-century leftism demonstrates an opposing truth. Whereas most leftists, like Bahro, continue to speak of socializing individualism, I argue that the more compelling need is to individualize socialism.

Here, Flacks's main argument is extremely helpful, which is that the American left, for all its substantial accomplishments, has never solved the problem of how to overcome the disjunction in American life between the commitments of everyday life and the needs of organizations committed to "making history." If most Americans tacitly express the nation's revolutionary heritage by asserting their *liberty* to shape their private lives through their daily commitments to dependents, to work, and to leisure, most American leftists construe that heritage as a *democratic* one, which conceives as central an actively engaged citizenry intent on reshaping the community. As Flacks observes, when the left has been successful at organizing large numbers (the Socialist Party and the IWW during the first two decades of the century, the Communist Party during the 1930s, and groups like SNCC and SDS during the 1960s), it has managed to integrate, however imperfectly, its history-making goals with those everyday activities of ordinary Americans meant to maintain and develop individuals and their loved ones—hence, for example, the Socialist Party's week-long summer encampments on the plains of the Midwest, to which families would flock for education, food, and entertainment; the IWW's network of meeting halls, which served as safe havens for itinerant workers; and the Communist Party's panoply of study groups, book clubs, lectures, debates, travel groups, summer camps, and social functions allowing for the self-development of workers.[3]

What, then, would it mean for a political movement to provide for a fuller expression of individual needs within the context of cooperative action? It would mean, first of all, the largest possible extension of the movement's intellectual and cultural life so as to enable individuals representing diverse experiences and opinions to reconsider the left's response to an exceedingly complex society. The refusal to engage in ideology—or, as Lionel Trilling put it, the ritual of showing respect to certain formulas for the sake of emotional safety—would serve to erode the Manichaean models and categories that have characterized so much leftist debate, familiar binaries such as progressive versus reactionary, reform versus revolution, individual versus group, politics versus direct action, confrontation versus co-optation, and "us" versus "them."

A political movement providing for a fuller expression of individual needs would also be a movement for which subjective experience—that of nonmembers as well as members—would be of crucial importance. The phrase "the personal is political" at one time represented such a mandate for the feminist movement, but, as Flacks observes, although it began by connoting the invigorating effect of the imaginative life, it degenerated into the axiom that ideology determines private, not just public life. As things now stand, leftists are usually encouraged to separate their personal lives and feelings from their political activities, especially when discrepancies exist (this is significantly less true of the feminist, gay, and lesbian movements than elsewhere within the left). Thus the left presents few opportunities for exploring (or validating) doubts, personal problems, aesthetic interests, or spiritual concerns. Activities are not carefully calibrated to reward all types and intensities of commitment, and moments of leisure within the time frame of the movement's unyielding schedule are rare, causing weary activists either to endure in a diminished capacity or to leave abruptly with feelings of bitterness or guilt. In responding to the needs of the whole self, the left has not substantially improved upon the dominant culture: it has not, for the most part, been kind to its members and has been particularly cruel to its leaders.

Finally, a political movement responsive to the minds and imaginations of its members would be in a better position to appeal to the goodwill of liberal democrats, moderates, and even conservatives among mainstream Americans, whose differences with the left are sometimes differences of opinion rather than differences in what Tocqueville termed "habits of the heart"—moral attitudes, values, sensibilities. A small local example serves to illustrate my point. Several years ago students at Princeton University mounted a campaign to expand the resources and purview of the campus program responsible for handling sexual assault and harassment. With eyes fixed exclusively on the pressure points of the administration, the students quickly opted for a dramatic takeover of the president's office, followed by a sit-in. In certain respects this was a perfectly reasonable choice: the program had been created in the first place in response to the militant tactics of campus groups whose members numbered no more than several hundred at most. But the priorities of the campus group reflected a more vital relationship with the administrative bureaucracy than with the majority of students making up its constituency, whose political and moral sentiments were largely overlooked in the activists' overattentiveness to the apparatus of power and to legal (and illegal) redress. In fact, many of the progressive students tended privately to express disdain for the average student, whom they regarded as hopelessly apolitical at best and downright reactionary at worst. After all, they maintained, many students came from wealthy families, many supported George Bush, and many displayed racist, sexist, and homophobic attitudes. But Princeton's student body, despite being an elite group, presented to activists a set of attitudes and beliefs not much different from those they would encounter off campus—indeed, in some respects they were more liberal. That these attitudes and beliefs were not progressive should not have prevented activists from taking up popular issues, finding a common style and vocabulary, and attempting to build a broad-based, democratic movement. Unfortunately, the barriers separating campus activists from mainstream students resemble those dividing leftists in general from much of the population. In spite of real benefits, one of the ill effects of the current decentralization of the left is that each group focuses on its own constituency to the relative neglect of the broad middle. The past two difficult decades have seen a disturbing attitude emerge among many leftists: a willingness to write off large constituencies based on the belief that the average American is hopelessly mired in false consciousness, whether expressed as a form of patriotism, racism, sexism, or homophobia. This view tends to overlook the possibility of unarticulated motives and commitments, which, once revealed, could provide a basis for fostering communication and even unity. If most Americans are consumed with the business of making significance of their private lives, it is, I believe, incumbent upon leftists to fashion vocabularies and a politics capable of contributing to the difficult tasks Americans face: living a moral life, constructing an identity, dealing with ambiguity and uncertainty, and finding pleasure.

II

Incredibly, it has been a quarter century since the publication of Fredric Jameson's *Marxism and Form* (1971), arguably the most important volume of Marxist criticism published in the United States in the postwar period. Issued by Princeton University Press at a time when Marxist criticism had little influence among American critics, the book at first did not gain a wide audience. Only gradually, as the 1970s wore on, did it become one of the most important critical texts to a generation of young scholars politicized by the antiwar, antiracist, and feminist movements of the 1960s. Jameson's book was instrumental in inaugurating a new phase of Marxist criticism in the United States, in which the orientation became continental, the subject literary form rather than content, the mode theoretical, and the manner complex and technical (yet at its best lucid and urbane). The contributions of this criticism have been substantial: it has revised our understanding of nearly every major category of traditional Marxist thought, including class, causation, base and superstructure, production, ideology, and form; and it has expanded other categories such as culture, gender, and race. This has earned for Marxist and materialist criticism unprecedented stature within the academy.

Compared with the 1930s, the last great heyday of Marxist criticism in the United States, the current efflorescence has already enjoyed a lengthier duration, although threats from inside and especially outside the academy continue to make its existence controversial if not precarious.

Ironically, one factor that has helped neo-Marxism gain respectability within the academy has been that it has forgotten earlier, indigenous efforts at Marxist criticism. In his prefatory appeal for his largely academic audience to consider Marxism anew, Jameson himself quite seriously misrepresented this earlier tradition. By ignoring its diversity, and the degree to which it was vilified if not suppressed by New Critics during the postwar period, he lost an opportunity to recover examples of accomplished, engaged, and indigenous left-wing criticism. Jameson is worth quoting in full:

> When the American reader thinks of Marxist literary criticism, I imagine that it is still the atmosphere of the 1930's which comes to mind. The burning issues of those days—anti-Nazism, the Popular Front, the relationship between literature and the labor movement, the struggle between Stalin and Trotsky, between Marxism and anarchism—generated polemics which we may think back on with nostalgia but which no longer correspond to the conditions of the world today. The criticism practiced then was of a relatively untheoretical, essentially didactic nature, destined more for use in the night school than in the graduate seminar, if I may put it that way; and has been relegated to the status of an intellectual and historical curiosity.[4]

Jameson's condescension, however auspicious it turned out to be for the subsequent institutionalization of Marxist criticism, nonetheless did a disservice by classifying an entire era of left-wing criticism on the basis of its most vulgar expressions. By dismissing American criticism with a close connection to actual political movements as tendentious and reductive, and by claiming elsewhere that Anglo-American criticism as a whole was mired in a myopic and reactionary bourgeois empiricism, Jameson's turn to Europe, with its own series of binarisms, did not escape the Manichaeanism he deplored in others. This turn seemed to preclude the possibility of comparative study, and certainly discouraged further exploration into indigenous left-wing criticism.

There was another problem with the new academic variety of Marxism. Aside from a suggestive, penultimate sentence about criticism's obligation to remain critical and "to keep alive the idea of a concrete future,"[5] Jameson remained silent on the question of commitment. Indeed, one of the interesting paradoxes of the dramatic politicization of theory and criticism during the past two decades is that it has often been confined to the politics of criticism alone, as if such an activity could be separated from the politics of the critic as citizen—as voter, as a community member with political beliefs and commitments, as activist. Until very recently, the highly politicized debates that have taken place in the academy over modernism and postmodernism, discourse and ideology, high and low culture, canons and conventions, and even issues of race and gender have occurred largely without crucial references to concrete political events, movements, causes, or constituencies outside the academy. Thus many academic critics who espouse a generally leftist or progressive perspective, have done so with little regard for the consequences their views might have for a democracy of informed citizens. Fortunately, the current attacks on multiculturalism and "political correctness," however cynical and hypocritical they often are, have at least presented the academic left with an opportunity to think seriously about the public consequences of its critical interests and take its case to the public.

As this process continues to unfold, academic leftists will, I hope, be increasingly drawn to the achievements and the limitations of the indigenous tradition of left-wing criticism, whose connections to nonacademic constituencies were relatively strong by comparison with current university-based criticism. As this book aims to show, many issues and circumstances faced by contemporary critics can be better understood once critics develop a historical perspective enabling them to compare their problems and solutions with those of former times. Thus, for example, the reader will find that current debates raise numerous issues that the American left has actually addressed before: the relation between ideological correctness (what used to be called "belief") and literary quality; the politics of modernism; changing and expanding the canon; the powers of Marxist and formalist critical modes; the relation between "high" and "low" culture; and the political and social responsibilities of the critic.

In the past, of course, the New York intellectuals have been the focus of a good deal of interest, to the point that not so long ago a reviewer, referring to the abundance of material produced by and about the group, referred to the New Yorkers as America's own Bloomsbury.[6] It is not difficult to find reasons for the intense interest. Many of the New Yorkers have themselves survived to write memoirs in order to set the record straight about the remaining controversies over Stalinism and anti-Stalinism, McCarthyism, the new left, and neoconservatism. Some have become quite conservative on cultural and political matters; thus Lionel Abel, William Barrett, Sidney Hook, Hilton Kramer, William Phillips, and Norman Podhoretz have all written memoirs or essays in which they seek to link their current conservatism to what they regard as the conservative cultural politics of the New Yorkers. (Howe, Kazin, and Macdonald have produced memoirs suggesting just the opposite—that the commitment to modernism and "high" culture they shared with others served their progressive politics.) And finally many scholars, who like myself have been influenced by the values of the 1960s, became interested in the New Yorkers because in the New Yorkers' early years they represented an example of intellectual dissent with their commitment to radical politics, literary experimentalism, and rigorous literary and cultural criticism.

I should add that most of these critics have argued, despite their enthusiasm for the early, heroic phase of the New Yorkers' enterprise, that these initial commitments were compromised and ultimately betrayed during the 1950s and 1960s, especially in light of the New Yorkers' success and power, and their encounters with McCarthyism and the new left. Thus an unusual confluence of opinion can be found between conservative and liberal or leftist critics, insofar as all agree that the radicalism of the New York intellectuals, whether compelling or not, whether fortunately or unfortunately, was short-lived and, moreover, not to be understood as part of the legacy of the American left. Thus on the right Hilton Kramer considers *Partisan Review*, the main organ of the New York intellectuals, to have been a precursor to the *New Criterion*, and Sidney Hook, in a piece entitled "The Radical Comedians," has lampooned *Partisan Review* as never having been serious about radicalism, even before it took to what he characterized as its comfortable middle-class surroundings.[7] The left-liberal view has similarly denied any real credence to the notion of *Partisan Review*'s continuing radicalism. James Gilbert's *Writers and Partisans* (1968), in part reflecting the strained relationship between the old left and the new, tells the depressing story of the magazine's gradual "embourgeoisement" and "failure of nerve." According to Gilbert, even by the late 1930s *Partisan Review* had become more than an organ of the New York intellectuals—it had become an institution whose requirements as a center of power caused it to transform its Marxist outlook into something innocuous. In Gilbert's view *Partisan Review*'s failure was the inevitable result of the magazine's "replacement of the proletariat by the intellectuals."[8] Likewise Serge Guilbaut's *How New York Stole the Idea of Modern Art* charts the trajectory of *Partisan Review* as revolutionary magazine to *Partisan Review* as aestheticist magazine. To this end he measures *Partisan Review*'s growing preoccupation with the intellectuals against the yardstick of an unstated version of "genuine" Marxism (For Guilbaut the process of "de-Marxization" began as early as 1935). In order to support his thesis that *Partisan Review* sold out during the 1950s and became an unwitting cultural tool of American global hegemony, he disputes the seriousness with which the magazine continued to address problems in postwar American society at the same time that it mounted a sustained critique of aspects of American leftist, progressive politics.

That *Partisan Review* and the New York intellectuals as a whole moved in the postwar period from the left toward the political center is an accepted fact. By the late 1930s and early 1940s the magazine had rejected or revised key components of doctrinaire and classical Marxism. Most significantly, the magazine rejected the Marxist belief in the revolutionary character of the proletariat, and accordingly emphasized the intellectuals as the body most likely to promote change. For many of today's readers interested in social change, *Partisan Review* may seem to have relinquished too much of its early radicalism. This may be so, but I would stress that the magazine's move toward the center entailed a careful assessment of political and cultural matters that should remain part of any leftist politics today.

In what follows I shall be arguing that many of the conclusions arrived at by *Partisan Review* and the New York intellectuals concerning both the limits of Marxism and postwar American society remain prescient. Confronting a series of social and political crises, this group eventually rejected Stalinism, then the dominant form of leftism, and more generally the utilitarian attitudes and overly rational, prosaic habits of mind that pervaded the left. Although their efforts included explicit critique of doctrine, frequently they explored the more subtle ideological manifestations that lurk in unexamined assumptions, attitudes, and values. In these efforts they addressed problems that have long plagued the American left and are hardly as yet extinct: axiomatic thinking, a penchant for a mechanical form of materialism distinctly out of touch with American realities, and a confusion of individuality with individualism. Thus they presented the left with the challenge to commit itself to rigorous self-scrutiny—to a reassessment of its traditional understanding of politics as fundamentally a matter of correct positions, tactical effectiveness, and instrumental change. The writers and critics I discuss wished to see politics transformed, so that it might give due emphasis to subjective experience, to the moral, spiritual, and cultural dimensions of political life, and to greater flexibility and experimentation with regard to a range of ideological issues. The New York intellectuals envisioned a politics that could address subjective experience, encourage diversity, accommodate spontaneity, and adjust to complexity and uncertainty. Theirs was a critique of the culture of American progressivism and liberalism in the best liberal tradition of self-reflexive inquiry.

In addition, New York intellectuals should be credited for being among the first in the modern epoch to realize it was no longer obvious that refusing to affiliate with official Marxist organizations meant repudiating leftism or dissent, that criticizing and finally rejecting Marxist theories of class struggle and revolutionary socialism, or striving to attain a middle-class rather than a working-class readership, or focusing on subjective experience rather than objective social forces, together constituted such a repudiation. For a time—roughly from 1937 to 1945—they hoped to build a new political culture based precisely on these things. Some, like Philip Rahv and Irving Howe, maintained an explicitly leftist perspective through the 1970s and (in Howe's case) well beyond. Most remained committed to the values of critical nonconformism despite profound disagreements with the counterculture and the new left of the 1960s. If by the mid-1940s *Partisan Review* was less committed to socialist revolution and the uncompromising critique of capitalism than it was in 1937, this does not mean it was not sharply critical of certain aspects of bourgeois society—sharp enough indeed to have helped sustain an audience of skeptics and critics of American as well as Soviet society. Although the audience did not include ordinary Americans, or encourage anxiety or regret over their exclusion, it was a relatively wide audience that was encouraged to relate cultural matters to public issues. One need not be nostalgic for an earlier age of public intellectuals and disdainful of subsequent academic scholarship to believe that an understanding of the New York intellectuals' critical projects can help to guide leftists today.

With regard to the limits of these critical endeavors, the following study aims to explore how the New York intellectuals' imperfect legacy might affect our views of contemporary criticism and politics. I recognize that the New Yorkers present us with certain perspectives that leftists will no longer want to sustain, such as their inadequate treatment of race, gender, the new left, and the workaday world of ordinary Americans. The 1930s and 1940s were decades rich with examples of rigorous, engaged criticism—indeed one might argue they saw the last great intervention in America of literature into politics. But the costs of the acrimonious ideological debates between Stalinists and anti-Stalinists were great. The valuable autonomy the New York intellectuals were able to wrest for themselves, and their ensuing prominence as cultural and political arbiters, came at the price of vital connections they might have developed with other Americans, particularly with dissenting and systematically marginalized groups. As a result, during the 1950s and 1960s they adopted a badly distorted picture of American life by exaggerating the pernicious influence of mass culture, Stalinism, and, later, the new left.

All the same, I wish to contend that these very weaknesses reflect the New York intellectuals' unique historical, critical, and political circumstances. Today's priorities are different for us precisely because we have absorbed some of the criticisms of the New York intellectuals. I see the priority of the left being to explore—rather than to rehearse the stale attack on—the values of individualism, liberalism, and the life of the imagination. I believe that, in addition, such an exploration would benefit by a closer and animating relationship between academic and nonacademic constituencies. Given the achievements and relative stability of the academic left at the present time, a more public perspective would make this a distinct possibility, provided it proceed with care and with the examples of past decades clearly in mind. For at the same time that the academic left presses on with innovative critical projects that lack reference to public politics, mass-based organizations press for changes in public policy without the benefit of new ideas and cultural forms that might be engendered through cooperative work with intellectuals. Today's left is badly in need of the kind of self-critical renewal strongly encouraged by the writers and critics whom I discuss in this book. Given the opportunities to break with old stereotypes presented the American left by the fall of communism, and given the extraordinary role that writers and artists have played in these events, one would hope to see a careful turn toward public politics in the work of academic critics.

III

I begin *Renewing the Left* with the 1930s because it was then that many of the ideological paradigms within which leftists continue to operate were solidified. Here I refer not so much to the overt Stalinist/anti-Stalinist and the Communist/anti-Communist polarities, whose efficacy has quite obviously been reduced, but to the binaries that fueled these ideological conflicts: self versus society, imagination versus politics, free will versus determinism, and gender, race, and culture versus class. As I have indicated, these polarities, combined with the historical tendency of left-wing organizations to discourage an open, democratic, and critical intellectual life, have greatly weakened the left's cultural life over the decades, causing many writers and critics wishing to maintain the vitality of their work to loosen or sever their ties to political movements. This was the case for most intellectuals at least until the 1960s, when various attempts to reintroduce a close, noncoercive relationship between politics and mind met with some success within the civil rights movement, the women's movement, and the counterculture. But the post-1960s left continues to struggle over the place of intellectual and cultural production within the movement. It is time to reverse the question traditionally asked by leftists (today it manifests as a demand for politically correct writing): How have writers and critics treated the left in their work? The more pertinent question today is: How has the left treated its writers and critics? Put another way, has the left created a hospitable environment for independent, critical, and creative intelligence?

In the first section of this study I examine in detail the politics and criticism of William Phillips and Philip Rahv, the founders and chief editors of *Partisan Review*. In chapter 1, entitled "The Antinomies of American Radicalism," I explore the schizophrenic character of leftist discourse during the early 1930s, when young critics such as Phillips and Rahv found it necessary to write highly reductive criticism for Communist Party publications while confining their more suggestive criticism to "highbrow" journals like the *Criterion*. Close readings of several key early essays show the precarious status of serious intellectual endeavor within the Stalinist movement, including Phillips's interesting attempt to revise reductive Marxism (his and others') by injecting into it not Lenin's but I. A. Richards's claims for scientific rigor and professionalism.

In chapter 2, entitled "*Partisan Review*'s Eliotic Leftism, 1934–1936," I look at the founding of the magazine and Phillips's and Rahv's unprecedented emphasis on literary criticism as the impetus for a proletarian literary renaissance. In order to transform orthodox Marxist criticism into a flexible instrument of analysis, the two critics systematically appropriated T. S. Eliot's early criticism (using concepts like "dissociated sensibility" and "objective correlative"), thereby anticipating later critics who claim to have discovered the dialectical impulse of Eliot's early criticism. I explore the debt to Eliot in greater depth than previous readers, making the polemical point that, despite Eliot's politics and damaged reputation at the hands of liberal and leftist critics today, earlier leftists managed to find something genuinely constructive in his contributions.

Chapter 3, entitled "Politics and the Autonomous Intellectual," examines the acrimonious split between *Partisan Review* and the Communist Party in 1937, and the former's subsequent interest in Leon Trotsky, then exiled in Coyoacan, Mexico. Although never developing official ties with the American Trotskyist movement, *Partisan Review*'s editorial board did enter into an important, two-year correspondence with the exiled revolutionary. These letters, until now never examined in detail, continue to illuminate the very difficult problem frequently faced by engaged intellectuals, namely of securing a degree of autonomy for themselves and their work while maintaining ties with active political movements and constituencies.

In chapter 4, "Modernism and the Autonomous Intellectual," I explain the process by which *Partisan Review* rejected the proletarian literary movement and attempted to form an extraordinary new union between political radicalism and modernism. Having first explored the possibility of a working-class movement allied with the avant-garde, a strategy outlined by Trotsky, Phillips and Rahv eventually despaired of such a possibility, pointing to the weakness of the working-class movement and the inability of the bankrupt Communist Party to sustain high culture. They opted instead for a highly defensive and custodial stance, in which, as Clement Greenberg famously put it, "Today we look to socialism *simply* for the preservation of whatever living culture we have right now."

The second section of my study looks more broadly at the New York intellectuals and the role they played in three crucial cultural battles, which took place before and during World War II. In chapter 5, "Modernist Renewal," I turn to the literary criticism of Rahv, Trilling, Dupee, and Troy, arguing that from the late 1930s to the mid-1940s, important differences notwithstanding, these critics were engaged in a project best understood as a radical appropriation of modernism for the purpose of renewing the culture and politics of the American left. By the "left" in the 1930s I mean the whole spectrum of political groups and constituencies from New Deal liberals to the Trotskyist sects of the far left. By renewal, I mean the attempt, in Trilling's words, "to form a new union between politics and the imagination"—such that the largely instrumental culture of the left might attain a new degree of subtlety, self-consciousness, and moral sensibility. Chapter 6, "Notes toward the Supreme Soviet: Wallace Stevens and Doctrinaire Marxism," examines the engaged poetry of Wallace Stevens, who wrote frequently for *Partisan Review* during the 1930s, and whose critique of doctrinaire Marxism roughly corresponded with that of the New York intellectuals. In chapter 7, "The Culture Wars of the 1940s: Literature, Popular Culture, and the Battle over a Usable Past," I look at two key cultural battles of the 1940s that apply to contemporary debates: the split between "high" and "low" culture; and the conflict between modernism and realism, the latter partially disguised during World War II as a debate over "literary nationalism." In both these battles the New Yorkers tended to polarize the issues in ways that continue to cause confusion. Thus like the Frankfurt critics Adorno and Horkheimer, Clement Greenberg in "Avant-Garde and Kitsch" (1939) and Dwight Macdonald in "A Theory of Popular Culture" (1944) allowed their dispute with Stalinism and their understanding of fascism in Europe to obscure their vision of popular culture in the United States. By equating the populism of the Communist Party with mass-produced, commodified culture in general, which they considered fascistic, they unrealistically saw the specter of advancing totalitarianism in a range of cultural practices, including publishing, moviemaking, and television.

Similarly, in Dwight Macdonald's and F. W. Dupee's polemics with Van Wyck Brooks, Archibald MacLeish, and V. L. Parrington over literary realism and a "usable" American past, certain high forms of culture—usually modernist—were narrowly considered to be the only possible sources of serious criticism of bourgeois values. Here the famous breakdown of the American literary tradition into the antitheses of "experience" versus "mind," and "paleface" versus "redskin" (along the lines of Brooks's highbrow/lowbrow dichotomy and Eliot's dissociated sensibility), were particularly damaging because they enforced exclusion, reduced texts to essences, and reversed the previous priorities rather than created new syntheses. In this chapter I point out the limitations of critical vantage points, which, for all their political sophistication, encouraged almost no sympathetic responses to examples of popular culture. Nor was there much favorable response engendered toward twentieth-century realist, naturalist, or minority literature. This said, we can decipher the political "hieroglyphics" within such influential texts as Trilling's "Reality in America" and F. O. Matthiessen's *American Renaissance* (Matthiessen was outside the *Partisan Review* circle but very influential) in order to uncover the salutary, and still relevant, correctives each offered to the progressive and Popular Front critics of the day.

The third section of the book focuses on the relevance of the New Yorkers' efforts to the current political and cultural situation, particularly where gender, race, and the idea of literary value are concerned. In chapter 8, "The 'Dark Ladies' of New York," I inquire into the experiences of four women–Tess Slesinger, Mary McCarthy, Elizabeth Hardwick, and Susan Sontag–whose work remains extremely important yet who have not received the attention that their male counterparts have. Despite obvious differences, these women produced fiction and criticism that presented the intellectual left with a new and critical vision of itself as profoundly patriarchal. I explore the ways in which these writers challenged the reigning notions of what politics were: impersonal, objective, public, masculine. Through their innovative narrative strategies, styles, and subject matter, they focused on experiences and subjects neglected by traditional leftism: pleasure and sexuality, marriage and divorce, childbearing and childraising–in sum, everyday life.

The New Yorkers' response to African American writing is the subject of chapter 9, entitled "'Their Negro Problem': The New York Intellectuals and African American Culture." (The title derives from the well-known essay by Norman Podhoretz.) Here I explore the critical reception of such writers as Ralph Ellison and James Baldwin, whose work the New Yorkers championed, as well as those black writers either received without enthusiasm or simply ignored. My purpose is to show how the complex politics of ethnicity and race from World War II through the civil rights and black power movements affected literary criticism. I argue that despite their strong support for certain black writers, the New Yorkers failed to respond adequately to the range of accomplished black writing. In part this was because they used an ethnic model of assimilation, often based on their own experiences as Jews, to understand a community whose epochal racial, national, and cultural differences demanded a different perspective. The New Yorkers thus imposed critical paradigms from the white literary tradition onto African American writing that was in many ways incommensurate with it.

During the 1950s and 1960s, with many of the New York intellectuals teaching in the academy, their increasing distance from marginalized groups manifested itself as a defensive, rather hidebound attack on the Beats and then on the counterculture. In chapter 10, entitled "'Preserving Living Culture': The 1960s and Beyond," after Clement Greenberg's well-known phrase, I explore the limitations of the New York intellectuals' reception of postmodernism, a term they themselves invented. In doing so I look closely at representative, influential essays by Norman Podhoretz, Leslie Fiedler, Irving Howe, and Lionel Trilling that attacked postmodernism. But beyond identifying limitations, my interest is twofold: to understand how and why the debate over postmodernism quickly became polarized, and to fix upon elements of the New York intellectuals' critique that remain compelling. Thus I apply some of the insights of the New Yorkers to several examples of contemporary academic left criticism.

In my last chapter, entitled "What's Left of Lionel Trilling?" I argue that in spite of Trilling's current reputation in certain quarters as a conservative critic, he is best understood as a dissenting critic of the left whose challenge to humanize radical politics has gone unanswered by later generations of leftists. Even his later work of the 1960s and 1970s, I argue, deserves more sympathetic treatment from leftists willing to acknowledge the excesses of the 1960s and the need to renew politics along the lines I have suggested.

I believe that the idea of the critic as both independent of and auxiliary to popular movements and aspirations, an idea whose exemplary American practitioners have included Emerson, DuBois, Gilman, and Dewey, and whose most persuasive recent expressions are Michael Walzer's *The Company of Critics* and Cornel West's *The American Evasion of Philosophy*, holds out the best model for the contemporary academic critic interested in carrying out his or her civic responsibilities in a democratic society. Surely if we are to explore such a role, the example of the New York intellectuals is valuable, both as warning and inspiration. As many have observed, they were, after all, the "last intellectuals" to reach a nonacademic audience with their accessible, rigorous discussions of literature, culture, and politics. But their public was neither the whole public, nor was it representative of the whole public. If academics are serious about intervening in the public sphere so that all sectors are democratically engaged, they will have to learn from earlier problems and find their own way.

PARTISAN REVIEW AND THE REMAKING OF RADICAL CRITICISM

1

The Antinomies of American Radicalism

I've thought a lot about the period. And I used to have a pretty good memory; I had something close to total recall of these events. And I found myself, as the years went on, suppressing it.

WILLIAM PHILLIPS, 1966

I

Following the severe repression of the 1919 Palmer Raids and the subsequent bolshevization of significant sectors of the American left during the 1920s, the circle of leftists surrounding *Partisan Review* emerged in the late 1930s as a major force for reestablishing a productive relationship between politics and intellectual life. As Lionel Trilling put it in a well-known phrase from *The Liberal Imagination*, *Partisan Review*'s major achievement was to have "organize[d] a new union between our political ideas and our imagination."[1] How far the magazine succeeded is one of the questions this book will attempt to answer. What is important at the outset is to acknowledge that the magazine's "new union" involved bringing the cultural avant-garde and the political vanguard into some form of productive mutual relation for the first time since the 1910s. The last time the American left had managed to maintain this kind of equilibrium—one might add the only time—was during the extraordinary period from 1911 to 1918 when radical writers and artists joined political activists to produce new cultural forms on behalf of actual constituencies struggling for social change. Not only did the *Masses* magazine provide a forum for innovative graphics, design, journalism, and, to a lesser extent, poetry and fiction, but it encouraged writers and artists to venture out of bohemia and into working-class communities (even if only passing through), to record living conditions, strikes, revolution (in Russia), and, in the case of the Paterson Silk Strike Pageant, help mount a major theatrical event, staged at Madison Square Garden, with a cast of thousands of working people. But however much we may still be moved to admire the experimental force of John Reed's journalism, Louis Untermeyer's poetry, Floyd Dell's fiction, or the graphics of John Sloan, Art Young, Robert Minor, Stuart Davis, and Boardman Robinson, it remains important to note that in terms of formal innovation they fell short of the more radical efforts of the postimpressionists or expatriates like Stein, Pound, H.D., and Eliot. Nonetheless, these were bold and controversial artistic initiatives linked to socialist, feminist, and anarchist movements. In this respect they were expressions of a momentarily tolerant and confident left, willing to give free reign to committed but autonomous artists and intellectuals.

Historians have attributed the demise of this happy set of circumstances to the postwar loss of innocence, although it would be more accurate to say the demise resulted from the tragic fact that lost innocence gave way to cynicism rather than maturity. Thus began a lengthy period in which many American radicals—independent and critical thinkers when it came to analyzing the ills of capitalist society—were willing to subordinate themselves and their movement to the example and the will of the Soviet Union. This was certainly the case for those who rose, or wished to rise, within the ranks of the newly bolshevized Communist Party. It was less true for the Party's rank and file, who in general were more capable of independent, effective, and courageous organizing work than were the Party leaders. But during the 1920s, doctrinal purity and organizational discipline gradually replaced the free-spirited, diverse leftism of the 1910s. Looking at the chief political and cultural magazines of the Communist left at the time, we note that although the *Masses*' successor, the *Liberator*, maintained an editorial policy that encouraged a degree of heterogeneity during the early 1920s, that policy gradually disappeared in its successor the *New Masses*. By the early 1930s political and doctrinal considerations largely shaped the magazine's artistic content, and this further eroded whatever autonomy and vitality was left of the movement's intellectual and cultural life. This has had indirect but baleful consequences for the fate of a democratic and culturally rich socialism in the United States, for it was the old left with its incapacities which shaped the imperfect responses of the new left during the 1960s, and these in turn provide the contemporary left with its main legacy. Given this genealogy, my hope in these opening chapters is that by tracing *Partisan Review*'s trajectory we can gain an understanding of some of the perennial difficulties that continue to characterize the left's political culture.

For young radicals coming of age in the early 1930s, openly heterodox cultural work was incompatible with the interests of the Communist Party, a democratic centralist, cadre-style political party leading a proletarian political movement under the rubric of Marxist-Leninist ideology. By heterodox I mean work with important and acknowledged sources outside as well as inside the official, orthodox Marxist-Leninist discourse of the international Communist movement,[2] whether these be bourgeois sources or sources derived from other radical traditions. Certainly before the more tolerant Popular Front period, inaugurated in 1935, the strict dichotomy between bourgeois and proletarian culture discouraged writers and critics from drawing too heavily upon the former. Those interested in maintaining their radicalism and, at the same time, pursuing their cultural work while transgressing this boundary had but three choices: they could negotiate a kind of division of labor whereby orthodox work would be expressed through official channels and heterodox work published outside the Party's milieu, they could furtively incorporate elements of heterodoxy within their apparently orthodox work, or they could operate on the fringes of the Party and suffer the consequent irrelevance. In this and the following chapter I show how William Phillips and Philip Rahv, the founders and chief editors of *Partisan Review*, gradually moved from the first to the second of these strategies. As I have indicated, this sort of discursive and institutional arrangement was certainly not new to the American left. The conjunction of politics and culture during the 1910s provided a brief respite from the usual situation in which little space was given in publications, and little attention paid generally, to the cultivation of a multivalent and diversely generated cultural and intellectual life. To be sure, the American left has had its share of dedicated, effective, inspiring political leaders: Daniel DeLeon, Bill Haywood, Emma Goldman, Eugene Debs, W. E. B. DuBois, and Norman Thomas come to mind immediately. But as many historians of the American left have noted, the tradition has produced no theorist of international scope, and only several of national importance (among them Daniel DeLeon, William English Walling, and Louis Fraina). Moreover, DuBois can be said to have been the movement's only figure of major intellectual stature, but of course his relationship to the major left organizations was usually a vexed one. Unlike the British left, which has harbored the likes of Morris and Shaw; the French left, with room for Aragon, Sartre, and Althusser; or the German left, whose leadership included Kautsky and Luxemburg, the American left has shown little interest in developing original minds. Hence whatever possibility there may have been to effect close and constructively critical relationships with figures such as Howells, Dewey, Hook, or DuBois, historically the American left has chosen instead to forsake intellectual debate and cultural challenge in favor of relatively narrow theoretical debates and instrumental political goals.

The early to mid-1930s looms as a moment in the history of the American left when the discrepancy between political and intellectual life stood out in particularly sharp relief, to the point that I think it not unreasonable to speak of a kind of cultural schizophrenia on the left at the time. Thus the environment in which Phillips and Rahv produced their earliest work, and produced *Partisan Review*, was substantially shaped by the antinomies of which I speak. Before they committed themselves to bringing out their new proletarian magazine, Rahv was trying to outdo the most dogmatic of critics in order to establish his Marxist credentials, while Phillips was producing cant for Party publications and quite sophisticated essays for the "bourgeois" little magazines. Both Phillips and Rahv began their careers in a milieu in which nearly everyone assumed the incompatibility between, on the one hand, Party life and official Marxism, and on the other hand, skeptical intellectual inquiry. Not until Phillips and Rahv were able to establish their own magazine did they feel confident enough to opt for a new approach, which was quietly to introduce nondoctrinaire, and even non-Marxist, criticism into a doctrinaire critical framework. Later, of course, when this strategy failed, they broke from orthodoxy completely and introduced still another strategy for maintaining a confluence of political and intellectual life.

II

William Phillips, born William Litvinsky in Manhattan in 1907, grew up in the Bronx, an only child of Russian Jewish parents. He attended public school, then City College of New York, where he was influenced by the charismatic professor of humanities Morris Cohen, and later did graduate work at NYU. At NYU he took courses in philosophy with Sidney Hook, and at the same time began frequenting Greenwich Village literary circles, where he made the acquaintance of several Communists. They in turn introduced him in the early 1930s to the John Reed Club, an organization of writers and artists affiliated with the Communist Party. With chapters in over a dozen cities across the United States, the John Reed Club was an organization in which youthful aspirants could contribute to the drama of a proletarian cultural renaissance whose goal was to sweep away the decadent art of the bourgeoisie and replace it with the vibrant, democratic art of the working class. Phillips was impressed enough with the organization to join, and he promptly became secretary of the New York City chapter, the largest and most important in the country.

It was here that he met Philip Rahv, who had arrived in 1932 from Oregon where he had worked briefly as a junior advertising copywriter. Like Phillips, Rahv was the son of Russian Jews (he was born in 1908), and his name was also an acquired one—at birth he was Ivan Greenberg. During the civil war in Russia his parents had emigrated to Palestine, where he left them in 1922 to live with his older brother in Providence, Rhode Island. His formal education ended with grade school; the rest of his education he acquired on his own: in an early article he proudly claims to have taught himself "Freud, Nietzsche, Proust, Joyce, Rimbaud, etc."[3] Upon arriving in New York from the West Coast during the Depression, Rahv began to translate the hardships of those years into compelling arguments for revolution and Marxism. It was at about this time—1931 or 1932—that he joined Jack Conroy's Rebel Poets' group, then based in New York, where he worked on the proletarian poetry magazine *Prolit Folio*. He also joined the John Reed Club, and began writing reviews and essays for various Party-oriented publications.

Two points need to be made concerning these brief biographical sketches. First, both critics have emphasized the degree to which exposure to modern literature—Phillips in college and Rahv through his own reading—helped shape their sensibilities. Thus, unlike many of their young peers, they were more or less grounded in modern bourgeois culture when they joined the proletarian movement. With their knowledge of modernism having preceded their education in Marxism, they were perhaps in a better position than most within the movement to approach Marxism critically and relatively independently. Second, as was the case for many young Jews of their generation, Phillips's and Rahv's political radicalization was ethnically and religiously overdetermined. Their radicalization meant access—though not necessarily privileged access—to secular Western intellectual traditions and to a distinctly American tradition of dissent. In the process they rejected traditional Jewish religious practices and beliefs yet simultaneously retained talmudic habits of study and the values of Haskalah (Enlightenment). Early in their careers, their political internationalism (to be distinguished from cosmopolitanism, a cultural phenomenon) provided a basis on which to reject nearly all things Jewish, including their own names. Yet at the same time they turned their backs on what they regarded as Jewish parochialism, they also criticized the society into which they were gaining entrance, thus easing perhaps their hidden feelings of anxiety about their own deracination.[4] It was only later, of course—once their internationalism subsided, once the reality of the Holocaust became known, and once the state of Israel was established—that the connection between their Jewishness and their politics became more overt and they even acknowledged it publicly.

As I have indicated, much of Phillips's and Rahv's earliest work exhibited the characteristics of political orthodoxy, not only in terms of style and habits of mind, but of ideology as well. Both critics defended the doctrines of historical and dialectical materialism, both supported the Party's strategic line as well as specific policies and attitudes; both evaluated literature on the basis of its conformity to Communist doctrine; and both tended to minimize the formal aspects of literature and the achievements of modernism. Rahv's first published works appeared in 1932, of which the earliest effort, "The Literary Class War," was a precocious and bombastic piece, which found its way into the *New Masses*. It began with Rahv characterizing proletarian literature as not yet having reached maturity: what was needed, according to the young author, was "a determined extirpation of all liberal, reformistic elements within itself,"[5] a task only begun by the October Revolution. For Rahv, proletarian literature was to be sharply distinguished from what he called "social or protest literature," which was bourgeois and "based on the premises of idealism." Proletarian literature must instead "formulate a clear dialectico-materialistic world-view," and thus contribute to "tearing asunder the last vestigial piece of bourgeois-esthete fancy-drapery . . . proclaim[ing] its position to be that of irreconcilable class-antagonism."[6] Rahv maintained that criticism could provide the "theoretical scaffolding" (a term of Lenin's, used in *What Is to Be Done* to describe the function of the planned party newspaper) for a literary movement that was expanding but in dire need of direction. Since proletarian theory and practice were seen as too elementary to provide such direction, it had to come through a systematic critique of bourgeois culture.

Rahv hastened to the task by turning to the classical concept of catharsis, an idea he claimed was extremely fertile, but which an impotent bourgeois literature could no longer seriously entertain because it had lost the power to arouse strong emotion in the reader. It therefore fell to the proletariat to reappropriate the concept, and Rahv undertook to do so. He began by defining Aristotle's idea as "a process effecting a purgation of the emotions through pity and terror." The problem with this notion, Rahv claimed, "is that in a slave-owning society such as Athens it is a passive, static conception," and consequently functions as a "mental laxative for a cultured leisure class." On the other hand, without catharsis artistic creation "loses all significance, loses that high gravity which is the most characteristic function of art."[7] To resolve this contradiction, Rahv proposed a new kind of catharsis, one that would constitute a new "implicit form" for proletarian literature. He described the new form as

> likewise a purgation of the emotions, a cleansing, but altogether of a different genus: a *Cleansing through fire*. Applying the dynamic viewpoint of dialectics, a synthesizing third factor is added to Aristotlean *[sic]* pity and terror—and that is militancy, combativeness. The proletarian Katharsis is a release through action.[8]

He went on to detail the way the form would work in proletarian drama, which, if executed properly,

> inspires the spectator with pity as he identifies himself with the characters on the stage; he is terror-stricken with the horror of the workers' existence under capitalism; but these two emotions finally fused *[sic]* in the white heat of battle into a revolutionary deed, with the weapons of proletarian class-will in the hands of the masses. This is the vital katharsis by means of which the proletarian writer fecundates his art.[9]

Rahv's use of Aristotle's notions was based on an interest in Aristotle that was not altogether uncommon among leftist critics during the thirties. Kenneth Burke had already begun an exploration of literary meanings and categories, which he would later call "neo-Aristotelian," and of course in Germany Bertolt Brecht, writing under the general rubric of official Marxism, was undertaking the critique of catharsis that would provide the foundation for his revolutionary epic theater. The interest in Aristotle was, to some extent, encouraged by the fact that within orthodox discourse Aristotle was referred to more often than any non-Marxist philosopher except Hegel. No doubt this solicitude came from the knowledge that Marx and Engels in their occasional allusions to Aristotle showed an appreciation for what was in their view his mastery of dialectical thinking and his elevation of philosophy to the highest limits possible within the framework of slave society.[10] It was this connection, in fact, between slaveholding Athens and certain of Aristotle's ideas that provided Marxist-inspired critics with what they considered a case history of how material conditions determine ideas.

Generally, leftist versions of catharsis invariably rested upon such a class-determined analysis. Whether the application was Rahv's, Burke's, or Brecht's, the premise was that catharsis was to be understood as a response that served an elite whose desire was to maintain the stability of the community. Catharsis was thus seen as an essentially conservative idea insofar as it was thought to encourage the return of a momentarily agitated audience to a state of harmony with things as they were. For leftist critics the purgation of pity and fear essential to Aristotle's definition of tragedy did not at all leave the audience in a state of higher awareness, certainly not with regard to their social and political situation; rather, the sense of satiation and acceptance that resulted from such purgation mitigated against effective political action and was thus viewed as a form of political quietism.[11]

Where Rahv's critique of catharsis differed from Burke's and Brecht's was in his insistence on forcing catharsis to fit an orthodox model. In doing so, he did significant violence to the original, so much so that his version bore little if any resemblance to it. Rahv's version depended on a radical, and rather arbitrary, redefinition of pity and terror as responses to exclusively proletarian forms of suffering and oppression. The result was that what in Aristotle were psychologically complex responses, in Rahv were reduced to the mere sentiments of sorrow and indignation. And Rahv's third factor, that of militancy, was equally arbitrary, so that it and the proletarian victory that implicitly followed would seem to have committed him to a different genre altogether—either proletarian epic or comedy. This seems to have been a problem that simply eluded the young critic.

Rahv next applied a portion of Aristotle's definition of tragedy ("an action that is complete and whole and of a certain magnitude") to examples of modern "bourgeois" literature. Needless to say he found this literature distinctly lacking. He condemned Joyce's *Ulysses* for failing to exhibit genuine magnitude, referring to Joyce's magnitude as merely "the magnitude of death." He also denounced Joyce's epic for failing to constitute an organic whole, employing a conceit so elaborate as to possess its very own quality of completeness and magnitude: "*Ulysses*," he exclaimed, "jumps at life like a cat at a canary, but the housewife arrives in the nick of time, and the disgruntled cat jumps out of the window and slinks down to the dungheap behind the gashouse by the bank of a slimy river, where it sinks into a fetid dream."[12] Generalizing from the assessment of Joyce, Rahv proceeded to repudiate literary modernism as a whole: "[Here is] a literature that is a rancid hotchpotch of mystic subjective introvert speculation, arbitrary and hallucinatory."[13]

In the essay's penultimate section, Rahv elaborated on his belief that literary forms and tastes are dependent on the dramatic shifts of classes in struggle. Such shifts, he maintained, involve the birth and death of comprehensive ideologies, which in turn transform group as well as individual sensibilities and psychological states. Drawing on a work entitled *The Economic Theory of the Leisure Class* (not Veblen's work of similar title, but Nikolai Bukharin's), Rahv attributed "the present asociality, blind anarchic individualism, and amorality" to a new ideology originating in the turn-of-the-century shift from entrepreneurial to finance capitalism (Rahv had again borrowed from Lenin, this time applying the major thesis of *Imperialism, the Highest Stage of Capitalism*). Whereas for Veblen (here he did allude to the somewhat better known *Theory of the Leisure Class*) "conspicuous consumption" was a trait of the entrepreneurial strata, Rahv argued that the new breed of banking capitalist imitated, adopted, and eventually expanded conspicuous consumption, so that by the 1920s it became dominant within the culture at large, earning for itself the new designation "pure consumption."

The literary manifestation of pure consumption was modernism. Contrasting

Frank Norris to T. S. Eliot, writers whose work for Rahv embodied the essence of these two very different epochs, he identified the ways in which transformation of one to the other had actually damaged literature:

> The heroes of Frank Norris' novels of industrial life are captains of industry, alive and buoyant with the optimism and vigor of a class still relatively young: they are in constant touch with the actual process of production: they are not coupon-cutters. This is no longer true of the literature produced during the period of finance capitalism. . . . Consider this statement by T. S. Eliot: "The arts insist that a man shall dispose of all that he has, even of his family tree, and follow art alone. For they require that a man be not a member of a family or of a caste or of a party or of a coterie, but simply and solely himself." . . . This statement offers us a concentrated expression of the asocial psychology of pure consumption.[14]

This sort of asociality, according to Rahv, was brought to the heights of absurdity in the work of the Transition Group. This group revolved around the Parisian little magazine *transition*, founded and edited by the Franco-American Eugene Jolas. Its pages included the writing of American expatriates and European modernists who were interested in applying the recent findings of psychoanalysis to literature. Their hope was to develop a wholly new and vital language that could replace the dead language of the bourgeoisie, embodied in the pervasive style of descriptive naturalism. Among those who published in *transition* were Hart Crane, Gertrude Stein, and James Joyce, whose *Work in Progress* (the early version of *Finnegans Wake*) was perhaps the magazine's most important contribution. Other Americans whose work appeared in the magazine included Djuna Barnes, Kay Boyle, Malcolm Cowley, Allen Tate, Matthew Josephson, and Horace Gregory. Rahv's view was that these writers all took refuge "in a flight from consciousness" and futilely sought "a haven in the subconscious." Fortunately for the proletariat, however,

> The Revolution of the Word can be explained from a Marxist standpoint. The bourgeois ideologues would like to think that they too are revolutionists, so the word-game is initiated, and we are treated to the ludicrous spectacle of grown-up people indulging in the most fatuous and infantile delusions. These experiments with word-dismembering are of no more value than the well-known experiment of children with flies. . . . In the ultimate analysis the Revolution of the Word is a pretext for indulging in psychopathological orgies; it represents a deep-seated craving for the prenatal stage, for non-being. The vagaries of Jolas & Co. and the necromantic method of producing literature through the immaculate conception of automatic writing are quite proper end-phenomena of a dying class, and of a crumbling hegemony.[15]

By attacking the Transition Group, Rahv was not only attacking the more advanced linguistic experiments of the modernist movement, he was also implicitly calling into question the whole phenomenon of expatriation and its self-imposed alienation from American life. For Rahv, the new vanguard of the proletarian movement would replace the old expatriate avant-garde, which in his view had played itself out with the coming of the Depression.[16]

Nor in his essay did Rahv have much praise for the literary and critical breakthroughs in the United States during the 1920s. Mencken suffered from "extreme individualism" and an "organic inability to think in socio-economic terms." Though Sinclair Lewis, like Mencken, had attacked "standardized philistinism," he was nevertheless responsible for helping to bring about "the individualistic philistinism of the people in *The Sun Also Rises*." Hemingway's protagonists were further described as "effete hypochondriacs, cataleptic individualists—the human dust of finance capitalism." In sharp contrast to writers such as these, the one writer who could still describe the plutocrat "with relish" and "with a certain amount of health" was Booth Tarkington.

The essay's final section dealt with the sudden spread of Marxist criticism. Praising writers such as Edmund Wilson, Newton Arvin, and Granville Hicks for incorporating Marxism into their work, Rahv described the proper attitude of Marxists toward fellow travelers in general. In a passage full of irony given Rahv's own fate at the hands of the Party three years later, he recommended what he called a lenient attitude "more or less": "Everything should of course be done to facilitate a fellow-traveler's assimilation, but once it becomes clear that his bourgeois class-roots are too strong, he should be neatly and rapidly dispatched on the road back, because he will only bring confusion into the ranks of the real militants."[17]

That Rahv's article was taken seriously by Party critics was indicated not only by its appearance in the *New Masses*, one of the most prestigious publications of the Communist movement, but also by the fact that it elicited a lengthy response from A. B. Magil, an influential Party critic. Magil immediately set to the task of rebuking Rahv for his "scholastic . . . unhistorical . . . and a little too 'original' use of Aristotle's concepts." Terror and pity, Magil asserted, may be responses appropriate for a capitalist spectator who pities his fellow capitalists and is terrified at the prospect of such a fate for himself. As for the revolutionary proletariat, however, terror is an emotion "absolutely . . . incompatible with militant class action."[18] Rahv's privileging of tragedy also provoked Magil, who asked rhetorically whether "this talk about 'high gravity' . . . the usual bourgeois professorial chatter that can be heard in any freshwater *[sic]* college," was supposed to rule out Chaucer, Swift, Rabelais, Cervantes, Molière, and the like.

Elsewhere Magil endorsed Rahv's use of Bukharin's epochal cultural distinctions as an explanation for the rise of modernism. But significantly, he rather surprisingly considered Rahv guilty of leftism for his hypercritical attitude toward modernism. According to Magil, Rahv had failed to appreciate the adversarial impulse within modernism that accounted for its profound critique of bourgeois society. This caused Rahv to display a sectarian prejudice against fellow travelers and potential allies in general. Instead of encouraging such writers, Magil claimed that Rahv's consuming suspicion had caused him to become exceedingly condescending toward them. Comparing Rahv's views with those endorsed at the 1930 Soviet Kharkov Conference (attended by influential American Communist Party members Mike Gold, Joshua Kunitz, William Gropper, and Magil himself), Magil accused Rahv of "mak[ing] demands of writers which not even the Communist

Party makes of its members."[19] The upshot is that here was an important Party critic pointing out Rahv's excessive dogmatism, and this suggests that at this early juncture in his career the future editor of *Partisan Review* was capable of a rigidity that even Party officials during a notoriously sectarian period looked askance at.

Another early piece shows the depths of Rahv's sectarian hostility to modernism. "An Open Letter to Young Writers" was a short statement published in Jack Conroy's *Rebel Poet*, the precursor to the better-known proletarian little magazine, the *Anvil*. In his letter, Rahv presented his audience with a choice, which he claimed was being forced upon them by the accelerated process of class division and antagonism:

> Shall we take on the coloration of the bourgeois environment, mutilating ourselves, prostituting our creativeness in the service of a superannuated ruling class, or are we going to unfurl the banner of revolt, thus enhancing our spiritual strength by identifying ourselves with the only progressive class, the vigorous, youthful giant now stepping into the arena of battle, the class-conscious proletariat.[20]

Rahv alluded once again to the primitive state of the proletarian movement: "[O]ur literary ideology remains opaque . . . due to ignorance of the foundational principles of revolutionary dialectic of Marxism-Leninism *[sic]*." Growing turgid, he warned that, nonetheless, the alternative was "the 'nice and waterish diet' of emasculated, unsocial writing, perennially engaged in futilitarian introspection and constipated spiritual lucubrations."[21] Reaching a peak of vituperation, he dismissed at once modernist writing and some of its most accomplished practitioners, not to mention an entire literary genre:

> 'There's nothing more up that street' someone recently remarked about lyrical poetry. The same sentence can be applied to all literature that pretends or misguidedly endeavors to rise above class-interests or to ignore this central issue by means of thematically immuring itself in what it falsely conceives to be neutral subjects like love, for instance (Lawrence), expeditions into the subconscious (Joyce), superficial social satire (Huxley, Mencken), bestiality and crime (Faulkner).[22]

If there was anywhere a hint that Rahv's hostility to modernism was less than all-consuming, it was in his admission that despite everything, Joyce and Lawrence were geniuses, although he never identified the nature of their genius. He did, however, have something substantive to say about the consistency of another's genius—T. S. Eliot's. This came in Rahv's next published piece, which he gave the simple title "T. S. Eliot," and which appeared in *Fantasy*, a short-lived proletarian magazine of poetry. In spite of the essay's dogmatism and brevity—it extended a mere thirteen hundred words—as a piece of proletarian criticism it was atypical for the specificity of its critical response, if only because few such essays were devoted to a single modernist writer.

Rahv opened by making a small gesture toward intrinsic criticism, but he quickly followed this up with the admonition that "technical-esthetic criticism" would be relegated "to a secondary position" behind the "socio-cultural background."[23] As to the purpose of the essay, it would be frankly polemical, even derogative: "Eliot has exerted an immeasurable influence on contemporary writing. But in view of his recent affiliations with all that is reactionary and sterile on the modern scene, one cannot but deplore this influence."[24] In an effort to neutralize Eliot's influence, Rahv first charted a trajectory of the poet's career that described a peak followed by a precipitous decline. The peak was the *Poems*, first published in 1919 and expanded in 1923; *The Waste Land* and *Ash Wednesday* represented the decline. No mention was made of Eliot's criticism, even though one would expect the immensely influential *The Sacred Wood*, published in 1920, to have figured in Rahv's assessment. Perhaps he was not ready to acknowledge publicly his debt to Eliot's criticism at the time, a debt that was substantial, as I will make clear in the next chapter.

In evaluating the *Poems*, however, Rahv was generous in his praise, expressing more of it, and with more conviction, than many in the proletarian movement would likely have thought necessary:

> [It] immediately became apparent that something decisive had happened in American literature. . . . His was a sensibility enormously superior to the reigning "talents" in American poetry. . . . and of course the taste of most readers was still too primitive rightly to appreciate the visions and twilights, the dawns and languors, the tears and the ice of the *Poems*. Much like Baudelaire . . . Eliot set out, with masterly metric and superlative technical virtuosity, to tear off the mask of piety and virtue from the smug bourgeois of New England. The absolute rightness of each and every word, the exquisite twist and turn of the psychic visage concretized in verbal diabolics of extraordinary efficacy—all proclaimed the master. . . . Here was . . . the intense dynamism of profound revolt.[25]

But if in Rahv's solicitude toward Eliot's early work we might be tempted to see any serious resistance to the formulaic treatment of Eliot found in most quarters of the proletarian movement, the following passage will quickly disabuse us, for Rahv's enthusiastic verbal flights crash-landed on *The Waste Land*:

> But . . . something happened to Eliot which cut short his creative progress and shunted him back to ally himself with the very elements of society he had satirized in his first work. The ancestral complex, the calvinist past, and the false evasions of the classical bourgeoisie had reasserted themselves. . . . At the present time Eliot must be discounted as a positive force in literature. His place is definitely with the retarders of the revolutionary urge.[26]

In *The Waste Land* Rahv detected what he felt were the telltale signs of Eliot's decadence. Quoting Ernst Robert Curtius, Rahv claims the work "'Breathes that despair, which is also the atmosphere of Proust and Valery': the end-of-the-world mood permeates its pages." Although Rahv admitted that in certain respects the work was more impressive than the *Poems*—"insofar as it condenses the values of the *Poems* it was a step forward"—this was not sufficient reason for him to alter his final judgment:

> The social realism evident in the *Poems* is here largely missing. . . . The Symbolist tradition is merged with that of Donne and Webster, a singularly tasteless mixture. In

> *The Waste Land* Eliot has slipped already, and fallen into the swamp of mysticism and scholasticism, a double damnation. Mysticism . . . robs the poem of the necessary precision, and the sinister obscurity, the arbitrary choice of allusion and evocation, render it a closed book to all but a few encyclopedic intellects. . . . Essentially, this poem represents a flight from reality, not a solution or even an authentic examination of contemporary time, but an obfuscation.[27]

As for *Ash Wednesday*, Eliot's next major poem, Rahv viewed it as an unmitigated failure and the nadir of Eliot's achievement:

> Throughout the poem Eliot refers to himself as the "aged eagle"; alas, it is all too evident that this particular eagle will fly no more. A poetry of aristocratic moods and ascetic ideas is neither possible nor desirable in an era of plebian revolt and materialist dynamics.[28]

For Rahv, Eliot's fall from grace had been swift and absolute. The poet's work demonstrated above all else the essential features of decadence, which defined the culture of advanced capitalism, and Rahv thereupon pronounced Eliot guilty of mysticism, obscurity, despair, and individualism. Such a verdict clearly placed Rahv within the bounds of official Marxism. By claiming that the early, subversive impulses of Eliot's poetry had succumbed entirely to the moribund culture of capitalism, he was reaffirming the quasi-official position on modernism that had emerged among the more doctrinaire critics of the Party and the Comintern.

Rahv's scorn for Eliot, Joyce, Hemingway, Mencken, Lewis, Jolas; his strict demarcation between bourgeois and proletarian culture; his intolerance of fellow travelers who did not toe the line; his general disdain not only for bourgeois but for all pre-Marxist culture, all was expressed within a doctrinaire Marxist framework. Within this framework, literature was judged according to the disposition of classes within the dominant productive mode, the particular needs of the proletariat, and the interests of socialist revolution in general. All specifically literary observations were subordinated to this larger schema.

III

Turning now to the earliest critical efforts of William Phillips, we are immediately confronted by the simultaneous publication in 1933 of two inaugural pieces that were in nearly every respect antithetical.[29] One of these was a review of Ortega y Gasset's *Revolt of the Masses*, published in the *Communist*, the theoretical journal of the Communist Party. The other was "Categories for Criticism," published in a highbrow literary journal, the *Symposium*. Whereas the review of Ortega's classic was little more than an exercise in formulaic criticism, "Categories for Criticism" was clearly a product of Phillips' academic interest in formalist criticism. Significantly it attempted to reconcile that discourse with a Marxist one.

Having a review of an important book published in the theoretical journal of the Communist Party was an extraordinary accomplishment for a young academic with no political experience. Surely it signaled a very bright future for Phillips as a Party critic, although he claims never to have joined the Party. "Class-ical Culture" began with a passage that hardly promised to be a well-reasoned assessment of Ortega's influential volume:

> In *The Revolt of the Masses* José Ortega y Gasset has rounded out most of the theoretical props of capitalism with a number of confusions and slanders of his own. It would require a volume to refute all of his fallacies and distortions, but this is made unnecessary by the fact that all of these fallacies are unified by his basic attitudes: an apology for capitalism and a condemnation of the revolutionary movement.[30]

Phillips promptly concluded that Ortega y Gasset's notable book was actually worse than an apology for capitalism. Applying the notorious strategic line of the Comintern's third period, he suggested that Ortega was actually a fascist: "The defenses of capitalism, common to fascist and social democratic theory, are assimilated into the vocabulary and turns of thought of a modern school of literary criticism (of which Ortega y Gasset is a leading representative)."[31] Elsewhere, Phillips was even more explicit:

> In this new program he calls for respect and enthusiastic support of "qualified" leaders by the people. This suavely phrased program is a thin veil for a plea for fascist dictatorship. . . . The book is symptomatic of the decay of bourgeois culture and of the attempt to enroll the forces of intellectual reaction under the ideology of fascism.[32]

Not only was the political line of the Party in evidence in Phillips's review; some of the favorite rhetorical tropes of doctrinaire Marxism were used in the article as well. He did not hesitate, for example, to employ the oft-used technique of guilt by association in order to condemn the book: "As might be expected, the book has been recommended by Wall Street and two of its chief agencies . . . the *Wall Street Journal* [and] the *New York Times*. . . . And the Catholic Church blessed the book by including it in its white list."[33] Another common gambit of doctrinaire Marxism, that of quoting from Marx as if his texts were the final, infallible arbitrator of all disputes, was also employed by Phillips against the Spanish philosopher. Quoting him to the effect that the rise of "the mass mind" (Ortega's phrase) will precipitate a return to barbarism, Phillips concluded that Ortega's observation was "supported by the trite idealist theory that culture is the product of great minds." But "the reactionary character of this idea is made evident by Marx's analysis of the relation of culture . . . to society." Whereupon Phillips the exegete maneuvered into position this weighty quote from Marx:

> "The sum total of these relations of production constitutes the economic structure of society—the real foundation, on which rise legal and political superstructures. . . . The mode of production in material life determines the general character of the social, political, and spiritual processes of life. It is not the consciousness of men that determines their existence, but, on the contrary, their social existence determines their consciousness."

As for Ortega's claim that "industrial technique" will disappear "when the masses revolt," Phillips met the challenge with the Marxist coup de grace:

> "At a certain stage of their development, the material forces of production in society come in conflict with the existing relations of production. . . . From forms of develop-

ment of the forces of production these relations turn into their fetters. Then comes the period of social revolution."[34]

Still others of Phillips's remarks concerning American critics demonstrated his fully doctrinaire approach. He compared *The Revolt of the Masses* to Joseph Wood Krutch's "reactionary, romantic criticism [which is] rapidly flowering into fascist theory." He referred to Irving Babbitt and Paul Elmer More as critics who provided "the core for fascist doctrine." What these critics and Ortega have forgotten, Phillips added in a characteristically upbeat conclusion, was that "'the revolt of the masses' is creating a revolt of many intellectuals against the culture which produces such reactionary tripe as Ortega y Gasset is trying to market, and that they are allying themselves in ever larger numbers with the workers."[35] It is ironic that Phillips saw the matter from the point of view of the intellectual given his own very obvious uncertainty about the value of intellectual inquiry in the first place.

In stark contrast to this dogmatic piece, however, was "Categories for Criticism," a sophisticated, if a bit abstract, attempt to theorize some of the problems faced by literary critics at the time. Actually more can be said for this remarkably prescient piece, which in some ways resembles contemporary examples of discourse analysis.[36] In addition, it demonstrates certain novel possibilities for Marxist criticism, which bear some resemblance to excursions in this direction by new historians and cultural materialists.

The essay addressed the central issue for critics during the 1920s and early 1930s, namely, how to lift criticism out of the morass of arbitrary opinion and personal impression onto a level where verifiable standards might prevail. Phillips wished somehow to translate the admittedly highly individual activity of criticism into something more collective, more accountable, and, although he did not use the word, more professional. He sought a model for the desired systematicity of criticism in the emerging ideas of a human science. Like I. A. Richards, a critic whose influence on Phillips was obvious, he set out to redefine criticism in terms that would make it responsive to a scientific mode of discourse. At the same time, however, he was influenced by modernism's antipositivism and especially by Sidney Hook's pragmatic Marxism, which Hook himself had shaped under the influence of his mentor John Dewey. These combined influences meant that in forging a more "objective" criticism, Phillips would have to avoid any reliance upon foundational, absolute truths. The difficulty would be to develop a notion of objectivity that was socially constituted rather than transcendent, and yet not so dependent on immediate experience as to reflect only the perceptions of the individual or the contingencies of the historical moment.

As a way into his problem Phillips briefly traced the notion of the "category" through Aristotle, Kant, Richards, Durkheim, Levy-Bruhl, Hulme, and Maritain. He extended Hulme's concept of the category, by which Hulme meant the web of ideas and emotions of individual experience within a given context, into the broader social dimension of "attitudes that form in the shift and flow of the history of ideas."[37] His concern was with the divisions, interactions, limits, and powers of these categories, which he emphasized were made and not found. In his description of the play of categories, he suggests that together they comprise a kind of discursive force field:

> Categories represent the divisions of methods and ideas in reason and in life. In science they are the fulcra of analysis; in the attitudes and judgments of intelligent men and in such independent fields of inquiry as criticism, the categories are constructs on the varying emphases and relevancies normal to life (thought). Like a centrifuge, each emphasis gathers its relevant material. The history of ideas is the record of the tension between, the interaction upon, the merging of these categories.[38]

We note the emphasis on the dynamics of power here: categories are imbued with forces that repel, attract, interact, hold in tension. Given the unpredictable and indeterminate character of this flux, and given that categories necessarily arise through their "forced separation from the waywardness of ideas,"[39] the central problem as Phillips saw it was somehow "to standardize categories—to bridge units of discourse that appear to be discrete."[40] With regard to the specific category of criticism, long considered essentially individual and subjective, he suggested something like an "interpretive community" to organize individual responses. "Criticism," he observed, "is achieved by a rich mind sensitive to a medium; but it is significant, intelligible and timely, insofar as it stems from the most intelligent groups."[41] Such a grouping would contribute to defining criticism's relation to other fields, and would both encourage and monitor theoretical advances.

Perhaps what is most interesting about Phillips's essay is the degree to which its own location outside the official discourse confirms its conclusions. Free from the ideological pressures prevailing within that discourse, Phillips proceeded to rethink historical materialism so as to purge it of determinist impulses that contradicted the principle of indeterminacy on which he premised his discussion. The result was a passage of impressive shrewdness, in which historical materialism was reconceived so that it could respond more fully to the reciprocal effect between active human agents and material conditions:

> Marxism predicts a range for future events against other ranges, and even this it does because a potential future event (unsuccessful proletarian revolution) is an end term in a causal series based on the historical analysis. The key values of the causal series are checked in that they hinge on present needs of the people. These needs and the actions that follow are shown to be such as to unfold this causal series toward the end term. In this way judgments on the basis of a future category are made, only because the future category is dialectically related to present ones, and because effective action to attain the future category is implicit in the judgments.[42]

Granted that the metaphysical status of this "causal series" is never thoroughly exposed here—"unfolding" assumes a preestablished shape awaiting favorable conditions to emerge. Nonetheless, Phillips has gone some distance toward pointing out that the "motor" of historical development—to use a favorite metaphor of doctrinaire Marxism—runs by the internal combustion of acting and judging agents who are conditioned by a particular set of historical variables. It is this reciprocal relationship between circumstance and human agency that accounts for change, not iron laws of historical development, according to which social development proceeds along a predetermined course laid out by the nature of productive forces and class contradictions.[43]

All of this, Phillips reminded his readers, was not to say that history might not be developing in a certain, even a prescribed direction, but that in whatever direction history may be heading, it must be continually ratified by people acting "effectively"—that is, according to judgments that are based on "present needs of the people." Although not entirely inconsistent with certain orthodox formulations that emphasized "the subjective factor" in history, Phillips went further. He had in mind some notion of a rational, self-created, and self-directed interrelationship of consciousness and action which is necessary for good judgment, for praxis. Literary criticism, as a rational, organized, and public discourse, would play an important role in providing for such judgment.

It could do so, suggested Phillips, by insisting upon a Janus-faced regard for history and human need; that is, it must manage to be both "forward-looking and backward seeing," assessing and balancing present and future needs with historical possibility so that judgments might be made to correspond to those needs.[44] His hope was that this critical correspondence would conform to what he called "the principle of relevance." By this he meant not topicality or immediate applicability but rather the adjustment of critical ideas both to the text *and* the overlapping context. Put another way, he wished "to construct a system of articulation" between the related but discrete categories of literature and history on the one hand and criticism on the other. In this model criticism functioned as a mediating field of inquiry among disparate personal and social experiences. Phillips seemed to be advocating a kind of capacious specialization if you will—a kind of cultural criticism:

> Successive correlations would build . . . toward a unified and consistent body of judgments and values, which would integrate more and more of the emotional and intellectual background. . . .
>
> On the basis of this analysis the function of the critic would be to translate the perceptions of art-works into current or new philosophical symbols, and to create thereby a system of judgments and values which shall retain the emotive unity of the art-work while subjecting its meaning to examination by all relevant ideas in terms of the philosophical symbols. Implicit in this system is the mechanism for criticizing and evaluating art.[45]

Phillips's interesting essay included a rethinking of conventional approaches to intellectual history, a reformulation of historical materialism, and an attempt to theorize a connection between historical and textual criticism, all of which were concerns central to the critical project of *Partisan Review*. But I have gone on at length here not only because I believe the essay to be anticipatory in this regard. Equally important is the fact that the appeal for discrete, rigorous, and interrelated categories with sources in diverse traditions had to be made outside the discourse of orthodox Marxism. The potential danger of this division becomes all the more apparent when we realize it was a division that existed inside, as it were, as much as outside critics like Phillips and Rahv. That is to say, critics interested in turning Marxism into a flexible tool of analysis were not only forced to write outside the orthodox movement, but in important respects they were forced to write against themselves. As we shall see, although *Partisan Review* made major contributions toward mending some of the fractures I have identified, once Phillips and Rahv left the orthodox movement, some old wounds, internal as well as external, reopened. Thus, for example, in spite of their courageous, trenchant, and justly celebrated criticism of orthodoxy following their 1937 split, the polarized discourse of the left discouraged them from distinguishing between the leadership of the Communist Party and their rank and file, which included many writers and artists who themselves had been resisting Party functionaries and Party formulas all along. If the orthodox left was tragically unable to provide its members with a diverse and critical culture able to support novel ideas and experiences, neither was it quite the nightmarish, totalitarian monolith Phillips and Rahv, among others, later maintained it was.[46] But no doubt habits acquired within a political discourse disfigured by conflict, coercion, and dogmatism, had much to do with their misjudgment.

2

Partisan Review's *Eliotic Leftism: 1934–1936*

Aesthetic criticism, if carried far enough, inevitably becomes social criticism, since the act of perception extends through the work of art to its milieu.

F. O. MATTHIESSEN

I

Had the twentieth-century likeness of Meyer Schapiro's urban promenader[1] walked the streets of New York in the winter of 1934, he might have happened upon one of the increasing number of newsstands in the city to carry leftist publications. There he might have noticed among the assortment of magazines the first issue of *Partisan Review*, then a tiny proletarian "little" magazine and only later, of course, the powerhouse of literary and cultural criticism of the 1940s through the 1960s. Had this wanderer not been possessed of some knowledge of leftist political circles, he would undoubtedly have been confused about this new magazine's outlook, for all indications were that it was assiduously avant-garde. In its physical makeup and design, it resembled a makeshift product of the Soviet constructivist style of the 1920s. Its cover was of poor quality coarse paper, pale white in color. Rising across the cover, outlined in pencil-thin blue ink and dynamically aslant to the left were the three-dimensional block letters "JRC," for the little-known John Reed Club. Superimposed over these was the magazine's name, urgently proclaimed as a banner headline. At the lower right the contents and contributors were listed in smaller typeface. Taking in these details, the wanderer would no doubt have concluded that here was another "highbrow" magazine offering difficult, experimental writing–had the term by then gained wide currency, he might have assumed he was looking at another "modernist" little magazine. Nor would the magazine's subtitle, "A Bi-Monthly of Revolutionary

Literature," have revealed that the new publication considered itself revolutionary in the literary *and* political sense, if only because during the first several decades of the twentieth century many magazines proffered themselves as revolutionary.

But any misconception about its outlook would have been immediately corrected had he bothered to read the magazine's opening editorial statement, for here its affiliation with the Communist Party and the proletarian literary movement was openly declared. The statement began by alluding on the one hand to the worldwide crisis of capitalism and on the other to the example of the Soviet Union, which was said to be embarked headlong on the epochal task of socialist construction. The Soviet Union was furthermore said to be in the vanguard of a worldwide revolutionary movement, of which the project to create a new and revolutionary art in the United States was a part. It was within this political context of intensified class struggle that the editors addressed their audience on the role of their new magazine:

> PARTISAN REVIEW . . . will publish the best creative work of its members as well as of non-members who share the literary aims of the John Reed Club.
>
> We propose to concentrate on creative and critical literature, but we shall maintain a definite viewpoint—that of the revolutionary working class. Through our specific literary medium we shall participate in the struggle of the workers and sincere intellectuals against imperialist war, fascism, national and racial oppression, and for the abolition of the system which breeds these evils. The defense of the Soviet Union is one of our principal tasks.[2]

The editors next turned to cultural matters:

> We shall combat not only the decadent culture of the exploiting classes but also the debilitating liberalism which at times seeps into our writers through the pressure of class-alien forces. Nor shall we forget to keep our own house in order. We shall resist every attempt to cripple our literature by narrow-minded, sectarian theories and practices.[3]

By now our flaneur would no doubt have long since left the scene, a fact that would not have fazed the new magazine's chief editors William Phillips and Philip Rahv in the least, for they intended their magazine for a proletarian audience, or at least a proletarianized audience, of which *les flâneurs* most definitely were not a part. Thus he could not have known that hidden within the formulaic language of Stalinist orthodoxy—in phrases such as "We propose to concentrate on creative and critical literature," "our specific literary medium," and "we shall resist . . . narrow-minded sectarian theories and practices"—was the beginning of a new critical project for the American literary left, one that has left its mark not only on the left but on American culture and politics as a whole.

If we attend to some of the details of the editorial statement, we will see just how completely Phillips and Rahv had tied themselves ideologically to the Party. First, all of the political ingredients of orthodoxy were manifested in the familiar litany of causes that adorned the statement: opposition to war, fascism, imperialism, and racism and support for the Soviet Union. The fact that these causes were so remote from the literary aims of the magazine showed the extent to which the editors were willing to saddle themselves with the full gamut of the Party's programmatic demands. Second, in the cultural sphere the editors adopted the standard assessment of bourgeois culture as, on the whole, decadent, and they pledged themselves to combat it by vigilantly guarding against its nefarious effects on susceptible progressive writers. In giving priority to the dangers of liberalism over those of sectarianism, they duly reflected the Party and the Comintern strategic line during the so-called third period, when capitulation to fascism was seen as a greater danger than fascism itself.[4]

Nonetheless, though the magazine was linked to the broad political program of the Party, the editors referred to their publication as a "specific literary medium." The ambiguity of this phrase deserves some attention. If their magazine was thought of as a specific literary medium, how then would it also "participate" in the political struggles alluded to in their extensive list of demands? Here we must keep in mind that despite the divisions between orthodox and "bourgeois" discourse, which we considered in the previous chapter, orthodoxy was not monolithic. We have already seen where A. B. Magil, a Party critic, assailed Rahv for his dogmatism[5]—this ought to warn us not to exaggerate the vigor with which Party functionaries pursued absolute conformity to doctrine within the cultural sphere. In fact, the degree of autonomy afforded literary magazines in the pursuit of literary matters was greater than is often thought, so long as Caesar was paid his due politically. What the editorial statement did *not* say was that *Partisan Review* or proletarian literature in general had to espouse specific political positions in an overt fashion. Rather, it said something quite different: that "through" their specific literary medium—that is, through publishing the movements' "best creative work"—the magazine would be a participant in political struggle. This distinction is important; if it is not made, we cannot account for the fact that during the first two and a half years of the early *Partisan Review*, Phillips and Rahv were able to pay almost exclusive attention to their "specifically literary" project while maintaining their standing within the overall political movement. Even as their criticism grew increasingly subtle and complex, they always managed to present themselves and their magazine as contributing to a political and ideological vanguard. This was not only because their idea of literature encompassed politics and ideology; it was also the result of a movement that made room, however reluctantly or cynically, for *Partisan Review*'s increasingly heterodox literary and critical project. This said, we shall see how swiftly and unconditionally *Partisan Review*'s autonomy was revoked as soon as political differences between its editors and the Party arose.

Finally, before getting to the main subject of this chapter—the literary program of the early *Partisan Review*—a word about the magazine's format, contents, and editorial composition is in order. The first issue was sixty-four pages in length, consisting of the opening editorial, five pieces of fiction, six poems, four book reviews, and a critical essay. In the course of the next two and a half years, which saw a total of fifteen issues (of which the seventh—the April–May 1935 issue—was brought out in combination with *Dynamo*, and the last six in combination with the *Anvil*), the format would shift, most notably in the direction of more critical essays and book reviews.[6] It was the emphasis on literary criticism—both theoretical and applied—that distinguished *Partisan Review* from the other proletarian little magazines, all of which focused almost exclusively on bringing out original proletarian poetry and fiction. Fully half of the average issue of *Partisan Review* was devoted to criticism, far more than other movement magazines. Aside from Phillips and Rahv themselves, who wrote a good portion of this criticism, the contributors were Party members or, probably more often, fellow travelers, and included critics such as Waldo Frank, Newton Arvin, Isidor Schneider, Harold Rosenberg, Alan Calmer (then an editor of *New Masses*), Edwin Seaver, Edwin Rolfe, and Granville Hicks. In addition, a significant number of European novelists, critics, and "men of letters" were published, among them André Malraux, André Gide, Georg Lukács, Ilya Ehrenburg, and Louis Aragon. As for the early *Partisan Review*'s offerings of poetry and fiction, nothing was particularly distinguished. As James Gilbert has noted, most of the contributors were already established as leftist writers, but were minor writers nonetheless. The best of the material contributed by Americans consisted of fairly inconsequential pieces by Tillie Olsen (Lerner), James Farrell, Nelson Algren, Josephine Herbst, Kenneth Fearing, and Richard Wright.[7]

The editorial board for the first issue, besides Phillips and Rahv, included Nathan Adler, Edward Dahlberg, Joseph Freeman, Sender Garlin, and Edwin Rolfe. All of the board members were active in the proletarian movement, and if they were not Party members they nonetheless maintained close ties. Phillips and Rahv have said that the board at first consisted mostly of "writers rewarded for political loyalty or paid off for their literary prestige by an honorary appointment to our editorial board." This has been only partially contradicted by Joseph Freeman, who has stated that along with Phillips and Rahv, he, Mike Gold, and Granville Hicks were active editors early on, but that they left to do other work, leaving Phillips and Rahv in charge.[8] In any event, the composition of the editorial board underwent continual change throughout the first two years of *Partisan Review*'s life, the only real continuity being provided by Phillips's and Rahv's presence. It was they, of course, who stayed, eventually fashioning a magazine that, to a significant degree, bore their particular intellectual and temperamental stamp.

Phillips's and Rahv's literary criticism of the 1934–1936 period naturally reflected their organizational and political obligations—but it did so in some unexpected ways. Indeed, what is most salient about their criticism is that, under the rubric of a proletarian aesthetic, they managed to combine a wide range of critical concepts from diverse sources. Phillips and Rahv were interested in redrawing the boundaries that then contained cultural discourse on the left. Their critical project consisted of the following elements: (1) a challenge to key notions of Stalinist orthodoxy and eventually a rigorous reevaluation of Marxism as a whole; (2) the attainment of an unprecedented degree of autonomy, tolerance, and rigor for literature and criticism—one could say intellectual work in general—within left discourse; and (3) the creation of a compelling, though highly unstable union between modernism and political radicalism. This union began even as *Partisan Review* championed the cause of proletarian literature, and, as we shall see, it was defended most strenuously during the late 1930s and early 1940s as the magazine shifted its allegiance from the Communist Party—which itself had abandoned the idea of proletarian literature in 1935—to "independent" socialism. Following several years during which the *Partisan* group moved toward Trotskyism, the difficulties for radicalism posed by the war led them to moderate and in some cases repudiate political radicalism in the postwar period. By the 1950s, of course, *Partisan Review*'s criticism had attained its greatest influence, contributing powerfully to modernism's canonization and the shape of American culture generally. By then the politics of the New York intellectuals' criticism was either liberal or conservative anticommunism, with some notable exceptions.[9]

In what follows I shall examine the changes in critical concepts, values and language that constituted *Partisan Review*'s subtle critique of doctrinaire Marxist criticism during the mid-1930s. I will look specifically at the appropriation of key concepts from T. S. Eliot's early criticism by Phillips and Rahv, who were attempting to turn their fledgling "little" magazine into a forceful critical and experimental center of the proletarian literary movement. Phillips's, Rahv's and the New York intellectuals' connection to Eliot is of course no secret: every historian of this circle has commented on it, and Phillips in his memoirs explicitly states that Eliot, and especially *The Sacred Wood*, served as his earliest introduction to modernism and "the new, complex questions of criticism."[10] Although Eliot was not the only major influence on Phillips and Rahv—Edmund Wilson, Van Wyck Brooks, and James T. Farrell were also highly influential[11]—nonetheless for them Eliot was the critic to be reckoned with, as he was for so many writers and critics from the 1920s through the midcentury. Yet it remains to explore in detail just how systematic Phillips's and Rahv's appropriation was, and what the results were for their renewal of Marxist criticism during the first two years of *Partisan Review*'s long life.[12]

To readers familiar with the many attacks on Eliot's crypto-fascism, his anti-Semitism, his misogyny, his traditionalism, and his mandarinism, attacks that have contributed to a discernible decline in his reputation, it may come as a surprise that his contributions had anything to do with renewing left criticism. But the politics of Eliot's early criticism, and by implication that of many modernist texts, have never been as transparent as many make them out to be. In truth their ideological valences are multiple and shifting, and although reactionary ideology plays no small part in helping to shape their politics, it is not necessarily determinate. A text's politics depends on a complex interplay between embedded ideological configurations and possibilities on the one hand, and groups of readers responding both to these textual pressures and the pressures and limits of an ensemble of social relations on the other. No text, I will be implicitly arguing, contains its own determinate ideological identity; it contains only dispositions, which are either developed or altered according to the desires and interests of contingent readers.

II

When *Partisan Review* made its appearance in the winter of 1934, its critical project was at best an incipient one. Within a matter of months, however, the editors had begun their attack on doctrinaire literature and criticism—what they called "literary leftism." In response to critical positions that demanded that literature make explicit appeals for socialist revolution, or present a dialectical-materialist world outlook, or render working-class life in a favorable light,[13] Phillips and Rahv offered a comprehensive reassessment of a range of critical problems and terms. They did so in numerous exploratory essay-articles on aesthetics and form, several of which they coauthored. In these essays they elaborated their most important critical ideas, and it was here also that Eliot's influence was most pronounced. "The imaginative assimilation of political content," "literature as a way of living and seeing," "specific content," form as a "mode of perception"—these were the critical phrases that denoted a major departure from doctrinaire Marxism. But the idea that informed each of them—Phillips referred to it as the "hub" of the problem of literature's relation to politics—was the idea of *sensibility*, also taken from Eliot.

The term "sensibility" has had a long history within the non-Marxist critical tradition, most notably as a term signifying literary effeminacy, as Janet Todd has amply documented with regard to the eighteenth century.[14] Eliot, of course, like Phillips and Rahv after him, chose to bury this particular meaning when he revived the term for twentieth-century criticism, thereby eschewing any attempt at feminization. In fact all three men, by suppressing this background and attempting to universalize the term, were significantly less thorough in their revisions than they might have been. As for American or European Marxism, neither made any serious use of the term before Phillips and Rahv employed it. Indeed the term seems to have held no special importance for either critic in their pre-*Partisan* criticism; only with Phillips's 1934 essay "Sensibility and Modern Poetry" did it acquire its strategic place in his rethinking of Marxist criticism. Phillips announced as much by concluding his essay with the unambiguous appeal for the proletarian movement to adopt as its slogan "Let sensibility take its course."[15] By giving the term such prominence he meant to counterpose it to the terms of the dominant reflectionist epistemology and corresponding social realist aesthetic of doctrinaire Marxism. The idea of sensibility represented a complex, mediated version of subjectivity in place of the reified antithesis of behaviorism and a spurious voluntarism that beset the Comintern's analysis of the "subjective factor" during its sectarian third period. Much was at stake according to Phillips: understanding the relationship between a poet's beliefs and the merits of his or her poetry, assessing the textual interaction between form and content, and responding productively to "bourgeois" literature, especially as it impinged on the shaping of the new proletarian literature.

As Eliot had before him, Phillips alternately referred to three types or manifestations of sensibility: that of the writer, that of the literary work or form, and that of a tradition or historical period. These corresponded, respectively, to different aspects of the literary enterprise: the creative process, the individual work's effect, and the development of literary movements based on shared sensibilities. As for the writer's sensibility, Phillips described it in Eliotic fashion as a medium or solvent by which specific opinions, programs, or ideologies are transformed by the writer into something more indirect, subtle, and affective: "We do not respond directly to a poet's beliefs for the simple reason that unless these beliefs are dissolved in his sensibility, the poet has not created a poem."[16] As a faculty that amalgamates both cognitive and affective experience, the writer's sensibility is the most important element in the creative process, not to be equated with either class consciousness, stressed by doctrinaire Marxism, or depth of feeling, emphasized by what he took to be the romantic view. Literary production demands a receptivity to all forms of experience, Phillips claimed, and it is precisely the sensibility that is able to assimilate these disparate forms. Without elaborating, he noted that in its turn sensibility is conditioned by a range of historical and biographical factors.

Like a writer's sensibility, a work's sensibility cannot be reduced to a single dimension, whether it be ideological or personal. This, I believe, is the idea behind the atmospheric passage "what we recognize as the quality of any poem is its timbre and its condensation of moods and attitudes which float in the world about us."[17] Thus, *The Bridge* by Hart Crane possesses "the sense of . . . the machine," and *The Waste Land* "conveys . . . a feeling of restlessness, tension, and futility" by attempting to "span our cultural traditions."[18] Significantly, Phillips was not concerned with Eliot's or Crane's politics, in or out of their poetry. Rather, as he indicated, his interest was with the feel of the poems, but the feel as it sublates the voice and the author through immersion in what would seem to be epochal subjectivities shaped by capitalism and class conflict.

Finally, groups of poems can share a sensibility if history seems to have the same feel to them. Thus Phillips referred to Auden, Lewis, Spender, and Horace Gregory as the "transition group" because their common "humanist" sensibility—their "sensibility of choice"—placed them in the camp neither of the bourgeoisie nor of the proletariat but rather "in transit" between traditional and proletarian verse. Employing language from Eliot, Phillips claimed this sensibility was the "poetic correlative of the intellectual's sense of conflict and the necessity of taking a position."[19]

The notion of literature "as a way of living and seeing" was another means of pointing out the irreducibility of literature to any single realm of experience, whether purely subjective or purely ideational. The phrase was first used on the occasion of Phillips's review of Henry Hazlitt's *Anatomy of Criticism*. Here Phillips took exception to Hazlitt's dismissal of proletarian literature because, in Hazlitt's words, it "attempts to enforce a specific article in the conventional moral code [and] bring about a specific reform." But Phillips effectively disarmed his opponent by defining proletarian writing in a rather unexpected way: "These are certainly not the kind of beliefs which Marxists advocate for literature. Proletarian literature does not 'enforce a specific article,'; it introduces a new way of living and seeing into literature. It does not enforce the new view; it embodies it."[20] "A new way of living and seeing" was a far cry from "the dialectical-materialist point of view"—it emphasized the necessary adjustment of subjective life to the new revolutionary situation, and, like the idea of sensibility, involved perceptions and values more than beliefs. Its particular resonance was to stress the necessary transformation of the entire individual as part of the transformation of a whole way of life.

The idea of literature as "a new way of living and seeing" encompassed many categories of experience that literature as ideology or literature as imagination at the time excluded: perceptions, attitudes, habits of mind, moods. Although Phillips and Rahv left out of their discussions of this period any reference to the unconscious, it should nonetheless be kept in mind that they assigned to literature's proper province a much wider range of experience than that encouraged by more doctrinaire critics. Thus literature might incorporate discursive, propositional, or even topical modes—as the orthodox critics tended to emphasize—but it should not be *expected* to do so. There was in fact an array of rhetorical modes, conventions, and forms by which writers might convey experience. The important thing was that these experiences be communicated so that their social significances are experienced subjectively by the reader. Felt experience must carry more than personal signification—it must bring the reader face to face with broader social contradictions. Viewing literature as a way of living and seeing was at once to call attention to the web of relations that connect literature to a historically constituted "whole way of life" and also to affirm the necessarily subjective way in which these connections are most effectively depicted. Here Phillips's and Rahv's views generally correspond with those of Raymond Williams concerning the individual/social nexus at the heart of culture. All three attempt to describe the play of personal and social, subjective and objective elements that go into the active making and reproduction of culture. Just as for Phillips and Rahv the notion of sensibility was meant to retain the quality of constitutive experience at both the individual and collective levels, so too, as I have indicated, Williams's concept of "structures of feeling" is meant to include the feelings, values, moods, and judgments of those actively engaged in making culture in the aggregate. All three insist that the affective dimension of collective life is part of the material dimension of collective life.

The idea of "specific content" was introduced by Phillips and Rahv in "Criticism," their second coauthored essay. Here they addressed the fact that in many works, especially modernist works, there existed little if any *direct* reflection of ruling class ideology. They made their point with reference to Faulkner's *Sanctuary*, and they based in on a distinction between specific content and ideology:

> Evidently the book does not present the author's "ideology" in the sense that it does not tell us about his general political opinions, his economic beliefs, his attitude to all the questions stirring the South; in fact, it does not even give us anything like a complete view of his esthetic opinions. What it does give us is not ideology directly, but *specific content* in the shape of attitudes toward character, painting of moods, patterns of action, and a variety of sensory and psychological insights. The patterns naturally contain within themselves the implications of a large world-view (ideology), which the critic may deduce. But the point must be made that *this specific content is not*

> *identical with any immediately recognizable reactionary or progressive non-literary program operating in the South today.*[21]

The editors concluded that it is "confusing" to attribute to a work a general ideology "with all its philosophical and political connotations." Its specific content is very much the product of an individual sensibility, historically conditioned to be sure, but exceptional in ways that matter to the reader and certainly to the critic.

Despite the fact that Phillips and Rahv, like nearly all leftists during the 1930s, accepted the narrow definition of ideology as an explicit, socially powerful system of belief, and despite their highly qualified and evasive definitions—"*immediately* recognizable"; "*all* its philosophical and political connotations" (it is unclear whether the authors would make the same conclusions about ideology minus the qualifications)—their remarks nonetheless spoke to the need for leftists to complicate the processes of literary analysis and valuation. As remains the case today, leftists were sometimes given to measuring the worth of a text largely on the basis of political conviction. As indicated, proletarian critics might demand that an author portray working-class life in a favorable light, call for socialist revolution, or exemplify the dialectical-materialist world view. Even sophisticated progressive critics like V. L. Parrington and Granville Hicks insisted that the best authors expose the abuses of capitalism and/or validate democratic traditions and values. Phillips and Rahv, in their remarks on Faulkner, called attention to the reductive outcome of such criticism. By focusing their attention on the textual force field as well as the social force field (to borrow and extend Adorno's term), they opened the way toward discussing the interplay between dynamic linguistic and formal elements that in successful texts are every bit as resonant and consequential as the complex social forces shaping them. One might say, of course, that diminishing the importance of systematic belief served to depoliticize criticism; I would argue instead that by suggesting that ideology may be deduced from a combination of textual and social phenomena by critics working through several intermediary levels of analysis, Phillips and Rahv anticipated critical modes that subsequently proved to be fruitful. They helped to create a situation in left criticism such that if critics were going to discuss ideology persuasively, they would in effect have to reinvent the term. It was in fact the similar critique of the limits of the orthodox understanding of ideology that informed the Frankfurt School's practice of *Ideologiekritik*, and later Althusser's significant expansion of the term in order to incorporate Lacanian and Gramscian insights. What is germane to my discussion of Phillips and Rahv is that they shared with each of these critics not only a belief in literature's (at least some literature's) relative autonomy from dominant systematic belief but also a belief in the power of this literature to destabilize dominant habits of mind.[22]

The complex, intermediary levels of textual analysis necessary to any adequate understanding of literature's ideological—or, better, *social*—function become clearer when we turn to Phillips's and Rahv's discussions of form. We have seen in the passage on Faulkner that they considered specific content and form to be inseparable: specific content includes point of view ("attitudes toward character"), modes of description ("painting of moods"), and plot devices ("patterns of action"). Not surprisingly, Phillips argued that form was actually a "mode of perception"[23]—that is, something intimately connected to a writer's and a work's sensibility. Addressing the form/content debate, the major conundrum of Marxist criticism during the 1930s, the editors further extended the idea of sensibility. Generally speaking, the Marxist view of the form/content problem has always been problematic. Its premises have been that form and content are dialectically related and that, in the last instance, content is the primary category. For those within the proletarian literary movement of the 1930s, the discrepancy between these views (which are either contradictory or at best partially incompatible, depending on one's definition of dialectical) was magnified by the repeated belittling of literary form and questions pertaining to it. For some critics "in the last instance" became "in each instance" as they diagnosed capitalism's cultural condition as decadent and perfunctorily labeled each modernist innovation in form as a particularly egregious symptom. And yet, inexplicably, these same critics insisted that the proletarian movement learn from past movements, including modernism, and not discard useful formal innovations. Phillips and Rahv, as did others, pointed out that this could not be an acceptably dialectical view of the problem. If form and content are dialectically related, how could Marxists justify preserving certain forms while rejecting their content? And what exactly was meant by these terms, so frequently, yet vaguely employed?

The most complete response to these questions was given by William Phillips in "Form and Content." He lost little time in rejecting the manner in which the question was being posed by critics within the movement—Phillips charged them with failing to define their terms and failing to explain what they meant by the "separation" of form and content. He then proceeded to critique the reified notions of form and content that contributed to their dilemma. He pointed out that "the shape or structure of a literary work" is often thought of as a kind of "mould" into which the writer pours his ideas. But "forms" in this sense, such as the sonnet or lyric, are better described as traditional "patterns of writing" rather than as receptacles. But even here, form as pattern, technique, or method speaks only to the "verbal surface," insofar as linguistic methods, though important in themselves, are nevertheless inseparable from the writer's "purposes and perceptions." Content, on the other hand, is depicted as "something solid, organized, and completely philosophical." "A more significant definition," hazarded Phillips, would reveal form and content "as two aspects of a unified vision," for if we try to point to the content of a particular work, then by the very act of identification we tear the object of our attention from its context. Consider

> Hamlet's soliloquy. As sheer content it asks whether it is nobler to yield to adverse fortunes or to resist them. As such, it is, of course, banal and silly. But the question takes its meaning from Hamlet's person and state of mind, in short, from Shakespeare's complete perception of the play of human motives and of the character of Hamlet. And this is given not only in the working out of the plot, in the innuendoes of action, but also in the very idiom of the soliloquy which imparts Shakespeare's

> grasp of behavior in his time. In saying this, the idea of form is included in that of content.[24]

Since "any suggestive idea of *form* would have to include the elements which give shape and quality to content," a more accurate way of defining form would be as a "mode of perception." Thus form serves to constitute content just as content shapes form—as is perhaps most readily seen in modernist poetry. Phillips added, however, that the more formally innovative modernist schools emphasize form to the point of "narrow[ing] the range of explicit meaning to the point of extinction."[25] After all, he continued, "while form should be regarded as a mode of perception, it cannot be forgotten that it is a way of perceiving a specific literary content." Perception, and ultimately the sensibility that underlies it, is thus the magnetic factor in the creative process: attracting, animating, and setting in tension elements of form and content, the intrinsic and the extrinsic, the existential and the historical.[26]

Phillips's and Rahv's dialectical, interpenetrating rendering of the form/content problem anticipated similar efforts by cultural materialists and new historicists. What was innovative about their handling of the problem was that for the first time they laid open a range of formal elements that previously had been considered beyond the pale of Marxist criticism in the United States. Moreover, they fashioned a way of discussing formal questions that avoided formalism, substituting for autotelism, a kind of modulated materialism of the text. We recall that for them a work's meaning is produced through the manipulation of its technical possibilities in rendering action, character, style, tone, and so on, according to the writer's perceptions. These in turn are products of his or her sensibility, the "agent of selection." And of course the writer selects from a larger range of possibilities set by broad social factors, including politics, economics, and ideology. Thus, a work's final form embodies both intrinsic and extrinsic phenomena. It is open to analysis on a local level, but the critical movement must ultimately be outward toward a broader social horizon. The positive effect of this approach, broadly shared by the New York intellectuals (Dupee, Hardwick, Howe, Kazin, McCarthy, Troy, and so on), and others such as Edmund Wilson and F. O. Matthiessen, can be fully appreciated only by rereading the work of these critics in context. Although in retrospect we are naturally apt to note telling absences and to dissent from particular judgments, alongside the doctrinaire Marxist critics and orthodox New Critics the work produced by these critics spoke, and continues to speak, with greater subtlety and insight regarding literature's social basis.

Not surprisingly, this approach to reading encouraged Phillips and Rahv to view modernism in general with more sympathy than most of their fellow leftists. In refuting the belief that bourgeois literature ought to be dismissed in toto, they strongly advocated that the cultural heritage be retained in all its diversity, Rahv going so far as to refer, somewhat capriciously, to the Marxist "principle of cultural continuity."[27] The editors argued for a continuing and close dialogue with modernism, always pointing out its positive as well as negative aspects. A typical example was a review of Hemingway's *Winner Take Nothing* in which Rahv strongly argued against rejecting bourgeois literature and for incorporating it into the new literature. In so doing he observed that "the principle that content determines form, if exaggerated, reduces itself to an absurdity, and withal a very dangerous absurdity, for it makes the proletarian artist insensible to those few—largely external—features of contemporary bourgeois art that are class determined in such a slender and remote manner as to render them available for use."[28] In spite of its qualifications, this was a bold theoretical statement to make at the time. By claiming this much distance from the class struggle, Rahv was granting to certain features of literature—albeit "few and external"—a degree of autonomy not normally acknowledged by orthodox critics. This allowed him to be significantly more receptive to modernism than were the majority of proletarian critics. Elsewhere, he underscored the profoundly adversarial role that modernist writing had played. He sharply contrasted "commercial" and "intellectual" art within bourgeois culture, maintaining that the former represented "the open instrument of propertied class interest in letters."[29] The literature of the "advanced intellectuals," on the other hand, had an alienated, dissident, and at times even subversive relationship to bourgeois society. Such art, which he gave the name "negative art," refused "the shallow optimism" and the "open cash-valuation of life" found in commercial art. Negative art "articulates despair . . . slashes certain forms of philistinism, and . . . even indulges in virulent social criticism (usually not stated in class terms but deflected through various crooked mirrors)." In spite of its vacillation, negative art has "served as an introduction and as a stimulus to social insurgence."[30] For Rahv, bourgeois culture was by no means homogeneous; nor was modernism to be understood primarily as an aestheticist and decadent source of counterrevolutionary ideology.

In the same issue that Rahv was developing the notion of modernism as bourgeois critique, William Phillips was providing a genealogy of proletarian literature that placed modernism squarely upon its family tree. Echoing Edmund Wilson's suggested synthesis of naturalism and symbolism, which he proposed in *Axel's Castle* (1931), the primer of modernism for so many critics during the 1930s, Phillips in "Three Generations" suggested that proletarian literature could be the fruit of a synthesis of two earlier, interacting literary generations: that of the native group (Dreiser, Anderson, Lewis, Robinson, and Sandburg) and that of the exiles (Hemingway, Eliot, Pound, Cowley, and Cummings). Given this genealogy, it was absurd, as Phillips pointed out, for leftists to repudiate the bourgeois heritage. Doing so would only cause them to "fall into primitive, oversimplified and pseudo-popular rewrites of political ideas and events." On the other hand, Phillips warned against the rightist mistake of "not complet[ing the] transition" from modernism and seeking to "assimilate the methods and sensibility of writers like Joyce and Eliot without a clear sense of the revolutionary purposes to which these influences should be bent."[31]

The effect of this analysis of modernism was twofold. First, it challenged the theoretical assumption of doctrinaire Marxism that the superstructure is a mere epiphenomenon of the base. Of course, most critics, doctrinaire critics included, were shrewd enough at least to pay lip service to the idea that the relationship was properly seen as a reciprocal one, so long as the base was finally determinate. Insofar as his remarks implied the ultimately determinate role of the base, there was nothing heretical, certainly, in what Rahv wrote. But his views on modernism went a step further by implying that the "dynamic interaction" between base and superstructure was characterized by reciprocity rather than determinism. As "negative" art, modernism was not "determined" by bourgeois relations and interests—many of its sources were to be found elsewhere, and many of its effects were either distinct from or inimical to bourgeois interests. This would be even more the case were modernism to be appropriated by an oppositional cultural movement.

Phillips's and Rahv's understanding of modernism was also bound up with their skepticism regarding official Marxist epistemology. Orthodoxy had for some time been committed to a reflectionist theory of knowledge, in which ideas come to the mind as phantoms, sublimates, or reflections of real objects. This caused many Marxist critics to adopt realist criteria for literature, the best of which was thought to achieve an accurate reproduction of real life. But Phillips and Rahv were ambivalent about socialist realism. They first of all rejected any simplistic notion of mimesis. Rahv explicitly rejected the idea of reflection and preferred to speak of "a deflection with crooked mirrors," which suggests that literature is not one but two removes from the object it purportedly "reflects." Not only is the original object "deflected" or altered by the act of mirroring it; the distortions of the mirror itself compound the discrepancy. In other words both the practice of writing and the material used in writing—one thinks of language itself—cause the original object to be reimagined. This revision of orthodox reflectionism compares favorably with that of Lukács, whose theory of realism did not so much reject reflectionism as alter it so that it applied to processes rather than to objects. Whereas the socialist realism of the proletarian movement called for the replication of objects, Lukács demanded that forces, trends, and contradictions be elaborated, all of which he thought comprised objective, verifiable historical processes. Whereas Lukács's aesthetic valued those works that most convincingly depicted the totality of objective historical forces, Phillips and Rahv tended to emphasize the ingenuity, resonance, and depth of a fully realized sensibility operating within the work. Their idea of realism, being less epochal than Lukács's, was also less insistent on nineteenth-century forms and methods that served to render society panoramically. They could afford to be more receptive toward literature of less breadth, perhaps, but of more concentrated focus and more various means. Their criticism was thus conceptually better disposed to modernism, a fact that invites comparison to the anti-Lukácian impulses in Brecht, Adorno, and Benjamin.

III

Looking at Eliot's early criticism, we can see that his famous comments on the creative process were an important source for the editors' criticism. I refer in particular to the well-known related discussion of the poet's impersonality in "Tradition and the Individual Talent," of sensibility in "The Metaphysical Poets," and of the "objective correlative" in "Hamlet and His Problems." In these essays Eliot attempted to find syntheses for several of the reigning antinomies of criticism: the "inspired poet" versus the "world"; the creative versus the critical, or intellectual, faculties; and, more generally, the subjective versus the objective world. Although Phillips and Rahv could not have known it at the time, Eliot's synthetic, or dialectical, impulse was grounded in his earlier studies of dialectical philosophy. In his dissertation, a critical evaluation of F. H. Bradley's neo-Hegelian idealism and its relation to the realism of Moore, Meinong, and Russell, he had already worked out a meticulously balanced view of the relationship between objective and subjective reality, between physical and mental worlds. It was therefore neither coincidental nor capricious that two young proletarian critics seized upon the dialectical aspects of Eliot's early criticism.[32] Following Eliot, Phillips and Rahv appealed to partisan writers to transmute their deeply felt opinions into persuasive literary emotion by laboriously mastering techniques and by striving for greater equanimity during the writing process. Thus Rahv praised Hemingway for his "impersonality of method,"[33] and the proletarian novels of William Rollins, Jr., and Arnold Armstrong because the solution to the class conflict was resolved "not externally, through the well-known device of preaching and finger-pointing, but internally."[34] Even Eliot's *Murder in the Cathedral* was rather courageously acclaimed (in very Eliotic language) because "it is precise, contemporary, sustained by a sensibility able to transform thought and feeling into each other and combine them in simultaneous expression"—this, remarkably enough, in the teeth of the controversy that followed the publication of *After Strange Gods* in 1934.[35]

Eliot's double-edge notion of sensibility provided Phillips and Rahv with a way of responding to critics of both the right and the left. In the face of those who in their fulminations against Marxism reduced literature to the immediacy of pure feeling, they could evoke the cognitive function of sensibility. To the doctrinaire critic who insisted on confusing literature with documentary or propagandistic writing, they could point out the need for felt ideas. In this new context, quite different certainly from Eliot's, his concept of sensibility served as a flexible tool for helping to fashion new forms of proletarian literature and Marxist criticism.

Several other of Eliot's concepts helped shape Phillips's and Rahv's thinking as well. Especially pertinent was a portion of Eliot's discussion in "The Possibility of a Poetic Drama" concerning the complex relationship that obtains between literary forms and society. To those who would simply manufacture a new form or copy an old one, he replied:

> To create a form is not merely to invent a shape, a rhyme or rhythm. It is also the realization of the whole appropriate content of this rhyme or rhythm. The sonnet of Shakespeare is not merely such and such a pattern, but a precise way of thinking and feeling. The *framework* which was provided for the Elizabethan dramatist was not merely blank verse and the five-act play and the Elizabethan playhouse; it was not merely the plot—for the poets incorporated, remodelled, adapted or invented, as occasion suggested. It was also the . . . "temper of the age" (an unsatisfactory phrase), a preparedness, a habit on the part of the public, to respond to particular stimuli.[36]

As we have seen, traces of this passage appear throughout Phillips's and Rahv's early criticism. Eliot's "precise way of thinking and feeling" became "a way of living and seeing," the dynamic understanding of form remaining the same. Form for all three critics became a "mode of perception" as opposed to a shape needing to be filled. And the public's "preparedness," its "habits of response," the "temper of the age," fairly well describe what the editors meant by the sensibility of an age.

In addition, there was the matter of *systems* of ideas in literature. As I have shown, this was no small concern for proletarian critics, many of whom demanded that literature reflect a dialectical-materialist viewpoint or support the Party's political program. Here Eliot was enormously helpful to Phillips and Rahv, who opposed placing the burden of systematic "correctness" on literature. Eliot had dealt with the problem of systematic belief in "Dante," the final essay of *The Sacred Wood*. He defended what he called "Philosophical" poetry against Valéry's claim that such was no longer acceptable as modern poetry. Eliot based his defense on the evidence of *The Divine Comedy*, which he maintained demonstrated that "Dante, more than any other poet, has succeeded in dealing with his philosophy, not as a theory . . . or as his own comment or reflection, but in terms of something *perceived*."[37] Dante did so, according to Eliot, by integrating his system of beliefs with the structure of the poem. The structure in turn was made essential to each of the parts. Eliot observed further that "the examination of any episode in the *Comedy* ought to show that not merely the allegorical interpretation or the didactic intention, but the emotional significance itself, cannot be isolated from the rest of the poem."[38] By "realizing" ideas rather than merely evoking them, by "inspecting" them rather than arguing them, the poet can indeed deal successfully with philosophical ideas. As Sanford Schwartz has perceptively observed, Eliot here reverses the emphasis on objectifying feeling found in his discussions of impersonality and objective correlative—the lone emphasis we have unfortunately come to associate with him. In the essay on Dante he provides the other element in his dialectical treatment of the subject/object problem by praising the poet for enlivening overly objective doctrine through transmuting it into sensation.

The relevance of these passages to Phillips's and Rahv's situation is clear: To the extent that late medieval religious belief and twentieth-century doctrinaire Marxist belief were codified and official, the issues at stake for religious and for proletarian poetry were the same. Like Eliot, the editors argued that for systems of ideas to operate successfully they would have to be thoroughly embedded in the structure and feeling of the work. Their idea that through the operation of a perceptive sensibility a proletarian writer could transform political doctrine into a work of imaginative power was a direct application of Eliot's insights into successful religious verse.

Finally, in assessing the direct and indirect results of Eliot's influence on Phillips and Rahv, we must not forget that their most passionate writing was that in which they defended criticism and theory against those within the movement who minimized their importance or in some cases rejected them outright. By the spring of 1935, not long after the appearance of their most important early essays, ominous allusions to "bourgeois aestheticism" and to "academicism" cropped up, and not long afterwards Mike Gold, the Party's best-known spokesperson, charged them with "mandarinism." These attacks prompted them to make a vociferous reply in "Criticism." Elaborating on comments they had made in their first coauthored piece, "Problems and Perspectives in Revolutionary Literature," in which criticism was likened to Lenin's notion of a vanguard,[39] the editors claimed that criticism was indispensable to the revolutionary literary movement. Only criticism, they declared, could "clarify the aims and premises" of the different schools and currents within the movement, and only it could relate these tendencies to one another and insure that the more promising among them succeed. To accomplish this, criticism would need to elaborate "a considerable body of aesthetic theory" in order to relate "the problems of art and literature to the body of Marxian theory." Without theory, practical criticism loses its bearings, giving itself over to "empirical observations, inverted estheticism, vulgar applications of political ideas to literature."[40]

Perhaps the most interesting part of their article was that devoted to refuting their critics' claims of academicism. Here Phillips and Rahv were unyielding in their defense of standards, maintaining that too much criticism was being performed by amateurs "who lack both the critical temperament and a knowledge of Marxism."[41] Rather, "recognized Marxists" with a command of both literary theory and Marxist theory ought to be writing the reviews and essays. Theory was not to be judged by its accessibility to a wide audience—that is a fair expectation for practical criticism, they allowed, which "simplifies for the purpose of daily reviewing." But more serious criticism, meaning theory, demands different criteria, and here they were defiant:

> [C]riticism is in the main a form of conceptual analysis, and is primarily directed at readers familiar with the problems of literature. "Criticism is not the passion of the intellect, but the intellect of passion" (Marx). It is to be judged by its validity, by its generalizing power, and not by its temperature, or by the number of readers who can easily digest it. Its effect is a slow one . . . it finally reaches its mass audience in an indirect form.[42]

A "cult of popularity," the editors maintained, had caused some to stigmatize theory as "academic." "But what is academic writing?" they asked. "It is not so much a quality of style as a content marginal to important literary problems."

And finally the editors addressed the problem of "the servile role assigned to [criticism] by many of our writers." For some, criticism is to be narrowed down to little more than "publicity for new proletarian novels and plays." To this they gave a stinging reply, in which they elevated criticism's status to a then unprecedented height. "Criticism," they proclaimed,

> is by no means obligated to herald each third rate poem as a boon to the proletariat. Its main concerns are with creating a new esthetic, with revaluating literary history, and with advancing proletarian art. The search for new creative methods so prominent in a new literature puts the stress on the critical faculty, not only in critics but in every poet and novelist. In this sense, the tasks of the critic in this particular period are perhaps even more complex than those of the creative writer.[43]

Within the boundary of a purportedly mass-based, popular movement, Phillips and Rahv had reaffirmed the need for a vanguard of critics—well-trained, expert even, whose profession was "criticism and not something else." For a moment at least, the possibility loomed of an unlikely alliance between Eliot's professional critic and Lenin's professional revolutionary.

IV

Phillips's and Rahv's Eliotic criticism of the mid-1930s constituted the most ambitious exploration of proletarian literary concepts undertaken within the proletarian literary movement, then under the auspices of the Communist Party and more broadly the Communist International. The very example of their dogged pursuit of literary and cultural questions represented a repudiation, in practice, of the prevailing view that culture, and specifically literary criticism, could not be a high priority for the revolutionary movement. Indeed Phillips's and Rahv's early work draws our attention beyond the limited confines of the proletarian movement toward other developments in Marxist theory contemporary with it and subsequent to it. Their early criticism can profitably be seen as a preliminary enunciation of important themes taken up by other Marxist revisionists. I emphasize *preliminary* for two reasons. One, their work represented an initial excursion into certain practical problems faced by an ongoing political movement with little theoretical rigor of its own—it did not offer finished theoretical arguments. Two, at this point in their careers Phillips's and Rahv's criticism was heterodox as opposed to unorthodox. That is, for every group of insights that seemed to show them embarking on a serious restatement of the dominant views of their movement, a seemingly discrepant opinion would reestablish them on conventional orthodox terrain. When it came to evaluating modernist texts, for example, ideological content alone sometimes shaped their critical response.

In other ways Phillips and Rahv failed to free themselves from traditional perspectives. As we shall see, after they rejected the idea of proletarian literature following their acrimonious split with the Communist Party in 1937,[44] they failed to maintain their earlier connections, tenuous as they were, with popular constituencies. Instead of developing new ties, they in effect substituted themselves for the masses, claiming that as adversarial intellectuals *they* were the revolutionary class. This encouraged them to put Eliot to new and often detrimental uses: professionalization sometimes became an excuse for elitism, and the dissociated sensibility was superimposed on the American tradition, creating several unfortunate binarisms, which have only recently been deconstructed.[45] With regard to the humanist ideal of a unified sensibility lending stability to a reader, writer, or text, there are grounds for concluding that Phillips and Rahv underestimated the destabilizing force of the unconscious, of desire, and of language itself, all of which, as we are now well aware, partially undermine even the kind of provisional, dialectical, and historicized equilibrium they sought.

Given their proximity to the public and their specific social and political obligations, however, I would argue that Phillips's and Rahv's critical humanist reformulation of Marxist criticism was remarkably resistant to essentializing the self—or any other category for that matter, whether it be class, revolution, truth, or beauty. Their relationship to a dynamic and popular political movement, with its culture in a state of flux, forced them to construct ideals and goals whose effects must be judged pragmatically and in context. That context was the attempt by writers and critics, often encumbered by the Party apparatus, to create a thriving proletarian literature that would be the center of a general cultural revival. Their commitment stemmed from their disdain for capitalism's separation of critical culture from the everyday lives of ordinary Americans. Proletarian literature was seen as a response to what was perceived as an elitist and increasingly specialized idea of literature within capitalism's class-determined organization of learning and acculturation. The editors rejected the view that the solution to the division of bourgeois society lay in a new arrangement in which the masses could passively enjoy classical or "serious" culture. Advocating a rather unique vision of proletarian literature, Phillips and Rahv prescribed a new kind of literature that would close the gaps existing within capitalist culture. They assumed that given access to culture, especially high culture, the masses would inevitably reshape it according to their own experiences and needs. Moreover, because these experiences and needs were inevitably linked to the struggle to make a living, their new culture could be expected to concern itself with material as well as spiritual dimensions of life. By combining a full treatment of objective factors with subjective ones—in the editors' words, the "habits" and "psychological relations" existing within the "unexplored continent of fiction" that was working-class life[46]—proletarian literature would become a basis for a whole new conception of literature. No longer would literature be thought of as imprisoned in individual consciousness, separated from its real grounding in social, even economic life. Literature would properly be seen as being deeply embedded in the historical process, yet not to such an extent that it would become a mere simulacrum of "real" social or historical forces. It would provide a common ground where an eminently social self might be explored.

This idea of a popular proletarian literature no doubt goes against the grain of Eliot's traditionalism and elitism—although Eliot's passionate love for such "low-brow" forms of culture as the music hall complicates matters. Yet Phillips and Rahv were not stymied by the acknowledged limitations of Eliot's views; their sense of urgency propelled them beyond the common leftist dismissals of Eliot in order to gain sustenance from others of Eliot's views. Thus the creation of a new, distinctly proletarian sensibility would be the means of achieving a new level of working-class cultural output. Unlike many of their colleagues on the left, they were willing to grant that an active individual sensibility capable of assimilating a range of social processes might arrive at a substantially new vision within the limits of what the social range makes possible. This was an advance, certainly, over criticism that overemphasized the reproductive aspects in creation. Yet against aesthetic theories premised upon the separation of the social and the personal and clinging to notions of "genius," "inspiration," or some innate talent to "get it right," Phillips and Rahv asserted the inescapably social nature of all experience, including that of literary creation. Raymond Williams's term "active reproduction" best captures the dialectical impulse behind their thinking.[47]

Why would Phillips and Rahv have appropriated key terms from a critic considered a class enemy by many on the left at the time? Certainly such systematic borrowing was not without its risks. One must remember that by the mid-1930s the disproportionate interest in reactionary politics among modernists had caused much of the American left to consider the entire literary and artistic movement anathema; those who lacked the political acumen to endorse modernism without the necessary and ritualistic qualifications were soon suspected of aestheticism, decadence, and petty bourgeois vacillation.[48] The deep skepticism toward modernism expressed by the doctrinaire left meant that young proletarian critics like Phillips and Rahv had to conceal their debt to Eliot in order to maintain their status within the proletarian movement. If the debt seems relatively obvious in the 1990s, this was certainly not the case for many mid-1930s Party functionaries and fellow travelers, many more of whom, one may safely hazard, knew of *The Sacred Wood* than had actually read it. It was thus rather remarkable that the two editors forged ahead with a critical strategy that, though it took its bearings from the literary and cultural life of an ongoing political movement, still managed to assert its independence from that movement's shibboleths. This was no small accomplishment, short-lived though it may have been. It offers some insights into what form engaged, worldly criticism might take today.

3

Politics and the Autonomous Intellectual

Our program is the program of Marxism, which in general terms means being for the revolutionary overthrow of capitalist society, for a workers' government, and for international socialism.

Partisan Review to *Poetry*, February 1938

I

We cannot understand the fate of William Phillips's and Philip Rahv's heterodox Marxism, or more generally the fate of *Partisan Review*'s leftism, if we confine ourselves to the analysis of ideas alone. Ideological, discursive, institutional, and class factors play their substantial roles as well, and these I should like to explore in this chapter. As I have shown thus far, Phillips's and Rahv's earliest critical work moved between the antagonistic discourses of bourgeois and Marxist criticism, and was thus marked by divisions in terms of subject matter, tone, and audience. Once the two critics passed over into the official Marxist discourse with the publication of *Partisan Review* in 1934, they urgently pressed for heterodoxy and Marxist renewal. In this and the following chapters I pursue their critical project as it evolved, first within the proletarian movement, second as part of the new anti-Stalinist socialist opposition, and finally as an expression of autonomous intellectual pursuit partially divorced from radical political movements. My intention is to demonstrate how *Partisan Review*'s critical project—its substance and rhetoric, and its failures and successes—were shaped by the pressures of impinging social forces.

I would begin by pointing out that for critics like Phillips and Rahv who believed in the possibility of a proletarian literary renaissance, the question of audience was crucial. This set them somewhat apart from the direction taken by most critics at the time, who were given to the analysis of ideological content (most proletarian critics), biography and social context (Edmund Wilson, Van Wyck Brooks, H. L. Mencken, Malcolm Cowley), or literary form and its moral and social implications (New Critics like Allen Tate, John Crowe Ransom, and R. P. Blackmur). Among critics still read, perhaps only Kenneth Burke, who was an early contributor to *Partisan Review*, paid explicit attention to the text's relationship to its audience. Although we can find examples of the New Critics' textualism in Phillips's and Rahv's work, as proletarian critics they were intent on adjusting their textual insights to the extraliterary experiences and interests of particular classes and strata of readers. This should not be surprising considering the fact that the proletarian literary movement took its name from its projected audience as much as from a specific political outlook.

But exactly who constituted this audience? Ideally, of course, it was the entire working class, replete with its range of progressive and retrograde beliefs. Most inside the proletarian literary movement, however, considered that reaching such a broad audience was a distant goal. So instead of appealing to average workers, the movement addressed, in the words of the two editors, the American proletariat "in the process of developing genuine class-consciousness" and "fulfilling its revolutionary goals." Nearly all of the writers and critics within the movement adopted a strategy, in other words, of winning an audience comprising the more progressive members of the class, the proletariat's "best self," as it were. Phillips and Rahv essentially endorsed this view, but with important qualifications.

The editors acknowledge that the anticipated literary renaissance, already underway, would radically redefine the relationship between writer and audience. As they put it in their editorial,

> The proletarian writer, in sharing the moods and expectations of his audience, gains that creative confidence and harmonious functioning within his class which gives him a sense of responsibility and discipline totally unknown in the preceding decade. . . . Indeed, it is largely this intimate relationship between reader and writer that gives revolutionary literature an activism and purposefulness long since unattainable by the writers of other classes.[1]

The editors appealed to writers to abandon the "spectator's attitude" and plunge into "the vast and complex background" of working-class life. As indicated earlier, what these writers would find was "the unexplored continent of fiction," the "working class with all its thousands of occupations, habits, psychological relations." After all, they added, "What Balzac did for many strata of the bourgeoisie, what Joyce did in Bloom for the lower middle class and in Stephen for the sensitive intellectual, still awaits the genius of proletarian artists."[2] The issue for proletarian writers was not one of taking a simple oath of fealty to a cause or a class. Their commitment must be to a new, broadly conceived and diverse literature, a literature of "an emerging civilization," rather than to an organization. It would include among its models the poetry of Crane and Eliot and the songs of Joe Hill. To those who were skeptical of such a mixture of high and low, the editors responded by warning that any attempt to conceive of one type of poetry to the exclusion of another "is to narrow the cultural expression of the revolutionary movement."[3]

As I have already indicated, the significance of this particular vision of proletarian literature is its ability to place Crane's and Eliot's modernism in the new scheme of things. Not only would intellectuals have to "remold" themselves and adopt working-class values, but the working class would also undergo a remolding process as it remakes values shaped by capitalism so as to become a more discriminating and ultimately enriched audience. The fact that for the editors the working class would always provide a "check" on the literary movement need not mean that serious literature would be threatened. Indeed, the proletarian renaissance would ensure that the writer would no longer experience the alienating and frustrating marginality that characterized his or her position in bourgeois society (they had the modernists chiefly in mind here). And no longer would ordinary Americans be isolated from books and works of art that would provide them with new experiences.

But Phillips and Rahv were acutely aware that it would be some time before such a harmonious intercourse between writer and audience would exist. In the meantime, they claimed, critics must acknowledge the distance between the two groups as a prerequisite for bringing them together—hence their singling out the problem of strata. They called attention to the fact that their audience was not at all homogeneous, that it contained constituencies with important differences in education and taste: "Workers who have no literary education prefer the poetry of Don West to that of S[ol] Funaroff, whereas intellectuals reverse this choice."[4] In admitting the divided and potentially bipartisan character of their audience, Phillips and Rahv were not alone among proletarian critics. Mike Gold, Granville Hicks, and others were aware of the fact that "petty-bourgeois" intellectuals "diluted" their movement. But for these critics, unity would be had only when the intellectuals "remolded" themselves and acquired class consciousness; those who could not, or would not, would simply drop by the wayside. For Phillips and Rahv, however, although they too at times endorsed such a view, a conciliatory tone made itself felt as they granted a place for "intellectual" poetry. Rather than require intellectuals to renounce their own conventions and traditions, the editors set as a goal for the proletarian writer the working out of "a sensibility and a set of symbols *unifying* the responses and experiences of his total audience."[5] Although they too could not finally sanction persistent divisions within their hoped-for audience, their solution was more protracted, and less hostile toward intellectuals.

As indicated, to achieve balance in their approach to the problem, the editors suggested that the development of a sensibility would have to "be constantly checked by the responses of [the writer's] main audience, the working class, even while he strives to raise the cultural level of the masses."[6] Exactly how such responses were to be gauged by writers and critics who had very little direct contact with the working class was a serious problem that remained unanswered. But it was Phillips's and Rahv's recognition of the various and changing cultural needs of the masses—as well as the recognition of their distance from the masses—

that caused them to envision a relatively long-term solution and thus, quite wisely, to settle for a hybrid literature in the meantime.

Looking around in 1934, they could not have observed many instances of intellectuals and workers sharing either work or leisure. On the whole, writers had relatively few opportunities to integrate with workers on a long-term basis. Although they had not yet become *Kathedersozialisten*—"socialist professors," who twenty years earlier had been scorned by Rosa Luxemburg and Karl Kautsky—neither did they have a relationship to the party that resembled Luxemburg's or Kautsky's, or for that matter Lenin's, Korsch's, Lukács's, or Gramsci's. These activist intellectuals combined intellectual work with Party work, and because their organizations were genuinely mass-based, their work brought them face-to-face with the working class. Aside from their responsibilities as Party leaders, they typically taught in Party-organized schools for members and nonmembers alike.[7] In the United States, because few serious Marxist intellectuals ever managed to join the Party and continue their intellectual work within that institutional framework, there was little chance of them coming into contact with the Party's mass constituency, which in any case was itself quite small compared with that of the major European parties.[8] Most intellectuals, whether Party members or Party supporters, made their contributions as intellectuals along the Party's periphery—that is, within its network of affiliated organizations and satellite publications.

As for Phillips, Rahv, and other literary figures, their work was organized around their respective publications, not around a union, caucus, mass organization, neighborhood, or some such entity in which workers participated. For the leftist writer or critic, the desk was the workstation, the office was the factory, and the local deli or bar the union hall. If they weren't putting in their time at the office, they were working at home or traveling from one strike to another, filing stories and moving on. There was little time to sink roots, and little opportunity to develop sustained relationships with ordinary Americans. At best, movement writers settled down in a locale long enough to complete an essay, short story, or novel. In short, writers and critics in New York rarely if ever came into sustained contact with working-class constituencies outside the subway, unless, in some cases, they were visiting family. An interest in workers' education did arise during the mid-1920s in New York, and a group organized in the Soviet Union calling itself the Proletarian Artists and Writers League did encourage leading leftist intellectuals to help coordinate workers' colleges and cultural groups, but nothing of any importance ever came of it.[9] Nor was the effort ever duplicated, with the possible exception of the John Reed Clubs, whose organizational history is worth reviewing, since for a time it was the only vehicle by which Communist intellectuals might come into regular contact with those from the working class.

The first John Reed Club was founded in New York City in 1929 by the *New Masses*. The purpose of the new organization was to develop a corps of writers and artists that would eventually provide the magazine with a network of reporters and experts extending to major industries and trades. According to Mike Gold's founding statement, each writer would focus on a single industry, spend several years studying it closely, and become an expert—in other words, write about it as an insider, not like a "bourgeois intellectual observer."[10] The *New Masses'* plan to organize the John Reed Club industrially never did pan out, although the club eventually expanded to almost twenty chapters and experienced some success in recruiting talented young writers from the ranks of the working class. In addition, a network of proletarian literary magazines modeling themselves after Jack Conroy's authentically proletarian *Rebel Poet* and *Anvil* carried their work. But gradually the character of the club changed as radicalized intellectuals increasingly asserted their leadership and as Popular Front policies caused the Party to court established writers rather than encourage young ones. No longer concerned to attach themselves directly to existing proletarian constituencies, the chapters developed their own spheres of activity—a likely enough thing for writers and critics to do who were serious about their craft. When the Party disbanded them in 1935, all pretense of developing working-class affiliations and writers was dropped in favor of the policy of making allies of established writers.[11] This did not mean that during the subsequent Popular Front period the Party abandoned all efforts to promote working-class culture: especially in theater, film, the visual arts, and music the Party's populist influence was in fact strongly felt and quite productive. But in the areas of literature and criticism, the onset of the Popular Front, with its interest "in authors instead of books," as one critic acidly put it, did not encourage a good deal of original work.

The demise of the John Reed Clubs was a political and institutional factor that had serious consequences for the problem of audience posed by the proletarian critics. Not that a closer proximity to their intended literary audience would necessarily have solved the difficulties inherent in their idealistic view of the potential alliance between working people and artists-intellectuals. But the failure of the Party to sustain any kind of invigorating literary connection to popular constituencies, and in fact accept the separate spheres of "high" and "popular" culture, encouraged critics like Phillips and Rahv to seek intellectual sustenance elsewhere. Increasingly they committed themselves more vociferously to ideas than to people, if I may put it this way, and to asserting their custodial functions over their equally valid obligations as citizens within a democracy to spread and promote diverse culture(s). As long as the masses lived in the minds of intellectuals as the most potent mythos of the period, rather than as a group to be known and understood firsthand within a democratic and diverse movement, the prospects for a proletarian renaissance remained dim. Ironically, what emerged instead were new possibilities for a strata of intellectuals to become a thriving, partially empowered, and politically more docile group than it had been during the 1930s.

II

In print now for nearly three years, *Partisan Review* had managed to diversify the proletarian movement and begin to redefine its relationship to bourgeois culture; however, in 1937 the political binarisms and intellectual schizophrenia it had tried to overcome reasserted themselves with a vengeance. These polarities would long remain an integral part of *Partisan Review*'s cultural project and, as I shall demonstrate, have in a broader sense been integral to American literature and cultural discourse ever since.

Given its reputation for defending culture and literature against their crass politicization by Stalinism, it may come as a surprise to some readers to learn that *Partisan Review*'s split from the Communist Party was caused by disagreements over political rather than cultural matters. Soon after the "old" *Partisan Review* was suspended in October 1936, largely for financial reasons, Phillips and Rahv began to question the Party's position on the three most crucial issues facing it at the time: its new Popular Front strategy, the Spanish Civil War, and the Moscow Trials.

The Popular Front strategy was officially promulgated at the Seventh Congress of the Comintern, which took place in Moscow from July 25 to August 20, 1935. As outlined in Georgi Dimitrov's address to the congress, the strategy included two major components: the United Front against Fascism (UFAF), and the Anti-Fascist Popular Front.[12] The UFAF was, at least in theory, a united front of the working class. It would include all working-class constituencies and organizations that opposed fascism and were willing to defend "the immediate economic and political interests" of the workers. Workers with allegiances to the bourgeois parties or to social democracy would be welcome, as would social-democratic parties and reformist trade unions. Joint action was now allowed with each of these constituencies, all of which the Communists had rebuffed during the previous period of sectarianism. Although it was hoped that the respective Communist Parties would assert their leadership within the United Front, demands for the overthrow of capitalism and for the establishment of the dictatorship of the proletariat were dropped. Further, the UFAF was to be the core of the Popular Front, a broader alliance between the proletariat, the peasantry, and "the basic mass of the urban petty bourgeoisie," which included the intellectuals. This multi-class alliance, led by the working class, explicitly rejected alliance with any strata of the bourgeoisie, even antifascist elements.

Yet Dimitrov's new definition of fascism as "the open terrorist dictatorship of the most reactionary, most chauvinistic and most imperialist elements of finance capital" invited Communists to begin making distinctions within the bourgeois class. Before long "tactical alliances" were being made with portions of the bourgeoisie. The Communist Party (USA) itself showed little compunction about proceeding along these lines. As dictated by the Seventh Congress, the Party's line initially was to build a farmer-labor party, but that effort was abandoned within a year due to Roosevelt's irresistible attraction and to difficulties with other left forces. It was replaced by a series of minor shifts in policy, which culminated in 1937 in the creation of the Democratic Front. This strategy was unique in that it involved an alliance with Roosevelt and the Democratic Party. The results in terms of the Party's popularity and influence were impressive; in terms of the Party's commitment to revolution and ideological consistency, however, they were disastrous, or so thought revolutionary intellectuals like Phillips and Rahv. Indeed, the bizarre presidential campaign of 1936 revealed to them the breathtaking political acrobatics that resulted from the new strategy. In that campaign the

Party nominated a Browder-Ford ticket, yet its fervent hope was that Roosevelt would defeat his Republican opponent. Consequently sections of the Party worked on the Browder-Ford campaign and other sections worked for Roosevelt. If the so-called collaborationism and reformism of the Popular Front were not enough for rigorous leftists like Phillips and Rahv to question the Party's commitment to revolution, these electoral vicissitudes surely fulfilled their worst expectations as to the Party's diminishing militance.[13] Phillips's and Rahv's doubts about the Party's commitment to revolution congealed around two major international events: the Spanish Civil War and the Moscow Trials. In 1936 most of the intellectual left was producing encomiums to the heroic Spanish loyalists, the American volunteers who were sponsored by the Party, and the Soviet Union, which alone was aiding the antifascist effort. A few faint and dissonant voices could be heard, however, with accounts of how the Communist-led brigades were withholding ammunition from, and even using ammunition against, the Marxist and semi-Trotskyist POUM (Partido Obrero de Unificación Marxista), a supposed ally. In addition, there were suspicions about the actual extent of the Soviet aid, and attacks were made on the Communist Party, mainly by Trotskyists, for refusing to condemn Roosevelt for his neutrality and for remaining content to blame his subordinates. According to James Farrell, then a close friend of the editors, the three of them agreed in conversation that neither the Soviet Union nor the European Popular Front governments had responded with alacrity to the situation. Stalin, they believed, had "ditched the proletariat of one country after the other."[14]

In August 1936, the Soviet government convicted and executed Zinoviev and Kamenev for allegedly conspiring with Trotsky (who was tried in absentia) to assassinate top Soviet leaders. In early 1937 Radek and Pyatakov were tried for collaboration with Trotsky and the Nazis.[15] The spectacle of these heroes of the Bolshevik Revolution being accused of such outrageous crimes struck many American intellectuals, Phillips and Rahv included, as highly suspect. As Harvey Klehr has trenchantly put it,

> If the confessions of his alleged co-conspirators were to be believed, Leon Trotsky had orchestrated the most extensive and diabolical plot in all of history. He had somehow united left-wing Zinovievites, right-wing Bukharinites, and disgruntled Stalinites with his own followers and then plotted with both Nazi Germany and Imperial Japan to overthrow the Soviet regime and install capitalism.[16]

Equally doubtful were the bizarre and unbelievable circumstances of the proceedings, in which all of the defendants testified unreservedly as to their own guilt. What sort of manipulation and coercion could have caused such a macabre spectacle in a supposedly liberated society? More to the point, in what kind of society could such an event take place, an event not only sanctioned by but produced at the highest levels of government? Although according to Farrell, Phillips and Rahv at first suspended judgment of the Moscow Purge Trials of 1936–1938,[17] the event ultimately became the most important symbol of the editors' disgust with the Party and the Popular Front.

By the time of the Party's well-orchestrated Second American Writers' Congress in June 1937, Phillips and Rahv were sufficiently disenchanted to join F. W. Dupee, Dwight Macdonald, Mary McCarthy, and Eleanor Clark in openly challenging some of the Party's key policies. In a move oddly prescient of current times, they made their challenge in a session of the congress devoted to literary criticism. To make matters worse for themselves, they recommended certain policies of Leon Trotsky, the archenemy of Stalinism, as alternatives to those of the Party. By October 1937, the differences had become irreconcilable, and given the Party's decision during this period to make the struggle against Trotskyism a major ideological campaign, there was little hope of any further dialogue.

Thus on September 14 the Party attacked *Partisan Review* in the editorial pages of the *New Masses*. In a short article entitled "Falsely Labeled Goods," the public was warned that it was being hoodwinked by the editors of the "so-called" *Partisan Review*, who had recently announced they would bring out a new version of the magazine. Comparing statements in Phillips's and Rahv's announcement with the editorial statement that had introduced the original, the authors declared, somewhat justifiably:

> The phrase "resumes publication" is, to say the least, a euphemism. A new publication is being founded, and its editors are taking an old name with slight regard for what that old name once stood for. Several of the editors have put themselves on record lately, and it does not require clairvoyant powers to foresee their policies. They have attacked the Communist Party, the people's front, the League of American Writers, and the Soviet Union. They have been extremely fond of Leon Trotzky *[sic]*, the P.O.U.M., and the Trotzky Defense Committee. No matter what attempts at camouflage may be made, there is no reason to suppose that the present activities of the editors do not clearly outline the future policies of the magazine. . . .
>
> The only thing that is necessary now is to make it quite clear that the new *Partisan Review* has no connection with the old.[18]

Almost a month later the *New Masses* printed a letter of protest signed by Phillips and Rahv in which they argued that in fact there existed sufficient continuity between the "old" and the "new" *Partisan Review* to warrant keeping the name. They claimed that the "old" *Partisan Review* had not followed the same policies throughout, but had experienced "numerous changes of policy and of editorial composition," thereby suggesting that the most recent change was not unprecedented. They also maintained that with the dissolution of the John Reed Clubs the magazine was "from then on . . . published by a group of individuals."[19] The editors reminded their readers that when the final issue of the "old" *Partisan Review* appeared, the magazine was owned and managed by three individuals: Alan Calmer, and themselves. Editorial and proprietary continuity was therefore preserved, they maintained, in their own persons, and by virtue of the fact that Calmer had been invited to join in the new venture, but had declined.

On the same day the *New Masses* printed Phillips and Rahv's letter (for which the editors were immediately and publicly rebuked by a high-ranking Party official),[20] the following article appeared in the *Daily Worker*, the Party's newspaper.

Entitled "Trotzkyist Schemers Exposed," and bordered in funereal black for good measure, it began:

> Philip Rahv and Fred W. Dupee, of New York City, have been expelled from the Communist Party as counter-revolutionary Trotzkyites, who associated themselves and voted with a Trotzkyist group at the Writers' Congress last summer, and who are now collaborating with other known Trotzkyites . . . in underhanded plans to dishonor the late "Partisan Review" and to start a Trotzskyist literary magazine under that name. The "Partisan Review" was started some four years ago by the John Reed Club, and after the latter disbanded in 1935, was continued along the same lines until 1936, when its publication was suspended due to financial difficulties. . . . They now hope to mislead in true Trotzkyist style, the former readers and supporters of the former "Partisan Review" in subscribing for *[sic]* their Trotzkyist magazine. They appropriate the name. They blandly state that it will be "Partisan Review . . . (which) not having appeared for a year, resumes publication." . . . They promise to honor subscriptions to the old Partisan Review. Their masquerading schemes will not stand the light of publicity.[21]

Of course, *Partisan Review* did stand the light of publicity—more than sixty years of it so far, making it one of the longest running little magazines ever. But in these early years the odds against survival were great as the magazine struggled to define an independent left position while being subject to the unrelenting abuse of the Party. What is important for us is that the wounds received during these battles never healed, and they remain a part of our literary discourse to this day.

III

In setting forth the broad outlines of *Partisan Review*'s new orientation, the editorial statement that introduced the December 1937 issue displayed two very distinct rhetorical registers.[22] The most pronounced was derogative, although given the lurid political environment the editors' tone was relatively restrained. They renounced their former affiliations with the Communist Party and condemned its interference in literary affairs. They claimed the Party subordinated literary matters to politics and was intolerant of differences of opinion. This was no mere example of overexuberance; it was a systematic demand for conformity over a range of literary issues. It was, in other words, a "totalitarian" policy, directing a movement that was fast becoming "an authentic cultural bureaucracy" as its ranks swelled with philistines, academics, and celebrities. "Every effort . . . will be made," they predicted with impressive accuracy, "to *excommunicate* the new generation, so that their writing and their politics may be regarded as making up a kind of diabolic totality."[23] The best that *Partisan Review* could do, they ventured, would be to deflate the effects of the Party's false claims by anticipating and then refuting them. But in the long run their main defense would be to counterpose a superior literature and criticism.

In this more affirmative mode, the editors attempted positively to define their own position. They announced that *Partisan Review* would be aligned with the magazines of the modernist revolt. Without enumerating which publications they had in mind—most likely they meant the *Dial*, the *Little Review*, the *Hound and Horn*, among others—the editors declared that henceforth *Partisan Review* would adopt the "exacting and adventurous" editorial policies of these earlier magazines. It would eschew, however, the "aestheticism" of the modernist revolt, and instead promote subsequent literature that has "look[ed] beyond itself and deep into the historic process."[24] Again no specific books or writers were identified.

A second assertion committed *Partisan Review* to an "independent" but nonetheless "revolutionary" political outlook. The editors claimed that "any magazine . . . that aspires to a place in the vanguard of literature today, will be revolutionary in tendency."[25] Once again, political ramifications of this commitment were not explained. All that the editors would allow was the cryptic reminder that *Partisan Review* "is aware of its responsibility to the revolutionary movement in general."[26] With regard to the magazine's independence, they curiously chose to give a literary rather than an organizational definition of the term. Insisting that the usual definition—independence *from* a group or movement (in their case the Communist Party)—was not what they meant, but rather "the conviction that literature in our period should be free of all factional dependence,"[27] they managed to evade the issue of their relationship to the anti-Stalinist left. If independence would allow the magazine to open itself to all literary tendencies ("conformity to a given social ideology of to a prescribed attitude or technique, will not be asked of our writers"),[28] what about political tendencies? Would they be equally latitudinarian when it came to Marxism or socialism? Here again they were noncommittal. The editors did reaffirm their commitment to Marxism, but in doing so they redefined it. No longer would it signify to them a simple and general theory or body of knowledge, and still less a specific set of beliefs. Rather, they regarded Marxism as "an instrument of analysis and evaluation," which must contend with other disciplines through the medium of democratic debate. *Partisan Review* was itself meant to provide such a medium.

These ideological and organization questions, unanswered in the editorial statement, can be boiled down to one fundamental question: What could it mean to be a committed socialist *and* a committed intellectual? Put another way, How does one remain politically engaged while maintaining enough distance to excel in art or criticism? Interestingly, the editors of *Partisan Review*, who now included Macdonald, Dupee, George L. K. Morris, and Mary McCarthy in addition to Phillips and Rahv, shaped their answer to this question most explicitly and comprehensively not in the pages of their magazine, but in their two-year correspondence with Leon Trotsky, then exiled in relatively nearby Coyoacan, Mexico.

Partisan Review had begun a quest for collaboration with Trotsky in July 1937 as the editorial board made its plans for the first issue. At that time Dwight Macdonald, on behalf of the board, drafted a brief letter to Trotsky requesting a contribution,[29] and Trotsky replied a week later wanting a more complete explanation of the magazine's purpose and political outlook. He granted that *Partisan Review*'s independence might signify an appropriate reluctance to become organizationally dependent, but he also believed that dependence on what he termed "certain fundamental principles" was also beyond question.[30]

Partisan Review's editors responded by briefly elaborating what they had already told Trotsky about the magazine. It would be "exclusively a cultural organ," they maintained, and it would not take positions on questions of political strategy "in the fashion of a political party or grouping." Nor would it take part in "*immediate* political controversies." Reminding Trotsky that as individuals the editors held various beliefs concerning the present political situation, nevertheless they proclaimed that "all of us are opponents of Stalinism and committed to a Leninist program of action. We believe in the need for a new party to take the place of the corrupted Comintern. But as editors of a literary periodical we cannot impose such ideas on the literary contents, although our political ideas do shape—in some ways—our work as editors."[31]

Trotsky wrote back a disapproving letter in which, under the rubric of comradely criticism, he castigated the editors for being "capable, educated, and intelligent people" with "*nothing to say*."[32] Independence was fine, he wrote, if it stood for a definite cultural value. If it did, *Partisan Review* would need to defend it "with sword, or at least with whip, in hand." After all, he observed, "every new artistic or literary tendency (naturalism, symbolism, futurism, cubism, expressionism, etc.) has begun with a 'scandal' . . . bruising many established authorities. . . . Artists as well as literary critics . . . fought and exactly through this they demonstrated their right to exist." Alluding to the approaching world war and the corresponding increase in ideological and cultural tensions, Trotsky suspected that the editors were "wish[ing] to create a small cultural monastery, guarding [themselves] from the outside world by scepticism, agnosticism, and respectability." And if any of this sounded sectarian to them, he concluded, this would simply provide further proof of *Partisan Review*'s desire to be a "peaceful 'little' magazine" content to drift lazily between major contemporary currents.

The editorial board responded with a four-page, single-spaced letter written by Rahv. It remains the most complete record available of the purpose, politics, and cultural significance of the magazine as the editors conceived them. The very comprehensiveness of the response, if it didn't exactly grant Trotsky the wisdom of his remarks, certainly did indicate how seriously the editors took his ideas. To be sure the letter mildly reproached him for his unencouraging attitude and for what the editors took to be his injudiciousness in insisting on measuring their magazine by the yardstick of world history rather than by the existing "determinate" situation. Were he to do so, they claimed, he would better understand the reasons for *Partisan Review*'s admitted limitations, and also better appreciate its modest contributions. Specifically, the letter acknowledge *Partisan Review*'s "uncertain line," its tendency "to deal somewhat gingerly and experimentally with issues that ideally require a bold and positive approach," its "lean[ing] over backward to appear sane, balanced, and (alas) respectable."[33] But, the letter explained, all of these "errors" were "to some extent conditioned by the uncertainty of the objective situation." The editors continued:

> Given the narrow social and literary base from which we are operating—the isolation of the magazine from the main body of radical intellectuals, and the unprecedented

> character of our project . . . encumbered with a Stalinist past and subject to the tremendous pressure of the American environment toward disorientation and comprise . . . it was inevitable that in the first few months of its resistance the magazine should grope for direction.

In addition to insufficient literary forces and an audience that was as yet unstable and politically unreliable, the letter alluded to further complications arising from the literary character of the magazine. Here Rahv raised key issues pertaining to the relationship between politics and literature that has always plagued the left, and he was uncommonly frank about the complexities and uncertainties involved:

> In literature . . . even under favorable circumstances . . . the problem of finding the precise relation between the political and the imaginative, the problem of discovering the kind of editorial modulation that will do damage to neither, is so difficult as to exclude any simple and instantaneous solution. In this sphere, in fact, the most elementary questions are still hotly debated; and a correct political line is altogether insufficient as a reply to the bitter and suspicious queries by means of which the writer challenges the claims of ideological systems and political leaders.

If these were not difficult problems enough, "totalitarian communism" had made matters discernibly worse by giving cause for the skepticism and agnosticism "that are at present making headway all around us."

Turning more fully to literary matters, Rahv made reference to the paucity of literary forces as the primary reason for *Partisan Review* being compelled to publish "purely formalist works."[34] Managing to sound extraordinarily defensive, the letter maintained that these works did nonetheless have value. They are "very good examples of a certain kind of discipline and intelligence" wrote Rahv, "to be preferred to . . . pseudo-radical shouting in verse and prose." All the same "it is obvious that an alliance with 'intelligence' per se opens no prospects to the magazine."[35] Here Rahv once again reminded Trotsky that his contributions, which the editors were still seeking, would be instrumental in allowing the magazine to transform this kind of formalist intelligence into genuine social knowledge. Trotsky's contribution, Rahv asserted, "would affect the character of the magazine in a drastic way."

The other means by which the editors hoped to sharpen the magazine's political edge was to publish a symposium under the title "What Is Living and What Is Dead in Marxism." Trotsky had taken strong exception to the venture in a previous letter, calling the title "pretentious" and "ahistorical," and questioning the credentials of the prospective contributors. Rahv's letter defended the symposium by alluding once again to the unfavorable historical situation that had in this case made a "re-valuation" of Marxism necessary (he hastily added that he did not mean a "revision"). Responding to Trotsky's complaints about the symposium, Rahv wrote:

> Unfortunately, to many people the successive defeats of the working class the world over and the moral abyss revealed by the Moscow Trials are tantamount to a theoretical refutation of the basic principles of Marxism. Surely this melancholy fact will

never be abolished by the refusal of Marxists to take it into account. After a defeat, one must often start over again. . . . We, for example, believe that the "basic principles" have stood the test of history; but in order to convince others that this is so we cannot approach them with the pride of knowledge.

Rahv's letter on behalf of the editorial board impressed Trotsky, and his response confirmed that in his opinion *Partisan Review*'s analysis was consistent enough with his own to merit friendly collaboration. But within the framework of this collaboration he stipulated a number of points on which he hoped the magazine would act. These included giving greater scope and sharpness to the critique of Stalinism, discrediting once and for all the *New Masses* as well as "the real sources" of Stalinism's respectability, the *Nation* and the *New Republic*; adopting a policy of "critical eclecticism" with regard to literary schools and methods; and shifting attention from former Marxists to the youth within the overall strategy of addressing the intellectuals. It is worth reviewing Trotsky's exact words with regard to this very important latter point, because *Partisan Review*'s special relationship to the intellectuals has been interpreted as evidence of its drift from authentic radicalism:

For the time being [my italics] I am not speaking about the workers. The laws of the working class movement are different, deeper, and more ponderous. One thing can be expected with certainty: a new wave of radicalism in the younger generation of intellectuals under the influence of those deeper processes which are at present occurring in the proletariat. At a certain stage these two processes will meet and the better elements of the intellectuals will fructify the workers' movement. At present the problem is in a preparatory stage. It is precisely in this preparatory period that the *Partisan Review* can play a very serious role.[36]

Partisan Review expressed essential agreement with Trotsky's stipulations in its return letter, the editors taking particular pleasure in his understanding of the need for "critical eclecticism." Following several abortive suggestions for contributions, Trotsky finally sent and *Partisan Review* published in its August–September 1938 number a lengthy letter to the editors entitled "Art and Politics." This was immediately followed by "Manifesto: Towards a Free Revolutionary Art," which Trotsky prepared in collaboration with André Breton and Diego Rivera.

Taken together, these documents succinctly expressed Trotsky's views on art, and their appearance in the magazine formalized a close relationship between Trotsky and *Partisan Review*, which had been growing for over a year. The views expressed in the manifesto can be summarized as follows. Because the bourgeoisie was in the throes of a crisis so severe that it could no longer tolerate cultural criticism, for the avant-garde to survive it would have to align itself with the political vanguard. The crucial difference between Trotsky and *Partisan Review* on this matter concerned the prospects for sustaining an avant-garde literary movement. Here the *Partisan Review* editors were not as sanguine as Trotsky. Although their analysis of the avant-garde paralleled his insofar as they too saw it gravely threatened by a panic-stricken bourgeoisie bereft of solicitude for its critics, they were not convinced that the revolutionary movement on its own was vital enough to sustain an avant-garde movement as well. This view, similar in its pessimism, by the way, to Adorno's, would have enormous implications for *Partisan Review*'s disposition toward literature and culture in the postwar period, for it caused the magazine to recoil from the critical task of aggressively criticizing established institutions of art (in fact, they helped to create these institutions) and from forging links between experimental, innovative art and emerging political movements. *Partisan Review*'s overall perspective was perhaps best exemplified in Clement Greenberg's widely influential essay "Avant-Garde and Kitsch," which ends with the classic formulation

> Capitalism in decline finds that whatever of quality it is still capable of producing becomes almost invariably a threat to its own existence. Advances in culture, no less than advances in science and industry, corrode the very society under whose aegis they are made possible. . . . Today we no longer look toward socialism for a new culture. . . . Today we look to socialism *simply* for the preservation of whatever living culture we have right now.[37]

The posture here is wholly defensive, the function prescribed for the critic wholly custodial, and in place of any reference to a constituency that might become an agent for social change beyond the avant-garde itself, we have merely the vaguely redemptive *idea* of socialism.

Finally, *Partisan Review*'s cooperation with Trotsky extended beyond ideological and critical considerations and involved organizational plans as well. The "Manifesto" was in fact a call and preliminary program for the International Federation of Independent Revolutionary Artists (IFIRA), an organization conceived by Breton, Rivera, and Trotsky. The plan was to develop an international network of radical writers and artists first by way of contact through press and correspondence, and then by convening local and national congresses before a world congress to mark the actual founding of the organization. *Partisan Review* both endorsed the plan and actively organized among intellectuals on behalf of the new group. As the editors announced on the first page of the Fall 1938 issue,

> An increasing number of writers, artists, and intellectuals are coming to realize that socialism offers the only permanent escape from the barbarism that is gaining ground so fast in capitalist society. We believe that these intellectual forces, hitherto scattered and isolated, should now draw together into some sort of organization for free discussion and for defense against their common enemies. We are, therefore, in complete agreement with the general aim of the IFIRA, and we are ready to take part in the formation of an American section of the federation.[38]

Partisan Review spent several months trying to get the new organization off the ground, but the results were disappointing.[39] This was not the end of *Partisan Review*'s quasi-Trotskyist organizational efforts, however. Later in 1939 the magazine took the lead in convening the League for Cultural Freedom and Socialism, with Dwight Macdonald as secretary. Once again *Partisan Review* printed a statement of principles defending working-class revolution and artistic freedom, and attacking fascism and Stalinism. It also warned intellectuals of the growing danger of regimentation and repression as an outgrowth of what they called "the new nationalism" sweeping intellectual communities. The entire editorial board signed the statement, but the magazine's relationship to the organization soon became problematic owing to disagreements over the issue of support for the Allied war effort. Nonetheless, for a brief period of time the organization provided a genuine alternative to the Popular Front's League of American Writers and its American Artists' Congress, both of which were mired in the mud of having to defend the Party's infamous double reverse of policy: first going from militant antifascism to defending the Soviet Union's nonaggression pact with Nazi Germany, and soon thereafter calling for all-out war against fascism when Hitler suddenly invaded the Soviet Union.

In spite of *Partisan Review*'s adoption of some of Trotsky's important ideas, the editors never allowed the magazine to become affiliated with the official Trotskyist movement. The issue, not surprisingly, was put to the editors almost immediately upon *Partisan Review*'s resumption in an article in the *Socialist Appeal*, the newspaper of the Trotskyist Socialist Workers' Party. Entitled "'Partisan Review': A Revolt against Stalinism among the Intellectuals," it praised the magazine for breaking with Stalinism and for rejecting its crude equation of literature and politics. But the article was not about to let *Partisan Review* off lightly. Once it finished with the somewhat perfunctory-sounding praise, it undertook a much lengthier critique. Although *Partisan Review*'s views concerning the Party's factionalism may have been accurate enough, allowed the *Socialist Appeal*, by proposing to remain independent *Partisan Review* had "swung over to the opposite extreme."[40] For the *Socialist Appeal* the only possible meaning of independence was neutrality and indifference toward the revolutionary politics of the labor movement, the only "real" politics. By refusing to decide "which among the tendencies struggling for supremacy within the ranks of the American working class most clearly and consistently fights for the ideas, interests, and aims of Socialism," *Partisan Review* was "ignoring, and thus denying in practice the close bond" between literature and politics.

Partisan Review's response to the *Socialist Appeal* article was first of all to note that regrettably the article had nothing to say about any of *Partisan Review*'s literary material. Calling the Trotskyists' charges "over-zealous simplifications" and their demands "ultimatist," the editors gave way to a lengthy letter that John Wheelwright, himself a Trotskyist, had sent to the *Socialist Appeal* hoping (in vain) to have it printed. Wheelwright's letter took issue with the official Trotskyist view of the relation between literature and politics, a view he considered to be at best inconsistent and at worst opportunist. He equated the *Socialist Appeal*'s criticism that *Partisan Review* had turned its back upon the working class with the Stalinist claims that the magazine had betrayed the working class. He questioned the Trotskyists' sincerity, referring to their call for openness to literary quality regardless of an author's politics as a "meaching plea" given their refusal to grant to *Partisan Review* its stated need for independence.[41] On the contrary, he declared, *Partisan Review* is very much concerned with the politics of its authors "in order to open up the best of contemporary literature, and . . . to bring proletarian politics to a Marxist position."[42]

Elsewhere, Wheelwright noted the incommensurability of statements such as "politics dominates everything in our world, including literature," and the demand for "complete autonomy for the workers in the arts and sciences." Indeed, in both cases, claimed Wheelwright, there was "special pleading"—in the one for politics and in the other for literature. In actuality, he asserted, literature and politics respond reciprocally and autonomously to the productive mode that underlies all culture. There is no transcendent, no essential politics that must provide literature with its center. What is needed instead, he argued, and what *Partisan Review* was correctly attempting to construct, is a "super-logical center [achieved] empirically, by methods of objective test, finding chords and diameters of the circles, spheres, and spirals of a reality beyond yours. It must do so without a political center if yours proves unfit."[43] He closed his letter in typical polemical fashion: "The brain behind the hand that wrote your editorial is the brain behind the tongue which would with *proletarian* vs. *intellectual* demagogy transform Trotskyism into a Stalinesque international slum."[44]

One can readily see why Wheelwright's letter so appealed to *Partisan Review*'s editors. It was passionately written, it argued for a more open Marxist understanding of literature, and it was an inside job. Yet the letter failed to address the *Socialist Appeal*'s most pertinent question, that concerning the revolutionary character of *Partisan Review*'s enterprise. From the standpoint of classical Marxist political strategy, how could the magazine justify its deliberate distancing of itself from all manner of working-class experience?

It was very much this kind of question which a very unlikely interrogator—*Poetry* magazine—was curious about. In announcing to its readers that *Partisan Review* had reappeared, *Poetry* got right to the point:

> The question arises . . . whether a magazine professedly revolutionary in character can avoid having some definite political program, either explicit or implied. Taken at its face value, the policy of *Partisan Review* seems to boil down to this: that literature, for the present, should lead not to action but to more literature. That may or may not be an excellent policy. But is it revolutionary?[45]

Indeed. For all of *Poetry*'s alleged "aestheticism," it was asking the same question the hard-nosed orthodox Marxists were asking. By asking as directly as it did, it forced *Partisan Review*'s hand:

> We answer *Poetry* as follows: our program is the program of Marxism, which in general terms means being for the revolutionary overthrow of capitalist society, for a workers' government, and for international socialism. In contemporary terms it implies the struggle against capitalism in all its modern guises and disguises, including bourgeois democracy, fascism, and reformism (social democracy, Stalinism).[46]

As to the role of literature in the revolutionary process, clearly, maintained the editors, it was not meant to supply political combatants with "weapons" in a direct sense. Marxism, as a guide to action, can also be a guide to literature. "But whether literature itself is, can be, or should be, typically a guide to action is one of the problems that *Partisan Review* is dedicated to explore."[47] What is certain, they observed, is that literature that leads to action and does not lead "to more literature, would not, we are convinced, be literature at all."

IV

Given the novelty of this perspective—uncompromising political militancy combined with a latitudinarian literary perspective—and given the magazine's disavowal of both the organized Stalinist and Trotskyist movements, where could the magazine turn for sustenance? How, and from which quarters, would it acquire an audience, contributors, and political allies? To answer this we need to examine the magazine sociologically. *Partisan Review* associated itself with a group of anti-Stalinist radical intellectuals who, by the mid to late 1930s represented a vocal and significant leftist presence in New York City. This formation, to use Raymond Williams's term, was where *Partisan Review* found a reliable base of operations, and before long the magazine became its unofficial organ.

For many of the intellectuals within this formation, an early association with the Communist Party provided a negative political and intellectual education. For a number of them—Elliot Cohen, Clifton Fadiman, Felix Morrow, George Novack, Meyer Schapiro, and Lionel Trilling—the break came as early as 1934 when they signed an open letter protesting the Party's brutal disruption of a Socialist Party rally at Madison Square Garden. Subsequent to their disillusionment, they drifted toward Trotskyism; several—Morrow, Novack, and Herbert Solow—actually joined Trotskyist organizations. When the Trotskyists then merged with A. J. Muste's American Workers' Party, this group joined with James Burnham, V. F. Calverton, Sidney Hook, and James Rorty to further their political work. This combined group founded the Non-Partisan Labor Defense organization in cooperation with the Trotskyist movement, and most worked for the Committee for the Defense of Leon Trotsky, formed in 1936 and led by John Dewey. That organization was instrumental in obtaining asylum for Trotsky in Mexico, and later provided him his solitary platform from which he made his defense against Stalin's charges of conspiracy and treason.

This loose formation was also responsible for the short-lived, but nonetheless formidable *Marxist Quarterly*, whose editorial board included James Burnham, Lewis Corey, Louis Hacker, and Bertram Wolfe. Others closely involved were Sidney Hook, George Novack, and Meyer Schapiro. The theoretical and critical output of these intellectuals was much more substantial than our literary histories have led us to believe.[48] Far from composing a group whose political work was separate from its intellectual pursuits, they often successfully negotiated an intelligent integration in their respective fields—whether philosophy, social science, or literary criticism. The result was that during the 1930s they and intellectuals close to them produced an impressive body of work, which provided a general theoretical and critical foundation for the reevaluation of Marxism to which they devoted themselves. Among the important preliminary texts that contributed to this project were James T. Farrell's *Note on Literary Criticism* (1936), which argued for what we have come to call the relative autonomy of literature; Edmund Wilson's

Marxist-informed *Axel's Castle* (1931), which, as I have mentioned, provided this formation its primer of modernism; Wilson's journalism dealing with Marxist criticism and politics,[49] much of which was incorporated into his own reevaluation of the Marxist tradition, *To the Finland Station* (1940); Sidney Hook's two full-length studies, *Toward an Understanding of Karl Marx* (1933) and *From Hegel to Marx* (1936), as well as his important article "Why I Am a Communist: Communism without Dogmas";[50] *Symposium* magazine's "Thirteen Propositions" (April 1933), being an outline for a new Marxism; James Burnham's "Marxism and Aesthetics"; and even, arguably, Kenneth Burke's *Counter-Statement* (1931). Together these works constituted an initial engagement with Marxism, which either implicitly or openly rejected the orthodox Marxist paradigm. Not only did they provide a general intellectual context for current and future magazines such as *Partisan Review, Politics, Commentary, Dissent*, and the *New York Review of Books*, they were also largely responsible for developing an ethos of engaged scholarship out of which a spate of classics arose: Lionel Trilling's *Matthew Arnold* (1939), Kenneth Burke's *Philosophy of Literary Form* (1941), James Burnham's *Managerial Revolution* (1941), F. O. Matthiessen's *American Renaissance* (1941), and Alfred Kazin's *On Native Grounds* (1942). These works compare favorably with any similar group of socially engaged works produced in the United States.

All of this is to point out that *Partisan Review*'s new environment and new affiliations were thick with ideas, intelligence, and opportunities for collaboration. *Partisan Review* may have been "triply marginalized" in terms of its Jewishness, its Trotskyism, and its advocacy of modernism, as Francis Mulhern has noted,[51] but that did not at all mean the magazine was isolated. The new *Partisan Review* was no doubt in an extremely precarious position at its birth due to the Party's intense hostility and attempted quarantine, but within several years it was well connected and thriving. Certainly in all respects save its proximity to actual social movements and working-class constituencies, its conditions of existence were far more favorable than were those of the old *Partisan Review*. The milieu into which the magazine had entered not only encouraged more flexible interpretations of Marxism but, more importantly, it exuded an enlivening culture. If the values of that culture were only in the process of being worked out, it was the informed, vibrant, and unencumbered nature of that process that mattered most. Here was a diverse group of individuals—some university-trained, some self-taught, a bona-fide polymath in Meyer Schapiro—of various ethnic backgrounds (though heavily weighted, of course, toward the Jewish), who took as a major tenet of their lives the idea that serious intellectual pursuit can profit by, and can in turn enhance, political commitment. In mastering the genre of high journalism, a genre with a rich cultural tradition in Anglo-American society (in need of defense only in response to certain academic trends whose arrogance is disguised as an appeal to professionalism and specialization), they sustained a dialogue with both pertinent contemporary issues and some of the major literary and cultural texts of the past two centuries. They were concerned, as none of their doctrinaire opponents were, with the nature of things and the nature of societies, and wished to arrive at some means for discriminating the best from the worst of these. They were concerned too with the quality of experience: their politics—when the ideological battle permitted—placed value on developing a cogent notion of the good life and the just society, rather than submitting to the politics of means.

To sum up the significance of *Partisan Review*'s break from the Party and its attempt in the late 1930s to reconfigure intellectual discourse on the left, it might be instructive to compare these developments with aspects of Antonio Gramsci's understanding of the "counterhegemonic" movement. For Gramsci the development of a counterhegemonic movement, although entailing a vanguard party, would of necessity possess a variety of political and especially cultural forms to rival those of the bourgeoisie. Such parallel institutions would comprise a much more heterogenous movement than the Bolshevik movement in Russia—this because a strategy determined by the hegemonic conditions of the Western capitalist democracies would involve a lengthy process of persuasion rather than an immediate, frontal assault on bourgeois power. *Partisan Review*'s project, although differing from Gramsci's in fundamental respects—it was not tied to a vanguard party and possessed increasingly tenuous ties to the working-class movement, it openly attacked Stalinism, and it did not include a sustained analysis of the bourgeois state—nevertheless in its dissent from orthodoxy showed some interesting correspondences with Gramsci's thought, particularly in its espousal of relatively autonomous intellectual labor, organizational diversity, and a willingness to engage in sustained cultural warfare in order to develop consensus around the critique of capitalism. Surely the editors' admission to Trotsky that the economic and political crisis had caused many to renounce Marxism altogether, and that under the circumstances "one must . . . start over again," raised precisely the issue of how best to persuade those who had wandered from the Marxist fold as a result of the apparent defeat of the working-class movement. This, combined with the pressure in American society always to compromise, forced the editors to see their struggle to win over the intellectuals as a protracted one, demanding a sophisticated, poised voice speaking to the range of political and cultural issues. The editors imagined their opponents to be exactly what Gramsci referred to as hegemonic: experts at winning the consent of those they exploited. For *Partisan Review* the chief means of unraveling the hold of Stalinism and nationalism on an important sector of the American public was to advance on the perpetrators culturally, using the most destabilizing and subversive intellectual "weapons" at hand—namely Marxism and modernism.

As I have indicated, however, there is a less sanguine aspect to *Partisan Review*'s new critical project. In 1941 the magazine's editors sent out a questionnaire to its 800-odd individual subscribers in order to get a more accurate picture of the magazine's readership. They received 250 responses, which represented less than a tenth of the total circulation, put at 3,500 for the recent May-June number. Although far from scientific, the survey did reveal some interesting facts about *Partisan Review*'s readers some three and a half years into their new project. First, they were young: fully 83 percent were under forty. Second, they lived in large cities, with a major concentration in New York City, where more than 35 percent lived. Third, and most important, they were almost all intellectuals or professionals. The occupations of the respondents broke down as follows: 19 percent were teachers, 12 percent students, 10 percent writers, 5 percent office workers, 5 percent physicians, and among the remainder business executives, journalists, and engineers were said to be common. In addition, 8 percent were unemployed or retired. Only 4 percent, it turned out, had working-class occupations. These were not concentrated in a particular locale or industry, but ranged widely. There was an ore sampler, a hosiery mill worker, a shipyard worker, a telephone repairman, and a Kansas dirt farmer.[52]

These figures, of course, reflected larger shifts taking place within American society. For several decades intellectuals had been incorporated in increasing numbers into key sectors of the American economy and state bureaucracy. Between 1920 and 1930, for example, the number of engineers increased from 134,000 to 217,000, the number of managers in manufacturing increased from 250,000 to 313,000, college faculty from 49,000 to 82,000, accountants and auditors from 118,000 to 192,000, government bureaucrats from 100,000 to 124,000, and editors and reporters from 41,000 to 61,000.[53] *Partisan Review*'s survey demonstrated that the character and influence of the magazine would have to be viewed in the context of its connection to these broad social, economic, and demographic changes—in particular, to the growing power of the professional managerial stratum, identified in James Burnham's *Managerial Revolution* (1941) and later in Barbara and John Ehrenreich's "The Professional-Managerial Class" (1976).

Thus in the late 1930s *Partisan Review* found itself in a paradoxical situation. To its editors and many of its contributors, serious cultural work simply could not be undertaken in or near political organizations or movements. *Partisan Review*'s strong distaste for democratic centralism and its desire for organizational independence implied that chronic bureaucracy and ideological repetition must give way to a freer form of intellectual inquiry. The result was that the magazine and its circle, rather than lean toward a party formation, gravitated first toward looser organizational forms such as IFIRA and the League for Cultural Freedom and Socialism, and later toward unaffiliated status altogether. To an extent thus far unacknowledged by leftists, this severing of ties was imposed by the constraints of the orthodox movement, Stalinist and Trotskyist. But it also represented a historic divergence chosen in part out of ambivalence and even antipathy toward popular constituencies. Although it provided a basis for dissent among individuals and groups of intellectuals, and although it fought for the autonomy necessary to produce compelling cultural criticism and political self-criticism much needed by the left, it also led, as I will demonstrate, to a neglect of ordinary Americans and the problems they faced living in a corporate liberal state. The *Partisan Review* circle faced a dilemma familiar to academic leftists today: how does one fulfill one's obligations as citizen within a democracy when political commitment often leads to simplification and autonomous intellectual labor often leads to elitism? It can be argued that other leftists who were contemporaries of the *Partisan Review*—F. O. Matthiessen comes most immediately to mind—provide us with better models of how to balance these obligations. Thus William Cain in his provocative and moving study of Matthiessen lists the extraordinary array of political and cultural organizations in which Matthiessen was active, including the Harvard Teachers' Union, the Boston Central Labor Union, the Massachusetts State Council of Teachers Union, the Massachusetts Federation of Labor, the American Federation of Teachers, the Progressive Party (Matthiessen gave one of the seconding speeches nominating Henry Wallace for president in 1948), the American Russian Institute, the American Committee for the Protection of the Foreign Born, Artists' Front to Win the War, American Youth for Democracy, Citizen's Committee to Free Earl Browder, Civil Rights Congress, Committee for Citizenship Rights, Committee for a Democratic Far Eastern Policy, Committee for Equal Justice for Mrs. Recy Taylor, Committee to Sponsor the Daily Worker, Committee of Welcome for the Very Reverend Hewlett Johnson, Conference on Constitutional Liberties in America, Defense of Communist Schools, Denunciation of the Hartley Committee, Educators for Wallace, Friends of Italian Democracy, National Council of Arts Sciences and Professions, National Federation for Constitutional Liberties, *New Masses, New Masses* Dinner Committee, Open Letter for Closer Cooperation with the Soviet Union, Samuel Adams School for Social Studies, Schappes Defense Committee, Sleepy Lagoon Defense Committee, Supporters for Samuel Wallach, Testimonial Dinner to Carol King, Veterans of the Abraham Lincoln Brigade, Win-the-Peace Conference, and Writers for Wallace.[54] These were mostly, of course, Popular Front organizations, few if any of which would have welcomed the participation of the *Partisan Review* renegades. Yet a list nearly as long could be compiled of independent progressive organizations active during the 1940s which *Partisan Review* as a magazine or its contributors as individuals might have affiliated themselves with but instead chose not to. This list includes the Artists League of America, a radical artists' organization active during World War II; the American Committee for Democracy and International Freedom (ACDIF), an antifascist group begun in 1938 by Franz Boas among others; the American Association of Scientific Workers, a radical scientists' organization started in 1938 whose members included J. Robert Oppenheimer; the League of Non-Violent Civil Disobedience against Military Segregation, headed by Bayard Rustin; the Council on African Affairs, run by Paul Robeson; the NAACP; the Independent Citizens Committee of the Arts, Sciences, and Professions (ICCASP), cofounded by Albert Einstein; the National Citizens Political Action Committee, an independent radical group that merged with ICCASP to form the Political Action Committee, one of whose divisions became the National Council of Arts, Sciences, and Professions; the Civil Rights Congress, which fought for civil rights rulings in the courts; Nature Friends of America, one of the nation's first environmental groups; the National Civil Liberties Bureau, which evolved into the ACLU; the Workers Defense League, a nonpartisan alternative to the mostly Communist International Labor Defense League; and the Young Negroes Cooperative League, which worked for black rights issues in the workplace.

In its quest for freedom from organizational interference in its intellectual and critical work, *Partisan Review* sundered all but a few ties to ongoing political and cultural organizations. Its aloofness is certainly understandable, and it cannot be denied that it provided more favorable conditions than did the Communist Party for open inquiry. Yet one cannot help but conclude that *Partisan Review*'s reluctance to align itself with any group—save the ill-fated Congress for Cultural Freedom (1950), which was revealed in 1966 to have been funded in part by the CIA, and the American affiliate the American Committee for Cultural Freedom (1951)—represented, finally, an excessive fear of affiliation with marginalized groups. One could argue that the members of the *Partisan Review* circle chose not to cooperate with some of the organizations mentioned here for specific political reasons; but in a more general sense the New York intellectuals chose to remain aloof from a whole range of organizations and thus a more intimate relationship with ordinary Americans and marginalized groups. This made it increasingly difficult for them to maintain a radical perspective, and when ideological disagreements marked their relationship with the new left and certain trends within feminism and the black movement during the 1960s, there was no prior history of cooperation or solidarity to prevent the unfortunate polarization that took place.

4

Modernism and the Autonomous Intellectual

One can note, in the case of the United States, the absence to a considerable degree of traditional intellectuals, and consequently a different equilibrium among the intellectuals in general.

ANTONIO GRAMSCI

I

If *Partisan Review*'s editors were surprised or disturbed by the results of their 1941 survey,[1] particularly by the absence of a significant working-class constituency, they gave no indication of it. Despite their clearly stated socialist politics, from the time *Partisan Review* reappeared in 1937 they made only symbolic gestures toward rendering working-class experience and attracting a working-class audience. All that remained of the old *Partisan Review*'s attempts to capture something of proletarian life was the short-lived column "Cross-Country," which featured reportage of strike scenes, poverty, and the inevitably sentimentalized offbeat. Otherwise, there was no proletarian literature, no coverage of the labor movement, nothing to indicate that the editors were unhappy with a working-class readership of 4 percent.[2] The "vast areas" of proletarian experience, which Phillip Rahv just a few years earlier had claimed could be the basis for a genuine literary renaissance, went unexplored in the new *Partisan Review*.

The magazine's distance from the working class was duly noted by fellow-traveling and Trotskyist critics. Among the former, Malcolm Cowley accused *Partisan Review* of retreating into a "red ivory tower,"[3] and several years later Milton Howard summed up these attacks by sardonically referring to *Partisan Review*'s political program as "socialism in one person."[4] However politically motivated and unjust these rebukes were, they did speak to a signal development in the magazine's shift away from political affiliation—the magazine's wholesale rejection of working-class literature and culture. This unfortunate development, coming as it did on the heels of the magazine's bitter political battle with the Communist Party, was less the result of a comprehensive critical evaluation of the literature encouraged by the proletarian movement than it was a result of the powerful ideological pressures of the battle itself. *Partisan Review*'s rejection of working-class literature and culture was thus among the very first casualties of a consequential political battle whose overall result was to end any hope of a constructive relationship between a generation of serious intellectuals and the cultural experiences of ordinary people.

The first public acknowledgment by Phillips and Rahv that they had given up on the idea of proletarian literature appeared in their coauthored 1937 essay entitled "Literature in a Political Decade." Included in a collection of essays entitled *New Letters in America* edited by Eleanor Clark and Horace Gregory, the piece outlined a new direction for the two critics and their magazine. It first of all rejected the earlier literary historical model which Phillips had offered in "Three Generations," in which proletarian literature was seen as a synthesis of the two earlier generations of realists and modernists. According to the new assessment, the literature of the previous two decades had played itself out. A vigorous modernism had passed with the demise of *Pagany* (1930–1933) and the *Hound and Horn* (1927–1934), and the literature of the left, with its emphasis on "the mechanics of liberation" and "sermons of proletarian virtues," was thoroughly discredited as "the eve of revolution transformed into a long vigil."[5] The revolutionary literature of the 1930s had failed to transform American literary consciousness because it continued to "reduce men to parts of themselves." What proletarian literature did was to substitute the "literary statistics of political man" for those of unpolitical man. Thus "the average proletarian novel is as much imprisoned within a pragmatic mold as the average bourgeois novel . . . [it] merely substitute[s] social behaviorism for individual behaviorism." The upshot was that "there is no revolutionary work written by an American which presents on the level of consciousness those moral and intellectual contradictions which appear in the struggle between the old and the new cultures."[6]

These notions concerning the failures of American literature would turn out to be crucial to Phillips's and Rahv's later use of modernism, and I shall return to them. Here I should like to focus, however, on their rejection of proletarian literature, a rejection that in Philip Rahv's 1939 essay "Proletarian Literature: A Political Autopsy" expanded into an unqualified attack on the culture of the Communist Party. In the essay Rahv focused exclusively on the degree to which the proletarian movement was produced and manipulated by the Communist Party, calling proletarian literature "the literature of a party disguised as the literature of a class."[7] Thus the movement's new emphasis on "politics and social relations," and its discussions of the connections between art, politics, and society, were not merely circumscribed by but directly reflected "the interests, needs, and attitudes" of the Party. The Party, according to Rahv, was the movement's "organizer" and "ultimate court of appeal." The proletarian little magazines "were published under its auspices" and in many cases "edited by its members," and the Party "appointed political commissars to supervise the public relations of the new literary movement and to minister to its doctrinal health."[8] Moreover, Rahv claimed, the Party provided the movement with an initial audience and an organizational base, and "it conditioned the writers that had come under its control to conceive of the Soviet Union, its own source of strength and seat of highest authority, as the living embodiment of their hopes for socialism."[9]

To the extent that Rahv's essay identified some of the glaring problems with the proletarian literary movement and the Party's manipulation of it, it was a salutary reminder to naive writers and critics that the Party's mode of politicization was dangerous indeed. But Rahv's essay exaggerated both the desire of the Party to impose its will on individual writers and critics, and the lack of resistance of many writers and critics to the control the Party did attempt to exert. It is thus not a little ironic that Rahv's essay could not account for the fact that he and Phillips had successfully operated first under the auspices of the Party and then within its milieu for well over two years. More to the point, however, in Rahv's haste to portray the movement as essentially monolithic, he barely mentioned that he and Phillips were not alone in avoiding orthodox strictures, a point I shall develop at greater length in chapter 7. Suffice it to say he only mentioned in passing writers such as Josephine Herbst, Grace Lumpkin, Robert Cantwell, and Kenneth Fearing, whom he admitted had managed to avoid the Party's "fantasies" by "sticking to what they knew."[10]

On turning its back on the Party, on the idea of working-class literature, and on an array of writers loosely connected to the proletarian movement, the editors of *Partisan Review* came increasingly to espouse modernism, or at least a highly selective version of modernism, as the only truly subversive literature and culture. In order to understand how they justified their new and rather exclusive interest given their political radicalism, we must return to Trotsky and his views of literature. Trotsky had, of course, written on art and literature long before his 1938 contributions to *Partisan Review*. His classic discussion of the cultural issues confronting the Bolsheviks, *Literature and Revolution* (1924), was well known to American radicals whose reading habits were not dictated by the Party. William Phillips reports in his memoirs that during *Partisan Review*'s brief hiatus in 1937, before he and others established contact with Trotsky, he was reading and studying his work.[11] As this polemical volume demonstrated, Trotsky took literary matters seriously, perhaps more so than any other Bolshevik leader. He looked on the latest European and modernist literary developments favorably, although his generally positive treatment of futurism and Russian formalism was far from uncritical. Overall, his perspective was to incorporate these important movements into the fabric of the revolution. His real animus was directed against those literary schools that sought inspiration from the past, and attempted to sink roots in backward peasant, rural society. All of this found its way into the essays of the *Partisan Review* critics, who preferred whatever was modern, urban, and cosmopolitan to what was backward-looking, rural, and provincial.

But the issue that most revealed Trotsky's influence concerned the possibility of proletarian literature itself. Trotsky's volume was largely a polemic against the Proletcult school and its notion of a distinctly proletarian culture and literature. He rejected this notion outright, claiming that it was based on a false historical analogy between the newly empowered proletariat and the rising bourgeois class of previous centuries. What made the analogy false for Trotsky was the fact that the bourgeoisie had centuries to develop its culture. The Russian proletariat, on the other hand, had no such lengthy gestation period, as it were, before it assumed state power. Trotsky believed the dictatorship of the proletariat would be ephemeral, and it would afford the working class little opportunity to develop a class culture in its own image. The new culture that would emerge would do so only gradually, and would constitute a truly socialist culture; that is, it would be a classless culture, unfolding a "fully human essence."

Moreover, for Trotsky any attempt to create a proletarian culture "by laboratory means" would of necessity fail because the proletariat had not assimilated even the basic elements of bourgeois culture. The main task for the "proletarian intelligentsia" arose out of this cultural impoverishment. Its mission, according to Trotsky, was not "the abstract formation of a new culture regardless of the absence of a basis for it, but definite culture-bearing, that is, a systematic, planful, and, of course, critical imparting to the backward masses of the essential elements of the culture which already exists."[12]

Phillips's and Rahv's thinking on these key issues largely paralleled Trotsky's. Certainly on the central issue of the viability of proletarian literature their opinions were identical. The following remark by Trotsky could just as easily have been made by Phillips and Rahv: "The 'worker poets' may have . . . [composed] significant cultural and historical documents," but "this does not at all mean that they are artistic documents . . . weak and, what is more, illiterate poems do not make up proletarian poetry, because they do not make up poetry at all."[13] It was with regard to the matter of "definite culture-bearing" that the editors and their magazine gradually came around to a view comparable with Trotsky's. They too came to see the proper function for the revolutionary critic, given the impossibility, or at any rate the absence, of proletarian literature, as in some sense bearing culture. But, of course, because they came to this conclusion in the United States during the late 1930s where revolutionary change was not the order of the day, their understanding of culture bearing inevitably differed somewhat from Trotsky's views.

As we have seen, Trotsky was far from a philistine—in a famous remark Leavis referred to him as a dangerously intelligent Marxist. He took art and literature much too seriously to exploit them for political purposes or to try to turn writers into political apologists. His sophisticated handling of these and other matters was reflected in his essay "Art and Politics," published by *Partisan Review* in 1938. Here he was very frank about the inevitable discrepancy between artistic innovation and the opinions and tastes of the masses, which often conform to the values of the dominant culture. Comparing the genesis of artistic movements with those of the great religious and political movements (including Marxism), he observed that

> Not a single progressive idea has begun with a mass base. . . . When an artistic tendency has exhausted its creative resources, creative "splinters" separate from it, which are able to look at the world with new eyes. The more daring the pioneers show their ideas and actions, the more bitterly they oppose themselves to established authority which rests on a conservative "mass base," the more conventional souls, skeptics, and snobs are inclined to see in the pioneers, impotent eccentrics or "anemic splinters" *[sic]*.[14]

To protect and preserve art's independence, and to insure that its emancipatory potential remained intact, Trotsky argued that its internal, aesthetic, and formal dynamics be respected by granting artists autonomy from outside interference in these matters. His underlying premise was that all authentic art is subversive of traditional values:

> Artistic creation has its own laws—even when it consciously serves a social movement. Truly intellectual creation is incompatible with lies, hypocrisy, and the spirit of conformity. Art can become a strong ally of revolution only in so far as it remains faithful to itself.[15]

As to why art that remained "faithful to itself" was necessarily adversarial, Trotsky explained that such art draws attention to the discrepancy between desire and reality, between the need for a connected and full existence and the materially and spiritually impoverished life offered by a society broken by class division:

> Generally speaking art is an expression of man's need for an harmonious and complete life, that is to say, his need for those major benefits of which a society of classes has deprived him. That is why a protest against reality, either conscious or unconscious, active or passive, optimistic or pessimistic, always forms part of a really creative piece of work.[16]

What is especially interesting about Trotsky's understanding is that in the end he placed a great deal of emphasis on the individual: for him the adversarial, even utopian role of art in the late bourgeois period manifests itself primarily in the contradiction between the individual and hostile social forces. In "Manifesto: Towards a Free Revolutionary Art," published in the subsequent issue of *Partisan Review*, Trotsky borrowed from Freud and suggested that at the heart of the artistic process is a dynamic that is psychological in nature, and only indirectly political. Sounding very much unlike what a revolutionary Marxist was supposed to sound like, and very much like a humanist (with teeth), Trotsky observed that

> The process of *sublimation*, which here comes into play, and which psychoanalysis has analyzed, tries to restore the broken equilibrium between the integral "ego" and the outside elements it rejects. This restoration works to the advantage of the "ideal of self," which marshals against the unbearable present reality all of those powers of the interior world, of the "self," which are *common to all men*.[17]

Thus, according to Trotsky, "true art, which is not content to play variations on ready-made models but rather insists on expressing the inner needs of man and of mankind in its time—true art is unable *not* to be revolutionary, *not* to aspire to a complete and radical reconstruction of society."[18] Quoting the young Marx on behalf of freedom of the press, Trotsky reached his crescendo:

> The free choice of . . . themes and the absence of all restrictions on the range of his explorations—these are the possessions which the artist has a right to claim as inalienable. In the realm of artistic creation, the imagination must escape from all constraint. . . . We repeat our deliberate intention of standing by the formula: *complete freedom for art.*[19]

Yet this was not quite the whole of it, because for the materialist Trotsky, art's revolutionary potential and even its autonomy were ultimately contingent on historical realities. In *Literature and Revolution*, he had made similar appeals for artistic autonomy and government noninterference, this time within the context of a bitter and highly politicized debate within the Soviet Union. By doing so he had intervened on behalf of a tolerant, ecumenical policy toward the arts. At the same time, it is well to recall that even Trotsky reserved the right to censor artists in the interests of the perceived needs of the revolution. The following passage from the enlightened revolutionary will no doubt sound ominous to today's reader, as indeed it must have to large numbers of artists and writers of the time, many of whom perished not only because Trotsky and others lost the struggle over the arts, but partly because he himself offered no principled opposition to suppression, and even supported it under certain circumstances:

> Our standard is, clearly, political, imperative and intolerant. But for this very reason, it must define the limits of its activity clearly. For a more precise expression of my meaning, I will say: we ought to have a watchful revolutionary censorship, and a broad and flexible policy in the field of art, free from petty partisan maliciousness.[20]

He continued:

> It is quite evident that the Party cannot, not for one day, follow the liberal principle of *laissez faire* and *laissez passer*, even in the field of art. The question is only at what point should interference begin, and what should be its limits.[21]

Had Trotsky changed his views on the matter of artistic censorship by 1938? His guarantee that the artist had "an inalienable" claim to his themes and explorations would indicate that he had, but we are nonetheless obliged by history to ask whether he meant "inalienable" in the current period or "inalienable" forever? What is interesting is that no one within the *Partisan Review* circle asked this question at the time, at least not publicly. Why not? Probably because these writers believed that the only threats to artistic freedom and openness arose from Stalinism. Whether their own cultural heroes (T. S. Eliot, for instance) or whether they themselves as powerful cultural arbiters of taste might then or in the future impose a direction on literature and thereby close off opportunities for certain writers was a possibility that seemed to escape them. Over the years *Partisan*

Review's vigilance concerning the repression of art was in many ways admirable, but too much vigilance emanated outward and not enough was directed inward so as to avoid the kinds of ideologically determined exercises of exclusionary cultural power, which, as we shall see, did mar the magazine's record in later decades.

In any case, in 1938 it was the immediate historical situation that was important both to Trotsky and the *Partisan Review* critics in their understanding of the role of culture. For Trotsky any new avant-garde would have to be created as a response to a swiftly deteriorating global political situation, the elements of which were the exacerbated contradictions of capitalism in crisis, the consequent rivalry of the democracies, the menace of fascism, and the absolute bankruptcy of the Stalinist movement. These factors defined what Trotsky called a "catastrophic" situation. Indeed the "Manifesto" began apocalyptically enough: "We can say without exaggeration that never has civilization been menaced so seriously as today."[22] The gravity of the political situation had radically altered the social position of art. As long as the bourgeoisie had been relatively stable, it was capable of controlling and assimilating the avant-garde: each moribund movement could be tolerably replaced by a fresh revolt in a vigorous dialectic of innovation and absorption. But given the current crisis, such a dialectic was no longer possible. The bourgeoisie was in the throes of an unprecedented economic, political, and cultural crisis, and the avant-garde alone could not withstand the new pressures placed upon it. Although the avant-garde movements of the past had provided numerous heroic models, Trotsky maintained that a politically indifferent or politically unsophisticated avant-garde was dead as a dynamic social and aesthetic force. If it was to survive at all, the avant-garde would have to position itself *alongside* the political vanguard.

In its broadest sense, *Partisan Review*'s literary program had a number of things in common with Trotsky's. We have seen that the editorial statement announcing the reappearance of the magazine on a new basis unambiguously declared the editors' support for artistic freedom: "Conformity to a given social ideology or to a prescribed attitude or technique, will not be asked of our writers." In addition, *Partisan Review*'s justification of modernist literature was largely based on the kind of argument Trotsky made in the "Manifesto": modernism is subversive because it exposes and invariably resists the social and psychological ills brought about by capitalism. Thus we can add to the major areas of agreement between *Partisan Review* and Trotsky—which we may list as the critique of Stalinism, the rejection of proletarian literature, the defense of cosmopolitan and internationalist values, and the view of Marxism as an analytic tool—a belief in the relative autonomy of art, the defense of artistic freedom, a focus on individual experience in art, and a belief that in the bleak political situation of the late 1930s all genuine art was subversive.

As I have shown,[23] the main difference between Trotsky and *Partisan Review* concerned the prospects for the development of a new avant-garde, which Phillips and Rahv felt could not be sustained through an alliance with the revolutionary movement. Rahv came to this conclusion in his eloquent and moving essay "Trials of the Mind" (1938), in which he commented on the tragic significance of the Moscow Trials for American radicalism. He began by observing that the defeat of the proletariat and the consequent dubiety of Marxism had suddenly become the fundamental condition of existence for contemporary radicalism:

> Ninety years have passed since the most subversive document of all times, *The Communist Manifesto*, injected its directive images into the nascent consciousness of the proletariat. We are not prepared for defeat. The future had our confidence, which we granted freely, sustained by the tradition of Marxism. In that tradition we saw the marriage of science and humanism. But now amidst all these ferocious surprises, who has the strength to reaffirm his beliefs, to transcend the feeling that he had been duped? One is afraid of one's fears. Will it soon become so precise as to exclude hope.[24]

But it was not the Moscow Trials alone, representing as they did the collapse of the socialist experiment in the Soviet Union, that led to Rahv's trepidation. His fear arose largely from his perception of the response of the Western intellectuals, particularly the liberals, to the Soviet debacle. As a group they were engaged in a "treasonable" charade of apologizing for the trials, "clamor[ing] for blood" like cheerleaders of the bourgeois war machine. "Much is being said and written about the 'moral collapse of Bolshevism,'" noted Rahv,

> but how little about the moral collapse of the intellectuals. Among them smugness has become the pseudonym of panic, and the more rapidly they abandon the values of culture the more sonorous their speeches in its defense. Everywhere they submit to the accomplished fact, everywhere they place themselves under the surveillance of authority, they rationalize, they explain away.[25]

For Rahv the appropriate metaphor for the intellectuals' pious subservience to bourgeois power was that of blatant incorporation:

> The masters of property . . . know that the "intellectual" of the family has no talent for practical affairs. He is outraged because the firm no longer treats its employees with the decency and kindliness that prevailed when business was flourishing. . . . But they are quite sure that in the end he will "come around," accept a minor post, and occupy himself with keeping the firm's books in order.[26]

This corporation was a far cry from that recommended by Julien Benda, to whose *Treason of the Intellectuals* (1928) Rahv alluded at several points in his essay. Benda had described the duty of intellectuals "to set up a corporation whose sole cult is that of justice and truth."[27] Part of Rahv's indignation at the "betrayal" of the intellectuals stemmed from his sympathy for such a notion of the intellectual vocation. For him the intellectuals, as the "guardians of values" within their assigned domain of "technical and spiritual culture," were "trampling on the very values they depend on for permanent sustenance."[28] Like Benda, Rahv passionately condemned the intellectuals for the unseemly services they were rendering the politicians. Unlike Benda, however, Rahv retained a classical Marxist analysis of the situation, in which the intellectual stratum, being a part of the petty bourgeoisie, gets pulled to and fro and is eventually torn apart by the proletariat and the bourgeoisie, the two major adversaries within capitalist society. Rahv was very clear on this point; for him it distinguished his position from Benda's ultimately idealist one:

> In what sense can we speak of the treason of the intellectuals? Marxism has taught us that the intellectuals are a special grouping within the middle class, as much infected with its unrest and ambition as with its fright and fantasies. This definition implies that when the working class is apparently beaten, the intellectuals veer back to their own positions, each time of course under the cover of a new set of rationalizations. The events of recent years have fully confirmed this definition.[29]

Rahv's class analysis corresponded to a somewhat different idea of vocation for intellectuals than Benda's, for Rahv believed in the compatability of proletarian partisanship and a reasonable, uncoerced search for truth. He argued for a rationally politicized intelligentsia, the example of the intellectuals grouped around *Partisan Review* providing an alternative to Benda's ideal of absolute detachment. Yet Rahv was willing to search for grounds that both ideals might share given the emergencies of the current situation. He put it as capaciously as possible:

> While it might be naive to expect [the intellectuals] to cleave to revolutionary ideas at a time when the proletariat is in rout . . . it does not require a revolutionary to be repelled by lies. It is not necessary to be a Marxist to be opposed to the perversion of historical facts.[30]

Under duress, then, Rahv's "minimal program" for intellectuals came to resemble Benda's "maximum program." Rahv was willing to welcome a stance of critical autonomy among intellectuals as long as the political situation made more direct forms of engagement dangerous to the intellectual enterprise.

II

The kind of interest in the intellectuals exhibited here very quickly grew among the *Partisan Review* critics, as we shall see. It was an interest largely caused by two factors. Politically, it grew proportionately as did the belief that the proletarian defeat was a more or less permanent feature of American society. Culturally, it took hold as most writers for the magazine began to understand modernism increasingly in the context of its social base within the intellectual stratum. Not surprisingly, these new interests were based on an evolving political and cultural analysis that entailed an increasingly critical view of Marxism.

The first overall assessment of the Marxist legacy to appear in *Partisan Review* was Rahv's commentary "What Is Living and What Is Dead" (1940). As the reader will recall, this was to have been the title of the symposium that *Partisan Review*'s editors had proposed to Trotsky.[31] As originally conceived, it was to have served as an important means to begin a full-scale reevaluation of Marxism by the anti-Stalinist left. But it failed, largely, as I have indicated, because Trotsky opposed it, owing to his belief that no one in the United States was qualified to answer the question, a question that in any event he considered presumptuous in the extreme. Thus Rahv's rather brief commentary was all that remained of what was initially meant to provoke a wide-ranging discussion.[32]

Briefly, the results of Rahv's search for life were as follows: still alive were the theories of the class struggle, the bourgeois state, and the capitalist economy, as well as the perspective of internationalism; dead were the theories of the vanguard party and the dictatorship of the proletariat; alive but in critical condition was the belief in the revolutionary status of the working class. He argued that this latter belief, hitherto held sacred by Marxists, needed to be reexamined in light of the failure of any Western proletariat to attain a significant degree of class consciousness. For Rahv events simply had not confirmed Marx's prognosis that the decline of capitalism would be accompanied by a deepening working-class consciousness. Assessing the international political situation of 1940, he declared "there exists in no country today a workers' movement of any size or influence which is carrying on the revolutionary tradition."[33] He asked why members of a supposedly revolutionary class "so consistently entrust their fate to non-revolutionary and even counter-revolutionary organizations and thus put themselves into the position of being betrayed."[34] Even so formidable a Marxist as Trotsky, according to Rahv, merely begged the question when he repeatedly blamed the Stalinists and the social democrats for faulty leadership. Is there not an implicit admission here, asked Rahv, that the proletariat "is intrinsically lacking in sufficient independence and political self-definition to check and control its leadership?"[35] Moreover, he continued, "it is time for Marxists to be rid of the absurd convention . . . that plain speaking about the proletariat is somehow indecorous, if not actually impermissible. In its own right the proletariat represents no ideals—it is only in its role as a dynamic and militant force in the struggle to liberate humanity that values can be attached to it."[36]

Unfortunately, Rahv's bracketing of the important Marxist axiom of the revolutionary character of the proletariat did not lead to "plain speaking" about the proletariat. Rather, it led to silence on the working class and an increasingly voluble discussion of the need to reevaluate Marxism in light of the growing importance to capitalism of the intellectuals. As we have seen, Rahv inaugurated the serious discussion of intellectuals in "Trials of the Mind"; he developed his views further in his important piece on cultural decline, "Twilight of the Thirties" (1939). As had Trotsky, Breton, and Rivera a few months earlier, Rahv portrayed in passionate terms the apparent disarray of the Western democracies in the face of the totalitarian threat, a threat he described as having both external and internal sources. He maintained that the democracies were actually following the same course as their totalitarian counterparts in destroying cultural values and "liquidating" literature—only the "form" and "tempo" were different. Whereas the fascists were burning books, in the democracies the intellectuals were "voluntarily subjecting themselves to a regimen of conformity." Obediently, and at times even with relish, they were adapting their work "to the coarsening and shrinking of the cultural market." This represented, according to Rahv, "the first stage in a process that might be called the withering away of literature."[37] As the critics effusively praised the likes of Steinbeck, Saroyan, Sandburg, and Caldwell, for Rahv the century-old tradition of the avant-garde literary rebellion was being buried. In its weakened state, capitalism had withdrawn "the privilege of limited self-determination hitherto granted its intellectuals. . . . Everywhere the vulgarizers and the big money adepts are ruling the literary roost."[38] For those few of conscience, the 1920s suddenly looked like a golden age, alive with bold experiments and independent minds.

Given the veritable stampede of the American intellectuals to enlist in the Popular Front and don the uniforms of dissent, Rahv saw new virtues in the relatively detached and apolitical stance taken by many modernist writers. Now, like Trotsky, rather than label as "escapist" the various apolitical means by which modernists had tried to transcend a corrupt social world, Rahv demonstrated a new appreciation for their diverse strategies of resistance. He insisted, also, upon seeing these strategies collectively, that is, less as individual effusions than as the products of what he called a "group-ethos . . . the proud self-imposed isolation of a cultivated minority."[39] In this way the modernist avant-garde represented an alternative to the catastrophic surrender of values by the intellectuals of the late 1930s. Sounding very much like his Frankfurt Institute counterparts, some of them now refugee scholars in Manhattan, Rahv observed that the isolation of the modernist "enabled the art-object to resist being drawn into the web of commodity relations. . . . In the sphere of art many producers still found it possible—through a valiant effort, certainly, and at the cost of much suffering—to remain the masters instead of the victims of their products."[40]

In order to justify this reevaluation of modernism, Rahv made several theoretical adjustments to Marxism. Declaring that "to speak of modern literature is to speak of that peculiar grouping, the intelligentsia, to whom it belongs," he judged that the intellectual stratum as a whole had been regrettably ignored by Marxism. Marxism, he asserted,

> has always laid too much stress on such terms as "bourgeoisie" and "proletariat." This is an error . . . because literature is not linked directly to the polar classes, but associates itself with . . . the life of society . . . by giving expression to the given biases, the given moods and ideas of the intellectuals. [Analysis of] the special role and changing status of the intellectual is therefore essential to any social examination of modern literature.[41]

In spite of Rahv's more positive assessment of the modernist avant-garde, he had become more pessimistic than Trotsky about the chances for a new avant-garde to arise. The prewar situation was too dangerous for the bourgeoisie to muster the necessary tolerance for the survival of an avant-garde. Given the climate of intense nationalism, shared even by the Popular Front, any vibrant, iconoclastic, truly adversarial culture such as the bohemian culture of the 1910s and 1920s was completely snuffed out. Certainly the revolutionary movement alone, in its weakened state, could hardly take the place of such a culture. This left *Partisan Review* with the task of insuring that the modernist avant-garde of the previous generation remain a vital source of social critique as well as a continuing inspiration to young writers. In this way Rahv and others justified a shift from an activist to a custodial function for the magazine. Rather than help to create a new culture, the function of *Partisan Review* would be that of "culture-bearing," to use the term Trotsky had employed to describe the main task of artists and critics in the Soviet Union during the 1920s. The reader will recall that Trotsky had then proposed, "The main task of the proletarian intelligentsia is not the abstract formation of a new culture . . . but definite culture-bearing, that is, a systematic, planful and, of course, critical imparting to the backward masses of the essential elements of the culture which already exists."[42] The best-known articulation of this project by one of the *Partisan Review* critics came at the close of Clement Greenberg's widely influential essay "Avant-Garde and Kitsch," which appeared in the issue of *Partisan Review* that followed the one containing Rahv's "Twilight of the Thirties." The reader will recall that Greenberg had warned, "Today we no longer look toward socialism for a new culture—as inevitably as one will appear, once we do have socialism. Today we look to socialism *simply* for the preservation of whatever living culture we have right now." However provisional this strategy was meant to be according to Rahv and Greenberg—as a kind of holding operation until such time as the revolutionary movement would strengthen itself so as to reclaim and reshape the avant-garde in its own image—it is clear in retrospect that neither *Partisan Review* nor the intellectuals or the American working class would fulfill the hopes even of this chastened perspective.

It is useful to compare *Partisan Review*'s understanding of the new and important role to be played by the intellectuals with Antonio Gramsci's influential insights into the function of intellectuals. Gramsci, of course, distinguished between *organic* intellectuals—intellectuals closely affiliated with a strategically placed social class, either hegemonic or emerging—and *traditional* intellectuals, those who have managed over the centuries to attain a certain distance from the immediate class struggle by virtue of a living tradition, established institutions, and specialized practices. In an illuminating passage from the *Prison Notebooks*, Gramsci applied these distinctions to the United States, noting especially the absence of traditional intellectuals:

> One can note in the case of the United States, the absence to a considerable degree of traditional intellectuals, and consequently a different equilibrium among intellectuals in general. There has been a massive development, on top of an industrial base, of the whole range of modern superstructures. The necessity of an equilibrium is determined not by the need to fuse together the organic intellectuals with the traditional, but by the need to fuse together in a single national crucible with a unitary culture the different forms of culture imparted by immigrants of differing national origins.[43]

Gramsci suggests that American intellectuals have developed an organic relationship to the bourgeoisie because historically they lack precapitalist, independent traditions, practices, and institutions; and because they have, by and large, cooperated in the overall cultural task of assimilating the various ethnic group into the hegemonic culture. It is interesting to observe how similar was Rahv's assessment of the function of American intellectuals within a hegemonic society; indeed Rahv's language bore an uncanny resemblance to Gramsci's as he criticized intellectuals for their lack of independence and their increasing subservience to a bourgeoisie increasingly more adept at winning consent than wielding force:

> Whereas in the Third Reich, for instance, all free and authentic creation, as well as thought, are forcibly suppressed, in the democracies no force is applied because its application is as yet uncalled for. Here most of the artists and "thinkers" are voluntarily subjecting themselves to a regimen of conformity, are "organically" as it were—obediently and at times even with enthusiasm—adapting their products to the coarsening and shrinking of the intellectual market.[44]

But whereas Gramsci called for intellectuals to develop organic ties to the working class, Rahv's and *Partisan Review*'s response was to advocate a strategy that would at least discourage intellectuals from developing organic ties to the bourgeoisie. Here the modernist avant-garde played a decisive role. Although certainly in no way organically related to the working class, the modernist avant-garde did share key characteristics—even if in incipient form—with Gramsci's traditional intellectuals. These included the avant-garde's intimate relationship to cultural traditions, its independent institutions in the form of its little magazines, its development of a new praxis emphasizing technical and formal proficiency, and its independent critical assessment of accepted values and taste. For *Partisan Review* the avant-garde provided the larger stratum of intellectuals with an example of how they might constitute themselves into a traditional intelligentsia. In concluding "Twilight of the Thirties" Rahv offered what amounted to just this strategy for moving the intellectuals leftward:

> If a sufficiently organic, active, and broad revolutionary movement existed, it might assimilate the artist by opening to him its own avenue to experience; but in the absence of such a movement all he can do is to utilize the possibilities of individual and group secession from, and protest against, the dominant values of our time. Needless to say, this does not imply a return to a philosophy of individualism. It means that all we have left to go on is individual integrity—the probing conscience, the will to repulse and to assail the forces released by a corrupt society.[45]

Compared with Gramsci's aggressive strategy, Rahv's modest hopes were conspicuously more defensive. Rather than join a working-class vanguard, Rahv elected one from the modernist writers of the previous generation and appealed to intellectuals to follow. It was not a "r-r-revolutionary" strategy, to borrow Lenin's word for ultra-leftism, but it was a leftist strategy that was consistent with Rahv's extremely pessimistic assessment of the state of the contemporary revolutionary movement.

Rahv's political analysis of the role of intellectuals was seconded and elaborated upon by William Phillips in his important essay "The Intellectuals' Tradition" (1941). For Phillips the traditional mode of Marxist class analysis, with its reliance upon the bourgeoisie/proletariat binarism, was unable to deal adequately with modern literature, whose tendency was "to withdraw from the conditions of its existence," both through recoiling from society's accepted practices and values and through striving toward self-sufficiency. One way of adjusting the critical framework, suggested Phillips, was to "locate the immediate sources of art in the intelligentsia . . . for the special properties of modern literature, as well as the other arts, are readily associated with the characteristic moods and interests of the intellectuals."[46] Phillips maintained that an emphasis on the intellectuals provided critics with an alternative to traditional criticism, which fetishized a writer's detachment, and Marxist criticism, which "overstressed the correspondences between the historical context and the work itself."

For Phillips the politics of modernism was to be traced to the development in Europe of stable groupings of secular intellectuals with their own practices and traditions. By virtue of sustaining a more or less coherent outlook, this intelligentsia had been able to provide literature with "a highly elaborate sense of its (own) achievements and its tasks, thus providing the creative imagination with a fund of literary experiences— a kind of style of work—to draw on."[47] Phillips argued that in contrast to Europe, American society had never sustained a well-established, relatively autonomous intellectual grouping:

> The outstanding feature—not to speak of the failures—of our national culture can be largely explained by the inability of our native intelligentsia to achieve a detached and self-sufficient group existence that would permit it to sustain its traditions through succeeding epochs, and to keep abreast of European intellectual production. One need hardly stress such symptoms in American writing as shallowness, paucity of values, a statistical approach to reality, the compensatory qualities of forthrightness, plebeianism, and a kind of matter-of-fact humanism. . . . To be sure, our cultural innocence has been practically a standing complaint of American criticism.[48]

It was Phillips's view, then, that the lack of a "traditional" intelligentsia had caused American letters to fluctuate wildly between the extremes of conformity and dissidence, self-effacement and confidence. Once again employing Gramsci's terms, we may say Phillips believed that *any* organic relationship on the part of intellectuals—whether with regard to the bourgeoisie *or* the working class—would cause them to relinquish their autonomy and hence their critical edge. Yet it is also true that Phillips was much more concerned about the dangers of an organic relationship with the working class, which to him would necessarily mean surrendering to Stalinism or some other less powerful but equally debilitating form of harassment and coercion. For all intents and purposes equating the Party with the entire working class, Phillips claimed it had been the intellectuals' "natural inclination to merge with the popular mind that has prevented any such lasting intellectual differentiation as has been achieved in European art and thought."[49] Specifically he had in mind the intellectuals' recent rush toward public acceptance by way of lending support to the Popular Front, whose mastermind, the Communist Party, had lately subordinated radical rhetoric to Main Street nationalism in its quest for acceptance by the American public. At this juncture—the late 1930s and early 1940s—it did not occur to Phillips, Rahv, or any other *Partisan Review* critic that their resistance to an organic relationship to the working class in favor of autonomous yet critical intellectual inquiry would require an equally committed resistance to an organic relationship with ruling elites.

WARTIME:
THE NEW YORK
INTELLECTUALS IN BATTLE

5

Modernist Renewal

To organize a new union between our political ideas and our imagination – in all our cultural purview there is no work more necessary. It is to this work that Partisan Review *has devoted itself.*

LIONEL TRILLING, "The Function of the Little Magazine"

I

Partisan Review's critical project from the late 1930s until 1945 is best understood as a radical appropriation of modernism for the purpose of renewing the culture and politics of the American left. I mean several things by this characterization: that *Partisan Review* offered incisive critique of some of the important political and cultural assumptions of the left; that in doing so it underscored the importance of literature and criticism as privileged sources of social knowledge; that it had a vision of a new and complex culture whose relationship to subjective experience and to history would provide the basis for a new politics. In the following pages I hope to elaborate on each of these themes as I examine representative critical essays by those in or near the *Partisan Review* circle.

I want first of all to be precise about what I mean by the "left" in the particular context of which I write. In one respect I mean the whole spectrum of political groups and their constituencies from the New Deal liberals to the Trotskyist sects of the far left.[1] As to the values these groups shared, and in many respects continue to share (minus the optimism), Lionel Trilling has made the best attempt to enumerate the more salient among them. Referring to Dos Passos's *USA* Trilogy as a series that embodied the cultural tradition of the intellectual left, Trilling summed up what he took to be this tradition's characteristic beliefs and attitudes:

> Briefly and crudely, this cultural tradition may be said to consist of the following beliefs, which are not so much formulations of theory or principles of action as they

> are emotional tendencies: that the collective aspects of life may be distinguished from the individual aspects; that the collective aspects are basically important and are good; that the individual aspects are, or should be, of small interest and that they contain a destructive principle; that the fate of the individual is determined by social forces; that the social forces now dominant are evil; that there is a conflict between the dominant social forces and other, better, rising forces; that it is certain or very likely that the rising forces will overcome the now dominant ones.[2]

Trilling supplemented these remarks several years later by observing that the left's attitudes included a suspicion of the profit motive, and a belief in "progress, science, social legislation, planning, and international cooperation, perhaps especially where Russia is in question."[3] The famous phrase that Trilling used to identify this particular set of attitudes was "the liberal imagination," and of course by this he meant to identify something broader than the left—an aspect of the American mind generally. This does not negate the fact, however, that for Trilling at the time the left was at once steeped in these values and therefore in dire need of undergoing a process of self-criticism that would complicate and humanize them.

The *Partisan Review* critics were as much concerned with the manner in which beliefs were held by the left as with the content of the beliefs themselves. What fell to them was largely the task of exploring the texture and the quality of the culture of the left—not that this culture was seen as anything but inextricably tied to politics, but that it came to be seen as shaping those politics and, therefore, as a reflection of the "condition of mind" of the left, something prior to and more fundamental than the articulation of politics in terms of positions and policies. In this respect the project of the *Partisan Review* critics was part of the general trend during the 1930s to discover a "usable past"—to define the meaning of culture and to determine what American culture might be.

This growing preoccupation with the culture of the left entailed an entirely new definition of politics. In "The Function of the Little Magazine," Trilling indicated what he hoped the essence of this definition would be: "The only possibility of our enduring [our political fate] is to force into our definition of politics every human activity and every subtlety of every human activity. . . . Unless we insist that politics is imagination and mind, we will learn that imagination and mind are politics, and of a kind that we will not like."[4] I have always been struck by the similarities in meaning and tone between this passage and Walter Benjamin's warning at the end of "The Work of Art in an Age of Mechanical Reproduction" that unless art is politicized, politics will be aestheticized. I take Benjamin's gnomic phrases to mean, despite appearances to the contrary, something close to Trilling's: that unless politics is predicated on man's desires as expressed aesthetically, desire and aesthetics will be predicated on politics. But this new basis for politics would necessarily entail a much broader definition of what politics actually is. Here is Trilling once again:

> If between sentiments and ideas there is a natural connection so close as to amount to a kind of identity, then the connection between literature and politics will be seen as

> a very immediate one. And this will seem especially true if we do not intend the narrow but the wide sense of the word politics. It is the wide sense of the word that is nowadays forced upon us, for clearly it is no longer possible to think of politics except as the politics of culture, the organization of human life toward some end or other, toward the modification of sentiments, which is to say the quality of human life.[5]

For the *Partisan Review* critics, this broadened definition of politics was accompanied by the corresponding notion that criticism ought to do something other than affect the outcome of the immediate political struggle. The opening up of politics to a broad cultural approach engendered the radical new insight (for the American left) that literature and literary criticism could provide knowledge essential to political understanding and practice. What was now needed was knowledge of the ways individuals and groups *experienced* politics and history, previously thought of as material, external, and objective. *Partisan Review*'s critical project was a major effort to reconceive political experience as significantly subjective—not unlike, by the way, Fredric Jameson's later appeal in *Marxism and Form* for a phenomenological criticism that could explain "not what a thought is, so much as what it feels like . . . not make statements about content, but describ(e) the mental operations which correspond to that content."[6] In forcing these considerations on a left political discourse that had habitually evaded the terms of personal responsibility and moral values, the *Partisan Review* critics looked to literature as a primary source of insight. As Trilling put it in *The Liberal Imagination*, "To the carrying out of the job of criticizing the liberal imagination, literature has a unique relevance. . . . Literature is the human activity that takes fullest and most precise account of variousness, possibility, complexity, and difficulty."[7]

This should not suggest that, in the hands of the *Partisan Review* critics, literature illuminated subjective experience and nothing else. For them literature was a means for wresting subjectivity from a political discourse that had crowded out the details of interior life with reductive abstractions. The reliance on abstractions such as science, progress, the masses, and individualism had effectively subordinated the self in an orgy of selflessness. By imposing a lifeless materialism and a pervasive utilitarianism on the already shrinking sphere of personal life, the left had suppressed that portion of experience where the moral sensibility might evolve into a source of political activity. But the *Partisan Review* critics were not given to separating the private self from the social self, even though their stress was increasingly upon the individual.

II

Having made this preliminary characterization of *Partisan Review*'s critical project, I want to turn to the literary criticism, for only by examining the magazine's appropriation of modernism, or portions thereof, can we appreciate the full force and subtlety of its critique. Of primary importance to this critique were Rahv's programmatic essays "Paleface and Redskin" (1939) and "The Cult of Experience in American Writing" (1940). In these essays Rahv argued that the left's preoccupation with unreflected experience—that is, its anti-intellectual propensity to advocate praxis without full consciousness—perpetuated the traditional American dualism between experience and mind, or, put another way, between mindless activity and mindful abstinence. For Rahv, and here he was representative of the *Partisan Review* critics as a group, the unifying antidote to this dissociated American sensibility had to be the proper relation to experience. Such a relation was not easily won according to Rahv: even Henry James, who among American writers was said to have come closest to possessing it, fell short because, an American in spite of himself, he was hemmed in by the national tendency to evade experience.[8] Rahv paid James the high compliment of referring to him as America's "Founding Father" of experience. Of *The Ambassadors* he observed "It is a veritable declaration of the rights of man—not, to be sure, of the rights of the political man, but of the rights of the private man, of the rights of personality, whose openness to experience provides the sole effective guarantee of its development."[9]

The issue to be raised here is not whether Rahv felt James could help to unify the dissociated American sensibility or internationalize American literature (I discuss these crucial issues in chapter 7), but rather Rahv's view of the impoverished cultural condition of the American left, which he felt was perpetuating the characteristic American tendency of separating "thinking" and "doing," or "being conscious" and "unconsciously being." What Rahv wanted, once again echoing Eliot, was "a person who thinks with his entire being, that is to say . . . a person who transforms ideas into actual dramatic motives instead of merely using them as ideological conventions or as theories . . . externally applied."[10] The point was that such people were simply not to be found on the left.

The kind of writing the *Partisan Review* critics felt could provide models for such individuals was rarely achieved by writers with liberal sympathies. In a well-known statement, Trilling dramatically called attention to this paradox:

> Our liberal ideology has produced a large literature of social and political protest, but not, for several decades, a single writer who commands our real literary admiration. . . . And if on the other hand we name those writers who . . . are to be thought of as the monumental figures of our time, we see that to these writers the liberal ideology has been at best a matter of indifference. . . . So that we can say that no connection exists between our liberal educated class and the best of the literary minds of our time."[11]

Contrary to the likely impression of current readers, this strong statement was not meant to suggest an inherent incompatibility between modernism and liberalism. Rather, it was a rebuke of liberals and leftists who refused the difficult task of making their culture diverse and capacious enough to assimilate the modernist critique of contemporary life. It was also implicitly an appeal to create a culture capable of sustaining serious writers. Here the function of criticism would be paramount, but it would need to go beyond the Arnoldian injunction to "create a current of true and fresh ideas" and more aggressively attempt to reshape modes of perception and experience. "Disinterestedness," therefore, was not quite the value sought, unless we are clear that we mean the kind of "disinterestedness" displayed by Arnold in "Our Liberal Practitioners"—in other words, circumspect engagement. In the best essays of the *Partisan Review* critics from this period, there is an attempt to bridge the gulf between modernism and liberalism by intervening in the politics of the left, but always with the understanding that both the crossing and the landing were extremely precarious operations. Hardened by political experience, possessed with only so much optimism as their times and immersion in modernism allowed, and laden with a full complement of irony and doubt, the *Partisan Review* critics proceeded to question the simple pieties of the left.

More than anyone else Lionel Trilling established the themes to which the *Partisan Review* critics returned again and again: the liberal-radical culture's false optimism; its blind faith in progress; its mechanical materialism; its diminishment of self, consciousness, and human agency; its impatience with moral and spiritual experience; its anti-intellectualism. "The America of John Dos Passos" (1938), a review of Dos Passos's *USA* trilogy, shows the extent to which Trilling shaped his ideas in response to the beliefs and attitudes of the Popular Front, here expressed by the critics Malcolm Cowley and T. K. Whipple. In typical fashion Cowley had taken exception to Dos Passos's alleged "despair," his failure to depict "the will to struggle ahead." Whipple complained that the class struggle was fatally absent in Dos Passos's so-called panorama of American life. What most troubled Trilling was the implication that despair was politically harmful, and he argued that unless one were entirely willing to rule out even the possibility of corruption or failure on the left—and only the most committed ideologue would—then one must acknowledge that Dos Passos's despair might be consistent with self-evaluative political work:

> There are many kinds of despair and what is really important is what goes along with the general emotion denoted by the word. Despair with its wits about it is very different from despair that is stupid; despair that is an abandonment of illusion is very different from despair which generates tender new cynicisms. The "heartbreak of *Heartbreak House*, for example is the beginning of new courage and I can think of no more useful *political* job for the literary man today than, by the representation of despair, to cauterize the exposed soft tissue of too-easy hope.[12]

As for the absence of class struggle in Dos Passos's trilogy, Trilling claimed just the opposite, that the class struggle was indeed decisive in *USA*, but was manifested in the "moral paradoxes" faced by marginal characters who felt the painful conflict between their "real and apparent interest." "If Dos Passos has omitted the class struggle," commented Trilling, "it is only the external class struggle he has left out; within his characters the class struggle is going on constantly."[13] By virtue of this shift in focus, Trilling meant to challenge the one-sided materialism of the left. He noted that for Dos Passos the great, solid things of modern life—the institutions, financial networks, conflicting classes—were to be thought of less as objective, determining forces and more as conditions affording opportunities for "personal moral action." According to Trilling, for Dos Passos the deepest tragedy of the social decomposition depicted in *USA* was that it led to "moral degeneration," not just "economic deprivation." Dos Passos's strong moral sensibility caused him to be concerned "not so much with the utility of an action as with the quality of the person who performs it. *What* his people do is not so important as *how* they do it."[14]

In justifying Dos Passos's moral approach, Trilling alluded to Dewey's belief that certain moral situations make it impossible to choose between ends. "As a result," remarked Trilling, "we make our moral choice in terms of our preference for one type of character or another: 'What sort of an agent, of a persona shall he be? This is the question finally at stake in any genuine moral situation. What shall the agent *be*? What sort of character shall he assume?'"[15] Significantly, Trilling admitted that this moral approach to political questions did not apply transhistorically. "Although dilemmas exist in every age," he observed,

> we do not find Antigone settling her struggle between family and state by a reference to the kind of character she wants to be. But for our age, with its intense self-consciousness and its uncertain moral codes, the reference to the quality of personality has meaning, and the greater the social flux the more frequent will be the interest in qualities of character rather than in the rightness of the end.[16]

Here we come to the crux of Trilling's argument concerning literature's relation to politics. In the wake of a diminished left, only literature, ushered in by criticism, retained the capacity for moral judgment based on a full consideration of the relevant social factors. Much of *Partisan Review*'s interest in the modernist novel was summed up in the following passage by Trilling:

> The modern novel, with its devices for investigating the quality of character, is the aesthetic form almost specifically called forth to exercise the modern way of judgment. The novelist goes where the law cannot go; he tells the truth where the formulations of even the subtlest ethical theorist cannot. He turns the moral values inside out to question the worth of the deed by looking not at its actual outcome but at its tone and style. He is subversive of dominant morality and under his influence we learn to praise what dominant morality condemns; he reminds us that benevolence may be aggression, that the highest idealism may corrupt.[17]

Through a logic unleashed by the defeats and complexities of the age, because of a historical conjuncture too changeful to be understood by any single totalizing paradigm, literature's moral vision becomes a political necessity. Trilling's argument, therefore, was as follows: often the consequences of political action cannot be known; ergo, the best indication of outcome is often the means employed; ergo, the means employed are known by virtue of knowing the character of those employing them; ergo, the mode that best prepares us to know character and therefore to judge action is the modern novel. The syllogism is completed in the recognition that the literary, with its concern for individual quality, especially as manifested in style and tone, returns us to the level of politics. As Trilling wrote in concluding his essay, "It is not at all certain that it is political wisdom to ignore what so much concerns the novelist. In the long run is not the political choice fundamentally a choice of personal quality?"[18]

One of the contemporary writers whose moral approach to politics coincided with that of Trilling and other *Partisan Review* critics was the Italian novelist

Ignazio Silone. His first novel, *Fontamara* (1930; English trans. 1934), told the story of impoverished peasants from southern Italy brutalized in their fight against the local fascists. The novel became an international sensation and Silone, a founder of the Italian Communist Party, became a celebrity of the Comintern's international antifascist movement. His next novel, *Bread and Wine*, was received with much less fanfare, in part because he had written a letter to Soviet officials protesting the Moscow Trials, and in part because the latest novel was less overtly political than the former in its abiding concern for moral and religious themes. Whereas in *Fontamara* Silone had depicted a heroic popular movement replete with examples of courageous resistance to fascism, in *Bread and Wine* the novelist told a story whose political impact was much more ambiguous. Here was a tale of an exiled revolutionary newly returned to fascist Italy disguised as a priest in order to bring together the scattered remnants of the former socialist groups. In this atmosphere of profound defeat, fascism was a fact of life and submission had become a daily ritual. Because revolution was unthinkable, instances of individual courage and integrity were rare, but they counted for that much more.

In a lengthy review of Silone's two novels, the playwright, drama critic, and Trotskyist Lionel Abel made it clear that he was not interested in the divorce between politics and ethics endorsed by most critics on the left. He dissented from the view that Silone had lost his critical edge with regard to the current political situation. On the contrary, the novelist had recognized that the gravity of the political situation had placed new demands on politics:

> In periods of reaction and defeat, when the class whose historical obligation it is to struggle and conquer is marking time or sunk in apathy—at such times the individual is severely limited; without personal resources of moral integrity he must inevitably fall in line and support the oppressor. . . . A political party which destroys the moral discrimination between values wants agents, automatons, not men.[19]

Silone's transmutation of a political situation said to require no individual choices into one requiring difficult moral decisions, offered the political insight that a movement that destroys the capacity for moral discrimination among its followers inevitably destroys "the spontaneous revolutionary enthusiasm" of the masses. Silone's value was not only that he implicity criticized Stalinism, but that he identified an important prerequisite for an alternative politics.

That T. S. Eliot's influence remained an abiding one for *Partisan Review* during the 1940s is evidenced by the magazine's publication of "East Coker" in 1940, "The Dry Salvages" in 1941, "The Music of Poetry" in 1942, and "Notes Toward a Definition of Culture" in 1944 (the latter was followed by a discussion of Eliot's ideas by William Phillips, Clement Greenberg, R. P. Blackmur, and I. A. Richards). In addition there was Trilling's "'Elements That Are Wanted,'" a surprisingly sympathetic discussion of Eliot's *The Idea of a Christian Society*, which, although objectionable to Trilling for its idealism and elitism, nevertheless raised issues that he thought the left incapable of handling.

Trilling began his piece defensively: he knew how controversial his peaceful visit to the enemy camp would be to his allies, and he needed to justify his transgression. He did so by referring to John Stuart Mill's sympathetic essay on the political reactionary Coleridge, in which this hero of liberal thought urged liberals to attend to Coleridge's capacity "to see further [than they] into the complexities of human feelings and intellect." Trilling claimed, not ironically, that Mill believed this would add "something practical" to Bentham's "short and easy political analysis." He quoted Mill telling his "radical friends" that they should pray as follows: "Lord, enlighten thou our enemies . . . sharpen their wits, give acuteness to their perceptions and consecutiveness and clearness to their reasoning powers: we are in danger from their folly, not from their wisdom."[20]

Trilling recommended Eliot's "religious politics" as a possible element in the "reconstruction" of the left. Specifically, Trilling pointed to the failure of the left "to ask of the good life, of the morality of politics, of the spiritual and complex elements in life."[21] The left, he maintained, had been much more interested historically in immediate rather than ultimate ends. The paradigmatic response to the question of ultimate ends had been Lenin's, who "postponed a choice as to what man might become until such time as he has a choice." But in the meantime a choice had been imposed, that of the Ideal Worker, a gravely diminished ideal to be sure. In order to counter the *imposition* of ideals by organizations and leaders of the left, Trilling sought clues in Eliot as to how ideals could be better constructed. Eliot should be read for an understanding of the elements that this culture, "to be adequate, must encompass." These elements, as we have seen, impel the left to inquire into the means by which it forwarded its various agendas for social change. What were the habits of thought, the condition of mind, being encouraged? How were the individuals involved understanding their roles as free and morally responsible agents of social change? What was there about the quality of their lives to recommend their beliefs to the unconverted? These were the kinds of questions Trilling was interested in, and he maintained they were raised with impressive consistency by Eliot, even if Eliot's own beliefs did not have genuine appeal.

What Trilling found in Eliot was an emphasis on the process of constituting the ideal and not on the ideal itself. Put another way, Eliot's ideal of critical intelligence was to be realized in the subtle details of a culture rather than in a society's public declarations. The point was not, in his words, "the conscious formulation of . . . ideal aims"—the real ideals of a nation are found in "the sub-stratum of collective temperament, ways of behavior and unconscious values" that go toward making up the formulation. Trilling's critique of the left—that it threw ideals and formulas at complex social and moral problems—had its positive side in the critical intelligence anchored in a moral sensibility. This critical intelligence was defined by its rejection of the various reductive models of the individual's relationship to society then available to leftists, and by its affirmation of those subjective categories alluded to by Eliot. For Trilling—and here he was again representative—what Eliot and other modernist writers seemed to be recommending to liberalism was

> the sense of complication and possibility, of surprise, intensification, variety, unfoldment, worth. These are the things whose more or less abstract expressions we recog-

> nize in the arts; in our inability to admit them in social matters lies a great significance. Our inability to give this quality a name, our embarrassment, even, when we speak of it, marks a failure in our thought.[22]

Eliot's challenge to the theorists of liberalism and leftism was to remind them that they could not have constructed their theories and programs without having made moral assumptions. Whether these were acknowledged or not, they were nonetheless practical necessities. The left had always said as much about its opponents, only it called moral assumptions ideology. To the myriad, anonymous individuals under the sway of the left, their opportunities for moral perfection and happiness were being sacrificed to the endlessly deferred ends of "the ultimate man." To them the left's moral assumptions were quite real. In this, claimed Trilling, Marx was as guilty as Lenin in believing morality was just an example of bourgeois false-consciousness. Although Marx "begins in morality, in the great historical and descriptive chapters of *Capital* . . . he does not continue in it. . . . He speaks often of human dignity, but just what human dignity is he does not tell us."[23] Trilling called this reticence "a practical, a political error." For as the idea of progress as never-ceasing improvement replaces the more demanding one of labored, self-reflective praxis, and as the future is worshiped and the past buried in ignominy, there will remain little left to prevent the underlying disgust with humanity from reaching ever more dangerous proportions. Eliot gave the left moral and political lessons, which Trilling hoped might prevent the drift toward totalitarian conformity within its ranks.

It was Kenneth Burke, a critic in close proximity to the *Partisan Review* circle, who was among the first to articulate the elusive qualities that needed to be grafted onto the culture of the left. In his essay "Thomas Mann and André Gide," included in *Counter-Statement* (1931), he defended modernism on the grounds that its tendency to challenge dogma was politically useful:

> Irony, novelty, experimentation, vacillation, the cult of conflict—are not these [authors] trying to make us at home in indecision, are they not trying to humanize the state of doubt? . . . Could action be destroyed by such an art, this art would be disastrous. But art can at best serve to make action more labored.[24]

Burke's early formulation—*labored* action rather than *inaction* or precipitous action—anticipated *Partisan Review*'s later attempt to redefine politics, and is at the root of his own understanding of reading and education and their links to social change.[25]

But among the New York intellectuals the new interest in complication, indeterminacy, and a needed ambivalence toward large questions was most obvious in Trilling. In the first of his three essays on E. M. Forster he referred to the novelist's "awareness not of morality itself but of the contradictions, paradoxes and dangers of living the moral life."[26] Because of Forster's refusal to adopt a systematic explanation of how the world which he knew so well worked, he was able to render the moral predicaments of his characters with due attention to the details of time, place, and character. The result was that notions of good and evil became enmeshed in a complex nexus of motive and circumstance which needed deliberate sorting out in the process of reaching moral judgments. Always in Forster was the understanding that too-certain judgments of virtue lead to evil. Similarly, Trilling regarded F. Scott Fitzgerald as another writer whose "moral realism" might provide political guidance: "We feel of him, as we cannot feel of all moralists, that he did not attach himself to the good because this attachment would sanction his fierceness toward the bad."[27] Trilling valued those writers whom he believed understood the relationship between self and society, or free will and circumstance, as dialectical—that is, as a tense, dynamic, and mutually defining play of opposites. No single factor, no one side of this relationship could necessarily be privileged. Fitzgerald may have linked Gatsby's fate to the social world he was enamored with but, in Trilling's words, "he never lays it upon that world." For Trilling, Fitzgerald's "moral and intellectual energy" issued from what the novelist called "the ability to hold two opposed ideas in the mind, at the same time, and still retain the ability to function."[28]

As I have indicated, Trilling's quarrel with liberalism was not confined to its understanding of self but took in related assumptions about reality. Perhaps the most effective articulation of Trilling's critique of the liberal view of reality was his polemical essay "Mr. Parrington, Mr. Smith, and Reality" (1940), better known under the bland title Trilling gave the politically expurgated version in *The Liberal Imagination*, "Reality in America."[29] As had Philip Rahv in "Paleface and Redskin,"[30] Trilling pointed out that it was a matter of historical record that the American mind was confused about its relation to its surroundings. It had persistently thought itself to be a wholly separate entity and free from social determination, from "conditions." This claim had a romantic, even a democratic sound to it, and received strong cultural support from the line of radical individualist thinking whose spokesmen were Emerson and Whitman. For Trilling, though, there was little that was emancipatory about this approach to mind, especially in the 1930s and 1940s, because in its own way it reproduced the same marginalization of mind that the materialism of American capitalism had produced. For both poles "reality" was "one and immutable . . . external . . . irreducible."[31]

According to Trilling, Parrington's celebrated survey *Main Currents in American Thought* expressed this "chronic American belief that there exists an opposition to reality and mind." As was true of so many cultural texts since Whitman, Parrington's advocated "that one must enlist oneself in the party of reality." Intellect is not only feared and avoided in this view, it is considered to be politically dangerous. Parrington "meets evidence of imagination and creativeness with a settled hostility the expression of which suggests that he regards them as the natural enemies of democracy."[32] One result of this disdain for consciousness of both self and society was that a terrible injustice had been done those American writers whose work was said to be insufficiently public or progressive. For Trilling, the left's notion of progress, symbolized in the image of the "current" (he objected that culture was not a current but a dialectical struggle), had caused many to look on the nation's most impressive literary achievements as so much debris washed up on the shore. Hence for Parrington and the culture he embodied, Hawthorne turned away from reality out of a "deficiency of feeling," and James out of "sinful pride." As a hopeless eccentric Poe would go with no current, and Melville's response to American life was said to be "less noble than Bryant's or Greeley's."

Trilling adamantly maintained, however, that as for Hawthorne, he had not turned away from reality; if he was "'forever dealing with shadows'" so be it: "shadows are also part of reality." Moreover "one would not want a world without shadows, it would not even be a 'real' world. . . . The fact is that Hawthorne was dealing beautifully with realities, with substantial things."[33] By virtue of his intense skepticism toward "the nature and possibility of moral perfection" and his heroic resistance to "Yankee reality"—that is, the reality of the marketplace, Hawthorne presented himself as the realistic writer par excellence.

Of all of Parrington's literary misjudgments, it was his unqualified preference for Dreiser over James that most dismayed Trilling. Here, once again, Parrington was said to speak for an entire culture and its assumptions about reality: "In the American metaphysic, reality is always material reality, hard, resistant, unformed, impenetrable, and unpleasant. And that mind is alone felt to be trustworthy which most resembles this reality by most nearly reproducing the sensations it affords."[34] Hence the extraordinary indulgence for Dreiser's clumsiness, stupidity, bewilderment, crudity—the words were those of Granville Hicks, Dreiser's advocate, for whom these qualities described "the sad, lovable, honorable faults of reality itself."[35] Trilling noted that regrettably, in this view of America as a raw and inarticulate democracy, James's "wit . . . flexibility of mind . . . perception . . . and knowledge" were thought to be synonymous with aristocracy and political reaction.[36]

The tendency of the culture of the left to take a simple view of reality by forcing it to conform to the pieties of a shallow philosophical materialism was not only responsible for a poorly formed literary canon. Just as upsetting for the *Partisan Review* critics was the fact that active and able writers were being led astray. Through the intense, high-sounding pressure of the Comintern-led international antifascist movement, eminent writers such as Thomas Mann, André Malraux, and Ernest Hemingway were being encouraged to make their work "socially responsible." In "Hemingway and His Critics" Trilling mourned the loss of Hemingway "the artist" and condemned the apotheosis of Hemingway "the man." For Trilling, whatever Hemingway's limitations as writer in the past, his fine receptivity to the demands of craft and personal truth produced an art Edmund Wilson had rightly called "'severe, intense, and deeply serious.'" With the publication of *To Have and Have Not* (1937) and *The Fifth Column* (1938), however, he had become a painfully self-conscious, thoroughly maudlin writer "with axes to grind." The blame for his decline was unequivocally directed at the Popular Front critics. Admittedly Hemingway's talent had always developed under close public scrutiny, but during the 1930s the pressure to use that talent in support of causes that Hemingway "the man" supported became irresistible. "Upon Hemingway," Trilling wrote, "were turned all of the fine social feelings of the now passing decade, all the noble sentiments, all the desperate optimism, all the extreme rationalism, all the contempt of irony and indirection . . . of the liberal-radical movement."[37]

In the face of the left's egregious abuse of Hemingway's talent, Trilling momentarily entertained the possibility that critics ought to claim literature to be separate from the surrounding social world. "One almost wishes to say to an author like Hemingway, 'You have no duty, no responsibility. Literature, in a political sense, is not in the least important.'"[38] But Trilling of course resisted this temptation, and instead used the remainder of his article to open up strategies in which the social relevance of Hemingway's most accomplished work could be discussed—without making it necessary that the critic demand open partisanship. Trilling made it perfectly clear that he considered Hemingway's best work to be "passionately and aggressively concerned with social truth." But not in the usual manner—not in terms of the content of his ideas. It was the politics of Hemingway's prose that most interested Trilling. He considered its celebrated spareness and precision to have manifested a linguistic rebellion against the fine Wilsonian rhetoric of the progressive era, which ended in the death and destruction of World War I. Hemingway's technique, his professional pride and cult of craftsmanship, were seen as attempts to rescue those rare and invariably individual examples of excellence that somehow survived the onslaughts of moribund traditions and the naive and therefore dangerous social engineering schemes of the liberals.

For Trilling none of this meant that Hemingway had attacked "mind," or self-reflexive rationality. What Hemingway attacked was "rationalization," or that combination of small motive and false feeling that went into the justification of large social projects. It was this "dull overlay of mechanical negative proper feeling" that constituted for Trilling the impoverished subjective self that was the creation of a materialist culture. Hemingway's best fiction exposed this modern encroachment into the subjective life by placing against it the primal emotions of characters facing the ultimate experiences of love, pain, and death. Hemingway's insistence on courage—on athletic proficiency and mental discipline in the face of death—could be best understood as an answer to a whole culture whose "reasonable virtue" and "mind" had wilted with such extraordinary ease when faced with the crisis of World War I, and again with the crisis of fascism.

Thus Hemingway, far from writing fiction whose implication for politics was nugatory—ironically this had always been the claim of the left critics before Hemingway was recruited to the Popular Front—was political in the broadest and profoundest sense. For Trilling there was no need to yoke artists to a political movement, for "Writers have always written directly to and about the troubles of their own time and for and about their contemporaries, some in ways to us more obvious than others but all responding inevitably to what was happening around them."[39] It was in a writer's complex relationship to his culture, "his acceptance of it (and by it) and his criticism of it" that his ultimate "political" effects can be measured.

The case of André Malraux as spelled out by F. W. Dupee largely paralleled Hemingway's. As politically progressive ideas became increasingly determinate in Malraux's novels, the quality of the novels diminished—to the point that his latest novel *L'Espoir*, was described by Dupee as "a work of the higher factional publicism."[40] According to Dupee, what was lost since Malraux's masterpiece

Man's Fate was that novel's complete assimilation of an idea—in its case the idea of mortality—into the character of the hero, himself immersed in politics. The result was "the death-ridden solitude of the individualist and the fraternal drive of the revolutionary collective" working themselves out in terms of complex motives, ideas, feelings, and actions. Thus in *Man's Fate* Malraux was able to weave "a pattern of human destiny" into "the pattern of class struggle." Death, although a metaphysical reality, was made "the common property of mankind."[41] In *Days of Wrath*, however, the transitional novel leading to *L'Espoir*, death was marginalized and placed outside the ethos of glowing fellowship shared by the antifascists—this was true even as the novelist pretended death was integral to life. In *L'Espoir*, a kind of revolutionary allegory in which Hope (the antifascists) confronts Despair (the fascists), the revolutionary novel that was *Man's Fate* had been transformed into a vehicle for "thinly-fictionalized reportage—vivid and readable . . . with only the meagerest human content."[42] Here the "imaginative conjunction of moral principles" achieved in the former novel was utterly absent. Seduced by People's Front propaganda, Malraux had written a novel that was a product of the "will-to-action" rather than the "literary will."[43]

Both Trilling and Dupee charted the decline of authors whose modernism was rubbed raw by the repeated caresses of the left. But what the left needed from modernism could no more easily be gotten through caresses than through the vituperative attacks of the third period which they replaced. What neither of these approaches provided for was a serious dialogue, a difficult but productive exchange. What modernism possessed, and what the left needed, was a sensibility highly receptive to doubt, ambiguity, limitation, even failure. Politically this meant, among other things, an appreciation for the fact that the dilemmas of praxis demanded solutions that took into account their own incompleteness. This kind of self-awareness was essential if the left was ever to adapt to objective and subjective conditions and thereby renew itself.

If through their affiliations with the Popular Front Hemingway, Malraux, and Mann (in his speeches and articles) had betrayed modernism's capacity for complex moral judgment, other writers stood out as unyielding opponents of liberal simplicities, both left and bourgeois. Among these members of *Partisan Review*'s literary pantheon, which included Kafka, Proust, Joyce, Lawrence, Dostoevsky, Mann, Gide, Stevens, and Yeats, Henry James deserves attention. By the late 1930s nationalist and Popular Front critics had done extensive damage to James's reputation. Although there were certainly many admirers of James during the 1930s, the judgment of Van Wyck Brooks in his *Pilgrimage of Henry James* (1925) and Parrington in *Main Currents* held sway over the vast liberal audience. Brooks's thesis was straightforward and unsympathetic: whatever James's talents so far as technician, his expatriation severed his relationship to the only material that could make his work interesting to Americans. Cut off from his roots, he shriveled as a writer, and had to sustain himself in the hothouse of European aristocratic society. As F. O. Matthiessen pointed out, Parrington managed in his three-volume study of American culture but two pages on James—a section entitled "Henry James and the Nostalgia of Culture." Like Brooks's, Parrington's unenthusiastic assessment of James's art arose out of the critic's disdain for James's expatriation and his cosmopolitanism.

The chief critical impetus of the Henry James revival as promulgated by critics in or near the *Partisan Review* circle was to oppose such evaluations. In the James criticism of Wilson, Rahv, Dupee, Trilling, and Matthiessen, there was a common claim for James's centrality on the grounds of his profound moral and social understanding, his unmatched technical proficiency, and his outward-looking Americanness.[44] The first piece on James to appear in *Partisan Review* was Edmund Wilson's "The Last Phase of Henry James" (1938). Although Wilson shared with the critics I have mentioned a favorable judgment of James, the basis for his judgment differed. Wilson maintained that in James's final works—*The American Scene* and the unfinished novel *The Ivory Tower*—one could observe both the reemerging American and the revival of "the old Balzac." With the "new kind of realism" of his last phase, James fixed his gaze on the "grotesque . . . America of the millionaires," and this America is "brilliantly observed." If, suggested Wilson, we add these acute observations of America to James's earlier interest in "all that was magnanimous, reviving and human in the Americans"—in the Milly Theales, the Lambert Strethers, and so forth—we come away with a new appreciation of James's Americanness.

The appeal of James for other *Partisan Review* critics was based on different criteria. Whereas Wilson seemed to accept realistic criteria in his judgment, others did not, arguing against the opinion that James was an aesthete by questioning the underlying assumptions of such a view. Hence in "The Heiress of All the Ages" (1943), Rahv argued that James's value for Americans should not be gauged by his immediate relationship to his native land but rather by his insights into the appropriate American response to Europe, which he expressed through his series of American heroines. What Rahv found when he examined James's depiction of his line of "passionate pilgrims" was, as James himself put it, a "magnificent" response to the beauty of Europe: "Even while keeping a tight hold on her native virtue," James's heroine "Liv[es] up to her author's idea of both Europe and America. . . . She is able to mediate, if not wholly to resolve, the conflict between the two cultures."[45]

After analyzing some of the salient details of James's depiction, Rahv noted that they hardly pointed to the novelist's alleged loyalty to Europe. On the contrary, they demonstrated that "his valuations of Europe and America are not the polar opposites but the two commanding centers of his work—the contending sides whose relation is adjusted so as to make mutual assimilation feasible."[46] Of course Rahv was not only speaking of James here; implicitly his remarks were a justification of *Partisan Review*'s own critical project of Europeanizing American culture.[47] James was so vitally instructive precisely because his amalgamating scheme enabled him to bring the American character face-to-face with Europe, the center of Western experience—of action steeped in thought, and of subtly rendered thought. And James also brought Europe face-to-face with the emerging American civilization. James freed his heroines from the grips of the limiting native dualism that divorced experience from mind, but he freed them not so much from their past as from their blindness to it. Part of their moral growth entailed giving their Americanness full expression, and this meant supplementing it, not disavowing it. "There is a world of difference," Rahv reminded the nationalist critics, "between the status of an ambassador and the status of a fugitive."[48]

In the modern period the Europe that is the center of Western civilization must be the place American goes to modernize, or else suffocate on its narrow nationalism. James's fiction demonstrated in meticulous fashion what Europeanizing might mean for the antinomies of the American character, and for establishing a new politics that would not shy away, in typically American fashion, from the subjective side of politics. His fiction represented to Rahv the fullest elaboration of what it might mean to pose the question of the quality of life, the question the left seemed obsessed with avoiding. Far from engaging in the endless refinement of precious sensibilities, as his critics had charged, James succeeded in depicting "the finer discriminations" of the self within the "envelop of circumstances" in which they are contained. By pressing the case for individuality (as opposed to individualism), he, and in their turn Rahv and the *Partisan Review* critics, were forcing open the instrumental categories of thought that seemed to them to be the dominant ones in America.

The Henry James revival meant other things as well for leftist criticism. Given the commanding role of content analysis in the criticism of the left, James more than any other novelist afforded the *Partisan Review* critics the opportunity to do formal analysis without becoming formalists. Although never as ingenious in dealing with form and technique as the New Critics, about whom they were always ambivalent, the *Partisan Review* critics incorporated formal considerations into their cultural criticism with occasionally impressive results. Much of "The Heiress of All the Ages," for example, was given over to formal analysis, as when Rahv compared the typically constricted structure of James's novels with the more expansive structures of Joyce and Proust. Rahv attempted to link formal and social phenomena by observing how the social was inscribed in the formal features of James's work—in this case, by seeing how it was that moral exigencies impelled James to constrict his focus and particularize. This kind of criticism was unprecedented among American leftist critics.

But above all it was F. O. Matthiessen, a critic outside the *Partisan Review* circle but nonetheless influential, who saw that James's social and moral insights could best be understood in terms of his formal technique and narrative strategies. As Matthiessen remarked in the preface to *Henry James, The Major Phase* (1944), "Aesthetic criticism, if carried far enough, inevitably becomes social criticism, since the act of perception extends through the work of art to its milieu."[49] Like the critics more closely associated with *Partisan Review*, Matthiessen positioned his criticism against the reigning simplifications of the left. In his article "Henry James' Portrait of the Artist" (1944), published in *Partisan Review*, he drew the overarching contrast between the "analytic subtlety" of James's inner world, and the "surface reporting" of H. G. Wells's outer world. This split was "a sign of one of the great cultural maladjustments of our age," namely the dominance of Wells's values over James's. Though Matthiessen's preference was for James, it was the split itself that he said damaged both types of writer and that therefore needed mending. Alluding to the dichotomy between mind and experience explored by Rahv, Matthiessen insisted that James was not typical of "mind" and the high-brows, but rather "bridged the gap" between mind and experience. Thus a major theme of his criticism also was James's ability to mediate between categories that the contemporary left stubbornly kept separate.

The criticism I have discussed thus far—of James, Eliot, Malraux, Silone, Dos Passos, and Hemingway—is representative of *Partisan Review*'s overall critical project, but it does not indicate the scope of the magazine's interest in modern literature. In order to make the reader aware of this breadth of attention, I would like briefly to enumerate several other important critical interventions made by the magazine's contributors.

From the 1930s to the 1970s Philip Rahv regularly returned to Russian literature in his criticism, especially to Dostoevsky (at the time of his death he was working on a book on the novelist). Rahv's first published piece on Dostoevsky, entitled "Dostoevsky and Politics" (1938), was a relatively detailed study of the politics of *The Possessed* and their implications for the American left. Since the Bolshevik Revolution, this novel had been roundly condemned by orthodox Marxists for its flagrantly reactionary politics and its vicious caricature of the Russian revolutionary movement. But for Rahv the emergence of Stalinism had made the novel seem very astute indeed. For him it was a thorough dissection of the ideas, motives, and actions of the Stalinist betrayers of the socialist ideal. Echoing Marx's evaluation of Balzac, Rahv added Freudian terms to suggest that the novel was counterrevolutionary only in its "manifest content," whereas in its "latent content" it was liberating. Through its exploration of complex and conflicting psychological motives it gave its readers an unequaled understanding of how and why revolutionary movements fail. According to Rahv, Dostoevsky suggested that they failed not for reasons of policy so much as for reasons that went to the heart of their metaphysical assumptions—their negation of religion, mystery, and sin; and their easy substitution of a faith in reason, in science, in unconditional free will, in progress. For Dostoevsky these were the conditions that led to the monstrous, Faustian character Verhovensky, the revolutionary who recognized no limit to his actions save what advanced the revolution. These conditions, maintained Rahv, given absolute hegemony, led to Stalinism. He suggested that the tragic course of the revolution proved Dostoevsky to have been prophetic, more penetrating even than the sacred leaders of the revolutionary movement themselves.[50]

Another writer in whom *Partisan Review* was intensely interested was Kafka. It published three of his stories: "Blumfield, an Elderly Bachelor" (1938), "In a Penal Colony" (1941), and "Josephine the Songstress" (1942); and several critical articles, including Max Brod's biographical study "Kafka: Father and Son" and Hannah Arendt's "Franz Kafka: A Revaluation" (1940). In addition, Rahv published two essays elsewhere on Kafka in the late 1930s: "Franz Kafka: Hero as Lonely Man" (1939) in *Kenyon Review*, and "The Death of Ivan Ilich and Joseph K" (1939) in

Southern Review. Kafka's importance for *Partisan Review* derived from two elements in his fiction: his uncanny ability to expose the machinery of modern bureaucratic society both in its impersonal and subjective workings, and his extraordinarily stubborn moral sense. The juxtaposition of these elements in the meticulously drawn nightmarish world of raw power where human purpose is systematically reduced to function, and in the small man of simple virtues, marked Kafka not as prophet, surrealist, or talmudist, but as sober analyst. "Only the reader," noted Arendt, "for whom life and the world and man are so complicated, of such terrible interest, that he wants to find out some truth about them . . . may turn to Kafka and his blueprints, which sometimes in a page or even a single phrase, expose the naked structure of happenings."[51]

But though valued for his "realism," Kafka was a realist of a different stripe than traditional realists. He refused to submit to society as such, to "life as it happens." His magical realism, if you will, was based on his desire "to build up a world in accordance with human needs and human dignities."[52] According to Arendt, Kafka accomplished this in a formal sense by distancing himself from the all-encompassing social fact of most realist and naturalist writing. Through his very abstractness—his concern above all for the overall structure of things and their functions, and for his heroes as "models" of goodwill—he was able to generalize both evil and the capacity for goodwill, in Arendt's words, "by anybody and everybody." She and the *Partisan Review* critics celebrated Kafka for this unique kind of realism, being at once an anatomy of a seemingly sourceless oppression and a documentary of the persevering moral sensibility.

Finally, I would like to mention the contributions of the unjustly neglected William Troy, a frequent contributor to *Partisan Review* whose prescient essays on Lawrence and Joyce bear closely on the subject at hand. Writing some twenty years before Northrop Frye, Troy introduced myth criticism as a new way of dealing with aspects of modernist literature—but not myth as an ahistorical cultural domain. Troy recognized that modernist myth was not a denial of or escape from history, but rather a means of forcing into one's understanding of history elements of unreason which, at least since the Enlightenment, have been tragically extirpated.

In his treatment of Lawrence, Troy departed from the usual ideological treatment and instead sought some realm of experience that mediated ideas and the primal subjective states so central to Lawrence's work. Troy granted the novelist his view of himself as atavist, and suggested that an understanding of his cult of blood and death would be more fruitful than the usual emphasis on his "formlessness" and "intellectual difficulties." Lawrence was chiefly concerned with what Nietzsche called "the agony of individuation"—the ontological crisis inherent in being part of and yet separate from nature. The value of Lawrence's Dionysian world, according to Troy, was that for all of its irrationalism it represented "the vicarious exhaustion of possibilities that are inherent in the human being in every time and place."[53] This utopian impulse had the powerful effect of negating bourgeois society, although, as Troy observed, to see Lawrence as primarily a social critic was to get him wrong; he was more a moral critic. His work was an implicit and sometimes explicit criticism of "mass-production in ideas, emotions, and men. . . . As a reflection of the formal and qualitative disintegration of human life at present, it is more compelling than the jeremiads of the reformers, the analyses of the psychologists, or the charts of the economists."[54]

Regarding James Joyce, what is particularly intriguing about the reception accorded him by the New York intellectuals was the disparity between their high praise and the surprisingly small amount of Joyce criticism they produced. Thus, for example, Irving Howe, in his influential essay "The Culture of Modernism," described the difficulty faced by the modern writer who wished to remain in opposition to established values and forms. He pointed out that few sustained the necessary "remarkable kind of heroism, the heroism of patience." Among the modernist writers, "only James Joyce . . . was able to live by that heroism to the very end. For other writers . . . there was always the temptation to veer off into one or another prophetic stance, often connected with an authoritarian politics."[55] But despite such praise, the New Yorkers as a group produced relatively few essays and no full-length studies of Joyce. Perhaps this was so because the complexities of Joyce's work simply overtaxed the limits of literary journalism; it is hard to say. It remains, however, a conspicuous absence among a group of critics who produced a goodly amount of material on Eliot, Kafka, James, Lawrence, Faulkner, and Mann, to name a few.

But what the New Yorkers did produce was of high quality. I am here referring to two essays, one of which was Troy's "Notes on *Finnegans Wake*," and the other was Lionel Trilling's controversial "James Joyce in His Letters." Eugene Jolas relates that it was Troy's essay, published in *Partisan Review* in 1939, that among all of the analyses that greeted *Finnegans Wake* was the one most admired by Joyce. Stanley Edgar Hyman writes in his introduction to Troy's *Selected Essays* that "The Joyce experts used to sit in on his lectures and have his insights in print over their names before the week was out." It is not difficult to see why. In his essay Troy was not so much interested in deciphering the detailed reappearances of the Odyssean voyage in *Ulysses*, or the myth of the Fall in *Finnegans Wake*; rather, it was the very existence of the myth as organizational framework that demanded explanation. Thus Troy was interested in the overall social significance of Joyce's polyvalent verse/prose, and in what he called the "method of simultaneity," which Joyce adopted in favor of chronological history. These interests fueled an exploration of *Finnegans Wake* as a synecdochal novel—that is, as a novel standing in relation to human history as part to a whole, rather than as primarily an ahistorical, essentially aestheticist text. For Troy, the novel's complex linguistic and narrative elements did not draw the reader out of history, but rather presented the reader with a radically relativized notion of history, one heavily indebted to Vico. Not unlike Fredric Jameson, in fact, whose 1982 essay "'Ulysses' in History"[56] makes for an interesting comparison to Troy's, the older critic considered the novel to be an implicit attack on modern instrumental thought, on contemporary social arrangements, and on political ideologies imbued with a faith in History as the unfolding of an idea of, or will toward, progress.

Troy implied that those detractors of Joyce who based their claims on his alleged failure to render history materially and in its general movement (critics like the early Rahv and the Soviet apparatchik Karl Radek, who at the 1934 Soviet Writers' Congress described Joyce's work as "a heap of dung, crawling with worms, photographed by a cinema apparatus through a microscope") themselves diminished the claim of the past over the most private, seemingly subjective and arbitrary elements of the human psyche and imagination. For it is often by way of inexpedient private impulses and egregious anachronisms, Troy reminded his reader, that individuals, not to speak of writers, most insistently resisted what he called "cultural confusion." For Troy, this resistance was very much a private and also a historical phenomenon. But whereas Jameson was largely content to demonstrate the ways in which *Ulysses* undermines critics who search for an achieved and codified symbolic order in the novel, Troy, while he too pointed out the futility of such efforts, sought a broader understanding of what Joyce's output might imply for the culture and politics of the left. As Trilling was later to remark in his essay on Joyce, "however else we read *Finnegans Wake*, we cannot fail to understand . . . that its transcendent genial silliness is a spoof on those figments of the solemn 19th-century imagination—History, and World Historical Figures, and that wonderful Will of theirs which, Hegel tells us, keeps the world in its right course toward the developing epiphany of *Geist*."[57] Trilling's remark merely highlights what was a key ideological component of Troy's essay: the muted but nonetheless telling critique of both orthodox Marxism and bourgeois progressivism, in 1939 comfortably aligned under the rubric of the Popular Front. Whereas Jameson was content to criticize mainstream scholarship—or what once was mainstream scholarship—Troy, like many of the New York intellectuals, extended his gaze so that it took in habits of mind found on the left as well as in academia. In addition to bourgeois critique, Troy and other New Yorkers offered rigorous self-critique as well.

Finally, a word about Trilling's 1968 essay "James Joyce in His Letters," which has been read by most commentators as an example of Trilling's growing quietism and disillusionment with the new left. The essay argued the incommensurability of Joyce's rebellion against the nineteenth century and the contemporary rebellion against liberal capitalist society. According to Trilling, for Joyce, himself saturated in nineteenth-century values, the act of rebellion was at once an aggression visited upon conventional society and a terrifying act of self-mutilation. For every thrust at the nineteenth century, which Eliot declared was killed by Joyce, there was damage done internally to tissues of concern and sentiment, belief and desire, fantasy and dream, all of which bound him to his hatred source. This onerous and immensely fatiguing "triumph," as evidenced by Joyce's letters, exacted the price of sociability, of civic-mindedness, of progressivism, according to Trilling: "Nothing in the way of 'humanness,'" he observed, "contradicts our sense that the letters of [Joyce's] years of fame were written by a being who had departed this life as it is generally known and had become such a ghost as Henry James and Yeats imagined, a sentient soul that has passed from temporal existence into nullity yet still has a burden of energy to discharge, a destiny still to be worked out."[58]

Trilling believed that, tragically, radicals in the 1960s had a much easier time rejecting their society than Joyce did his. "It may well be that never before," he remarked,

> have so many people undertaken to live enlightened lives, to see through the illusions that society imposes, doing this quite easily, without strain or struggle, having been led to the perception of righteousness by what literature has told them of the social life. Whatever we may *do* as getters and spenders, in our other capacity, as readers, as persons of moral sensibility, we *know* that the values of the world do not deserve our interests. We *know* it: we do not discover it . . . it is a thing taken for granted.[59]

Thus, for Trilling, Joyce's life combines with his writing to deliver a message not of conservatism, but of considered, self-conscious rebellion. Joyce was an enemy of liberal capitalism, and equally a burden to its opponents.

III

It has often been observed that the New York intellectuals came late to modernism—that, far from constituting a corps of advanced critics, not to speak of an avant-garde, they were unique only because along with the New Critics they helped to canonize the modernist classics a generation after they first appeared. It is further observed that their tastes, although more adventurous than the tastes of mainstream and academic readers, fell short of the radically subversive preferences of the modernists themselves. Yes, they championed Eliot, James, Hemingway, Malraux, Joyce, Yeats, Gide, Mann, and Lawrence, when large numbers of the educated public were still perplexed by them and unconvinced of their ultimate value. But where were the other experimentalists, those whose work was radically new in the 1930s: Djuna Barnes, Sterling Brown, Gertrude Stein, Nathanael West, Louis Zukofsky (who labored in New York City under the very nose of *Partisan Review*, which managed to publish but a single poem by Zukofsky)? *Partisan Review* clearly was not on the cutting edge of literary innovation.

I do not wish to quarrel with this assessment: there is something to the charge that the avant-garde that *Partisan Review* represented was at least fifteen years old in 1940, and in many cases was older than that. Moreover, *Partisan Review*'s rather exclusive focus on nineteenth- and twentieth-century white males of European-American descent meant that many writers were not taken seriously who should have been, a point I develop in chapters 8 and 9. But if the reader recalls the cultural and political predicament that *Partisan Review* found itself in in 1940, he or she will perhaps appreciate the magazine's accomplishments in spite of these limitations. For approximately seven years, roughly from 1937 until the end of World War II, *Partisan Review* attempted to create a self-conscious and self-critical culture of dissent that might correct some of the long-standing problems experienced by the American left, itself part of a broader liberal tradition. The magazine failed. It failed because, as we shall see, it was not careful enough in discriminating between the strengths and weaknesses of American radicalism. It failed because, as an organ for intellectuals who were part of the middle class, it did not sufficiently explore its relationship either to the lower or the upper classes. With regard to the working class and ordinary Americans, the *Partisan Review* critics possessed a great store of sympathy but maintained few connections and little knowledge; with regard to the comfortable and the powerful, these critics possessed little sympathy but maintained increasingly close connections. This did not mean, as some critics believe, that *Partisan Review* was fated for conservatism, but it did mean that their understanding of social change and cultural critique did not become part of the discourses of dissent that shaped the major movements for social change during the 1960s.

But finally, *Partisan Review* failed because the American left failed. At the time of course the left demonstrated little capacity to incorporate *Partisan Review*'s critique. As I argue in chapter 10, despite the movements of the 1960s, much remains the same, and much, therefore, might still be gained by reconsidering this critique in light of current circumstances.

6

Notes toward the Supreme Soviet: Wallace Stevens and Doctrinaire Marxism

Theoretically a period of attempts at a world revolution should destroy or endanger all stationary poetic subjects and words and be favorable in the highest degree to the recording of fresh experience. But the vivification of reality has not yet occurred in spite of the excitement. Only the excitement has occurred.

WALLACE STEVENS, 1948

I

No canonical modernist writer appeared in *Partisan Review* more often and more prominently than Wallace Stevens. If over the years Elizabeth Bishop, Robert Lowell, and William Carlos Williams appeared slightly more frequently, this was offset by the fact that Stevens's contributions were strategically placed and thus very highly regarded. He was sought after by William Phillips and Philip Rahv for the inaugural issue of the revived *Partisan Review*, which carried his poem "The Dwarf" as a powerful indicator of the magazine's new literary direction. Later, during the height of the struggle against Stalinism, the editors published "Life on a Battleship," Stevens's withering critique of totalitarianism. In all, between the years 1937 and 1948 *Partisan Review* published no fewer than eight of his poems, three pieces of prose, and five book reviews.[1] In addition, it later reviewed his *Collected Poems* (in 1955) and his *Letters of Wallace Stevens* (in 1967). In this chapter I explore one of the important reasons for this very interesting convergence between the poet from Hartford (who spent a good deal of time in Manhattan) and the New York intellectuals. It was of course not a simple relationship. There were obvious differences of background and style between the comfortable lawyer-poet with firm roots in his native land and the children of immigrant Jews. And there were doubtless anxieties that arose from the mutual quest for prestige and recognition, which each could deny or provide the other. But these factors did not prevent the two parties from developing a genuine respect and admiration for one another. Indeed, for his part Stevens considered *Partisan Review* the best literary magazine in the country, and he held immense respect for Philip Rahv, whom he recommended in 1940 to his friend Henry Church as the critic who should inaugurate the lecture series Church was proposing for Princeton University. One of the reasons for the mutual admiration of which I speak was a shared political perspective. This is not to suggest that their views were identical on the array of political issues and possibilities that confronted them in the 1930s and 1940s—indeed there were many differences between the rather confused politics of the poet and the carefully worked-out radicalism of the New Yorkers. But as James Longenbach has persuasively demonstrated in *Wallace Stevens: The Plain Sense of Things*, one major political and cultural perspective shared by the two was anti-Stalinism. In exploring this perspective as it was manifested in Stevens's poems of the period, I hope to shed light on Stevens's poetry and also on the nature of the New York intellectuals' attraction to Stevens and to modernism in general.

The subject of Stevens as political poet has not inspired many critics over the decades, except perhaps to provoke resistance. Certainly Frank Lentricchia was right to have called attention to Stevens's critics' lack of comfort with the idea, not to mention Stevens's own.[2] Indeed it is only relatively recently that critics have begun to look carefully at Stevens's relationship with the political world,[3] and one of the questions being asked is what the word "political" might mean in reference to his poetry. Clearly Stevens is no political poet if we insist on the traditional meaning of that word, which stresses matters of public policy and governing. But it is equally true that were we to apply a more contemporary understanding of politics, in which issues broadly pertaining to questions of social power and authority are involved, Stevens's poetry still would not be well served, for even his most socially engaged poems of the 1930s were more broadly political than this. In fact these poems implied that both these views of politics share a fundamental instrumentalism that obscures the relation of politics to essential areas of subjective and intellectual experience.

Despite recent efforts to the contrary, no doubt many readers still see Stevens as an intensely personal poet who, by and large, remained aloof from politics. If at times he did address himself to the pressures of social actualities, as in the poetry of the early and mid-1930s, it is stubbornly held that such responses were reluctant and self-distancing. To the degree that Stevens explicitly engaged with social problems and even social movements, as in "Owl's Clover," many readers continue to think that his poetry suffered as a result, its subtle meditative tones giving way to stridency and its minutely textured meanings succumbing to "ideology." But I submit that such judgments, although alive to recognizable formal and aesthetic issues within Stevens's engaged poems, often neglect the central question of what politics, and hence what political poetry, could and should be. We will see later that this is itself a subject about which Stevens had much to say in some of the very poems of the 1930s most widely considered to be inferior.

I have already touched on another reason critics have neglected Stevens's strengths as a political poet: their unacknowledged or unconscious assumption that politics must be something largely instrumental—that, above all, it must be objective and expedient, and sometimes narrow, hard, and crude. That is to say poetic acts and political acts have for some time now been thought to be mutually exclusive, save in those extremely rare cases when a poet's genius is conjoined with a sensibility unusually equipped to survive politics intact. As for the leftist politics that Stevens chose to discuss in the 1930s, it was accused not only of evincing these characteristics, but was "ideological" as well—that is, was uniformly based on systematic and utopian beliefs coercively maintained.[4]

Such assumptions about the corrupting effect of politics had strong precedents in nineteenth-century American cultural life, but the cold war's containment of political life in this century very much strengthened these tendencies. In short, many postwar realities, like Stalinism itself, combined to foster the belief that mainstream politics must be realistic, tough, and pragmatic (therefore antipoetic) in order to survive the constant threat of "ideology." Few literary critics or political analysts were ready to admit that their notion of authorized party-politics might also be "ideological" in some sense, and fewer still were interested in the possibility of reestablishing the politics of the left so that it might express more open and democratic values. Critics left and right writing the two decades after the war were simply not well disposed toward Stevens's political poetry of the 1930s, in which he attempted to revise leftist political discourse through his trenchant but patient and relatively sympathetic critique of doctrinaire Marxism. Academic critics and doctrinaire Marxists alike responded to postwar conditions by assuming that since politics was war by another name, no poet could by definition explore politics in poetry, especially a poet so insistently individual and self-reflexive as Stevens. Where Stevens wandered onto political terrain, as in "Owl's Clover," his critics agreed that the results made for poor poetry.[5]

If, in contrast to the generation of critics and analysts I describe, we acknowledge that "ideology" may inform all political discourses (possibly, but not necessarily equally); if we admit that the "ideological" does not contaminate so much as describe modern society; and if we entertain the possibility of forming a union between politics and the imagination, to use Trilling's phrase, by making politics respond to philosophical, cultural, and aesthetic issues, then we can, I believe, reassess our understanding of Stevens's political poetry. To show how far Stevens was able to carry out this project in the poems of the 1930s, let us focus on the undervalued "Owl's Clover."

In "Owl's Clover" Stevens has preceded those who currently wish to revitalize our understanding of politics in general and leftism in particular. Indeed, others have noted the extent to which doctrinaire Marxism has been superseded by significant sectors of the left in the West. For example, Stanley Aronowitz writes in the opening paragraph of his aptly titled *The Crisis in Historical Materialism*,

"We live in a time when all the old assumptions about politics and history appear enfeebled. Throughout the Western industrial societies, both the capitalist and the state socialist types, the theory and practice of workers, intellectuals, women, and ecologists have, in different ways, questioned the adequacy of Marxism as a theory of the past and present and as a guide to the future."[6] Writing nearly a decade ago, Aronowitz could not have known that the crisis in Marxism would soon go well beyond the West and include Marxist governments and movements in the second and third worlds. For many, these developments represent the triumph of capitalism and the official death of an idea that, for all intents and purposes, expired long ago, the idea of socialism. But this view fails to account for the continued and quite extraordinary vitality of leftist political thinking, which has drawn not only on feminism, theories of race, and environmentalism, but also on the critique of doctrinaire Marxism coming from within the Marxist tradition itself—for example, in the work of Lukács, Korsch, Gramsci, Adorno, Benjamin, Brecht, and Bakhtin. The assault on doctrinaire Marxist views of nature, history, causality, class, race, gender, and culture has been both intricate and productive, and one can only expect this critique to gain in influence. It is against the background of revision of classical Marxism that I wish to examine Stevens's insights into politics in general and Marxism in particular during the 1930s.[7]

None of this is to suggest that Stevens, though named after a politician, possessed political beliefs we could call remotely Marxist or even progressive. We need only examine some of his often inconsistent, sometimes rash, and sometimes offensive political statements to know this and to sympathize with Irving Howe's suggestion that modernist writers would have been better off had they kept clear of practical political issues. The inconsistencies and ambiguities of Stevens's politics have been conveniently and succinctly reviewed by Milton Bates in *Wallace Stevens: A Mythology of Self.* In his letters Stevens variously referred to the leftist program as "a magnificent cause" and a "grubby faith"; he claimed to be "headed left" and "extraordinarily stimulat[ed]" by his encounters, yet maintained that "the whole left nowadays is a mob of wailers" (unaccountably he attempted to resolve this contradiction by adding that "I do very much believe in leftism in every direction, even in wailing"). Stevens also described "Mr. Burnshaw and the Statue" as a "poetic justification of leftism," albeit a "vague" one, yet went on to remark that "to the extent that the Marxians are raising Cain with the peacocks and the doves, nature has been ruined by them." He referred to fascism as "a form of disillusion," but nonetheless informed Ronald Lane Latimer that he was "pro-Mussolini." He hardly cleared up this last matter in a letter to Latimer three weeks later where, commenting on the Italian invasion of Ethiopia, he wrote "While it is true that I have spoken sympathetically of Mussolini, all of my sympathies are the other way: with the coons and the boa-constrictors."[8] This sort of thing is painful to recall, especially because there is no evidence that Stevens was playing devil's advocate or was in any way testing Latimer, as Joan Richardson has claimed.[9] Painful too are the passages that reveal a disdain for ordinary people: the "butcher, seducer, bloodman, reveller" of "Ghosts as Cocoons," the childlike and insectlike pallbearers of "Cortège for Rosenbloom," the "sudden mobs of men" of "Sad Strains of a Gay Waltz." And there are the passages that patronize women, to which Sandra Gilbert and Jacqueline Brogan, among others, have called attention.

It is nonetheless worth recalling that as questionable as these responses to the historical crisis of the 1930s seem, Stevens's *description* of this crisis—he called it "the drift of incidents"[10]—could be surprisingly discerning. In several rarely quoted passages of "The Noble Rider and the Sound of Words" (1941), for instance, he offered a physiognomy of social life by describing, among other things, "the intricacy of new and local mythologies, political, economic, poetic"[11] arising out of the deadly combination of ubiquitous force—threatened and real—and the absence of widely recognized authority, which could limit that force. In this connection Stevens briefly discussed the importance of mass education, mass housing, mass communication, narcissism ("the generally heightened awareness of the goings-on of our own minds, *merely as goings-on*"), and alienated labor ("[workers] have become, at their work, in the face of the machines, something approximating an abstraction, an energy").[12] According to Stevens, what was unique about the era he described was the degree to which these "incidents," or this "weather," failed to elicit the expected reciprocal and creative response of its subjects. The "news"—of capitalism's collapse, of new societies being constructed, of war—was exerting so great a pressure on the consciousness of individuals that it tended in Stevens's view to exclude "any act of contemplation."[13] It is worth observing that here Stevens anticipated a central issue in the current debate over postmodernism by suggesting that the era beginning in the 1930s was threatening to end the more or less consistent ability of artists to transform the recalcitrant material of history into imaginative art. He spoke of the pressure of reality being "great enough and prolonged enough to bring about the end of one era in the history of the imagination and, if so, then great enough to bring about the beginning of another."[14] He also remarked—and this is relevant to the subject at hand—that it was this very achievement of art, suddenly jeopardized, that had hitherto belied the strictly materialist viewpoint.

It would seem that whatever one may think of Stevens's miscellaneous and frankly inconsistent political positions, his meditations on the consequences of an increasingly administered society for subjective experience and art remain compelling. Leftists or anyone else may wish in the end to dissent from Stevens's politics as he expressed them overtly or implied them, but this ought not to deflect attention from his perceptive identification of crucial social developments that have shaped the postwar era; nor should it prevent us from taking seriously his related critique of doctrinaire Marxism. Stevens's ideological vicissitudes cannot be the last word; if we insist that they are, we merely repeat the mistakes of the dogmatic leftist critics of the 1930s whom Stanley Burnshaw, in his famous review of *Ideas of Order*, quite properly assailed some sixty years ago.[15] This is not to suggest, however, that Stevens's intervention onto the political scene was independent of his own political and ideological positions—it is only to say that these positions were not necessarily determinate, nor in any way an equivalent to what resulted from what I believe was a productive encounter with the left. One final word: to the extent that we take advantage of the fact that the cold war curtain between culture and politics has been swept aside only by insisting on carrying politics into culture, we will have little of value to say about writers like Stevens who wished, for a time at least, to carry culture into politics.

II

Far from evading, fleeing, or otherwise resisting politics in the poetry of the early and mid-1930s (from *Ideas of Order* [1935] to *The Man with the Blue Guitar and Other Poems* [1937]), Stevens made his way steadily to the heart of the political crisis which by the mid-1930s had become global. Having said his farewells to Florida, he sailed his shaky craft north, not merely in order to relocate the solitary artist or argue at closer range for the exemplary power of the transcendent imagination, but to undertake the more formidable and infinitely more difficult task of intervening in the political struggle and determining how a place within it might be made for autonomous poetry and the unencumbered imagination. In part the difficulty can be measured by the extent to which Stevens was forced to interrogate his own understanding of the imagination, his relationship to romanticism, and ultimately his role as poet.

In *Harmonium* the poems of starkness, "The Snow Man" and "The Man Whose Pharynx Was Bad," had been offset by Hoon's affirmation of empowered imagination: "I was myself the compass of that sea: / I was the world in which I walked, and what I saw / Or heard or felt came not but from myself; / And there I found myself more truly and more strange."[16] In the poems of the 1930s, however, Hoon fought to be heard, and was heard above the din, but only fitfully. It is his diminished voice which speaks at the end of "Sailing after Lunch," when, laboring under "this heavy historical sail," he finds it sufficient "to give / That slight transcendence to the dirty sail" (*CP* 120)—a far cry from the confident, unqualified transcendence of "Tea at the Palaz of Hoon." In "Sad Strains of a Gay Waltz," Hoon is explicitly invoked, and here too his world is altered. He is as much the outcast as the chosen isolate, his powers apparently undermined:

> And then
> There's the mountain-minded Hoon,
> For whom desire was never part of the waltz
> Who found all form and order in solitude,
> For whom the shapes were never the figures of men.
> Now, for him, his forms have vanished.
>
> (*CP* 121)

It is not that for Stevens Hoon is expendable—far from it—only that he can find no form of expression because suddenly his isolation is more burden than boon.

A key toward understanding the poetry of the 1930s is that Stevens both refused to turn away from Hoon *and* refused to turn away from political actualities. He was intent on resettling Hoon, finding him a voice, and placing him within earshot of a hostile audience so that they would be forced to hear. Of course, Hoon too would have to adapt. Stevens's poetry of the 1930s offers a fascinating record of a poet struggling to find a form of expression that could argue for—or rather embody—this necessary arrangement, in which the poetic imagination has a home within the everyday world. At times a self-mocking Stevens despaired of ever contributing to making such a community. In the poignantly repetitive lines of "Anglais Mort à Florence" he laments the loss of the old solitude:

> A little less returned for him each spring.
> Music began to fail him. Brahms, although
> his dark familiar, often walked apart.
> His spirit grew uncertain of delight,
> certain of its uncertainty . . .
>
> (*CP* 148)

And later in the poem:

> He stood at last by God's help and the police;
> But he remembered the time when he stood alone.
> He yielded himself to that single majesty;
> But he remembered the time when he stood alone,
> When to be and delight to be seemed to be one,
> Before the colors deepened and grew small.

Yet in the 1930s Stevens persisted in confronting his world, sometimes quite sharply and directly, as we see in "Owl's Clover."

The directness and specificity of Stevens's struggle to find a place for poetry amid the social dislocations of the 1930s becomes evident when we focus on his critique of doctrinaire Marxism, one of the most important motifs in "Owl's Clover." Also worthy of consideration are the sometimes obscure passages of "Owl's Clover" in which he takes on orthodoxy for its historic failure to deal productively with modern subjectivity by neglecting to base its social diagnoses and prescriptions on an understanding of imagination, desire, sensuality, the unconscious, the appeal of authoritarianism, and of course poetry, in either its social diagnoses or prescriptions.[17]

"Suppose," wrote Stevens of the future revolution, "instead of failing, it never comes, / This future, although the elephants pass and the blare, / Prolonged, repeated and once more prolonged, / Goes off a little on the side and stops." Here, in "A Duck for Dinner," Stevens supposes a scenario that has, of course, come to pass, in which the planned procession of History leading to communism loses its way, sputters, and terminates in disarray. This is the historic failure of socialism in the West, in which the revolutionary process ends or is at least delayed by virtue of the fact that the agent of revolution, the working class, has opted for reform rather than revolution. To quote Stanley Aronowitz once again, the "spectre haunting Marxism since the first world war" has been the likelihood that "the practices of the workers' movement in reforming capitalism already constituted a new configuration for a future without revolutionary consequences"[18]—spectral because for many during the time of the Second and Third Internationals Marxism meant orthodox Marxism, whose doctrine promised the inevitability of capitalist crisis and its ineluctable collapse. The failure of the revolution to occur was no small miscalculation—for many it meant that the entire philosophical and intellectual edifice of doctrinaire Marxism was severely shaken if not destroyed, because the future was itself to be the ultimate justification of a teleological scheme of history rife with the strong foundational tendencies of determinism, rationalism, positivism, and scientism.

Stevens opposed these systematic views in the name of continuous change, stressing desire's responsive and constitutive counteraction upon reality and upon schemas designed to remake reality according to a totalizing vision. "It is not enough," replies the normative voice in "Mr. Burnshaw and the Statue," when faced with this future, "that you are indifferent, / Because time moves on columns intercrossed / And because the temple is never quite composed, / Silent and turquoised and perpetual, / Visible over the sea. It is only enough / To live incessantly in change."[19] This particular Marxist revolution will change little, Stevens suggests, or certainly will not change enough. The indifference it engenders results from its fatalistic belief in eternal truths; its telos, representing a modern equivalent of the division of heaven from earth, endlessly defers joy. "Everything is dead / Except the future," proclaims Stevens's ideologue, "Always everything / That is is dead except what ought to be" (*OP* 46). Doctrinaire Marxism's unmediated materialism promises a future of things, things that compose "Parts of the immense detritus of a world . . . that moves . . . out of the hopeless waste of the past / Into a hopeful waste of the future."[20] In addition, the insistent positivism that shapes this vision of the future betrays a palpable fear of art, the imagination, and the unconscious: "The statue seems . . . a thing / Of the dank imagination, much below / Our crusted outlines hot and huge with fact" (*OP* 47).

In this last passage Stevens's critique settles on what was for him the most egregious flaw of doctrinaire Marxism: its historic indifference to the problems of human subjectivity and culture, the result of a long-standing and not unproductive emphasis on objective social structures. This issue in my view informs a good deal of "Owl's Clover," each poem dealing with a different aspect of it. Indeed in remarks to Latimer concerning the overall theme of "Mr. Burnshaw and the Statue," Stevens raised the general issue of subjectivity by implying that Marxists must become fully aware of the degree to which imagination and pleasure redound upon the structural changes wrought by revolution. "You will remember," he wrote, "that Mr. Burnshaw applied the point of view of the practical Communist to *Ideas of Order*; in 'Mr. Burnshaw and the Statue' I have tried to reverse the process: that is to say, apply the point of view of a poet to Communism."[21] Several days later he added:

> The . . . question is whether I feel that there is an essential conflict between Marxism and the sentiments of the marvelous. . . . My conclusion is that, while there is a conflict, it is not an essential conflict. The conflict is temporary. The only possible order of life is one in which all order is incessantly changing. Marxism may or may not destroy the existing sentiment of the marvellous; if it does, it will create another.[22]

It should be pointed out that Stevens's relatively sanguine assessment of what the future holds for Marxism's relation to imagination is, in part, based on his own claims for material reality, claims that are much more generally acknowledged today than they were in 1935: "I am what is around me," he declares in the poem "Theory." "All of our ideas come from the natural world," goes the adage (*OP* 163). But, of course, included in the natural world is that most transformative of powers, imagination, whose marriage to physical reality results in a planet replete with endless argument—especially where the imagination is abused—and only partial recuperation. Ironically, within the Marxist tradition the reciprocal relationship between being and consciousness had been adumbrated nearly a hundred years earlier by Marx himself in *Theses on Feuerbach*, specifically in the first, third, and fifth theses.[23] But the Stalinist version of Marx that Stevens confronted produced nothing so subtle, insisting on subordinating both willful and spontaneous acts of the imagination to "the recognition of necessity" (Engels's phrase)—necessity defined at once as material necessity and as the necessary laws of historical development. In the orthodox version revolution is always "two-staged," wherein the problems of the quality of labor, personal relations, sexuality, imaginative life, and culture are endlessly deferred as their "solution" awaits the defeat of capitalism and the transition to the new society.[24] For doctrinaire Marxists the new "order" entailed all of the austere limitations upon self-expression which its equation of individuality and bourgeois individualism implied. In contrast, Stevens's emerging idea of order amid the economic and political ferment of the time would invest poetry with the authority to guarantee the provisionality of any order arising out of this chaos.

"Politic man ordained / Imagination as the fateful sin," wrote Stevens in "Academic Discourse in Havana," and in "Owl's Clover" he interrogated a politics whose suppression of desire, spontaneity, difference, and imagination was only much later acknowledged by the left. Marchers in the parades of the masses are "morbid and bleak," Stevens sadly observes in "The Drum Majors in the Labor Day Parade." "The banners should brighten the sun. / The women should sing as they march. / Let's go home" (*OP* 37). Similarly, in the playful and quite serious "The Revolutionists Stop for Orangeade" (1931), Stevens has his soldiers appeal to their leader to stop subordinating song to the rigors of the class war: "Ask us not to sing standing in the sun, / Hairy-backed and hump-armed" (*CP* 102). Instead, they claim, he must realize the radical incompatibility between music and what is narrowly defined as the real, according to which "There is no pith in music / Except in something false" (*CP* 103). Transform your notion of struggle, they implore him, open yourself, open politics to frivolity; let your altruism arise out of the wish to spread pleasure, not destroy injustice:

Hang a feather by your eye,
Nod and look a little sly.
This must be the vent of pity,
Deeper than a true ditty
Of the real that wrenches,
Of the quick that's wry.

(*CP* 103)

The Bulgar in "A Duck for Dinner" is another leader—a labor leader and likely either a fellow traveler or Party member—who makes the mistake of trivializing culture by legitimizing it only as an ancillary part of the overall movement toward liberation. Thus in the poem's first section he quite wisely puts forth a case for the gradual improvement of the quality of life for workers: "after all, / The workers do not rise, as Venus rose, / Out of a violet sea. They rise a bit / On summer Sundays in the park, a duck / To a million, a duck with apples and without wine" (*OP* 60). Yet as the poem soon makes clear, the idea of culture in this appeal for acculturation is much too narrow in scope—in fact the idea has an uncanny resemblance to the idea of culture as defined by the society that needs to be replaced, with its emphasis on distraction, relaxation, and creature comforts:

They rise to the muddy, metropolitan elms,
To the camellia-chateaux and an inch beyond,
Forgetting work, not caring for angels, hunting a lift,
The triumph of the arcs of heaven's blue
For themselves, and space and time and ease for the duck

(*OP* 60)

What results from this conception of culture is socialist realism, the commercial art of communism, later satirized in the poem as Basilewsky's "'Concerto for Airplane and Pianoforte.'"

But the poem is much more than a simple condemnation of socialist realism; Stevens acknowledges the profound material changes that have taken place in America since the nineteenth century, causing the older myths to become superannuated: "O Buckskin, O crosser of snowy divides, / For whom men were to be ends in themselves," he queries of the American Adam, "Are the cities to breed as mountains bred, the streets / To trundle children like the sea?" (*CP* 61) Clearly not, replies the normative voice: "For you / Day came upon the spirit as life comes / And deep winds flooded you; for these [i.e., the masses], day comes / A penny sun in a tinsel sky, unrhymed, / And the spirit writhes to be wakened" (*CP* 61). Stevens echoes Emerson here even as he updates him. "Go out of the house to see the moon," Emerson tells those who see fit to hunt natural beauty down, "and 't is mere tinsel; it will not please as when its light shines upon your necessary journey."[25] In the remainder of "A Duck for Dinner" Stevens speculates on what it might mean for beauty to accompany the masses on their journey, rather than become the desperately pursued object of their well-earned leisure time. Clearly the relationship between audience and work of art, audience and artist, must be integral, dynamic, and transformative for all three. "They [the masses] see / The metropolitan of mind, they feel / The central of the composition, in which / They live. They see and feel themselves, seeing / And feeling the world in which they live" (*OP* 64). The artist is given privileged status within the movement by Stevens, but in an interesting twist on Lenin's justification of the vanguard based on its standing above the masses and seeing further into the future, Stevens's vanguard artist, though something of a seer himself, is somewhat shorter, as much impresario as prophet, more magician than militant—in brief, to Lenin's scientific socialist a "worshipper of spontaneity," a quack:

> Exceeding sex, he touched another race,
> Above our race, yet of ourselves transformed,
> Don Juan turned furious divinity,
> Ethereal compounder, pater patriae,
> Great mud-ancestor, oozer and Abraham,
> Progenitor wearing the diamond crown of crowns,
> He from whose beard the future springs, elect.
> More of ourselves in a world that is more our own,
> For the million, perhaps, two ducks instead of one.
>
> (*OP* 64–65)

Put another way, the difference between the Bulgar and the artist with his statue is that the one organizes the masses to seize a future already divined and the other organizes the masses to fashion one. Unlike the Bulgar's, Stevens's idea of order can be said to describe a state of mind, or rather a process of becoming, in which change is expected, encouraged, absorbed, and sought after anew. Radicals must realize that even after the revolutionary negation of bourgeois society, if it should come to pass, unruly desire rises once more, intermittently perhaps yet irrepressible, a constant reminder to the negation of the negation that it must always affirm change. In the postrevolutionary scenario described in "Mr. Burnshaw and the Statue," desire provides the basis for permanent revolution. Moreover it involves shades of red not normally attributed to socialism:

> There even
> The colorless light in which this wreckage lies
> Has faint, portentous lustres, shades and shapes
> Of rose, or what will once more rise to rose,
> When younger bodies, because they are younger, rise
> And chant the rose-points of their birth, and when
> For a little time, again, rose-breasted birds
> Sing rose-beliefs.
>
> (*OP* 49–50)

The eroticism of this passage clearly indicates that for Stevens sexuality lends its energy to change, and I think it fair to say that passages such as this implicitly rebuked doctrinaire Marxism for its puritanism. More than puritanism was at stake, however, in other portions of "Owl's Clover," and in a poem such as "Life on a Battleship" (published in *Partisan Review*), where Stevens rather courageously challenged the doctrinaire Marxist cult of virility. As Paula Rabinowitz has amply demonstrated in her introduction to *Writing Red: An Anthology of American Women Writers, 1930–1940*, the ideology of gender pervaded the proletarian literary movement despite the Communist movement's attempts to politicize and empower women, especially in comparison with the conventional alternatives available to women at the time. A brief quotation from Mike Gold, one of the "founding fathers" of the movement, should suffice by way of example. In his 1926 article "America Needs a Critic," Gold wrote, "O Life, send America a great literary critic. . . . Send a soldier who has studied history. Send a strong poet who loves the masses . . . a man of the street. . . . Send no coward. Send no pedant. Send us a man fit to stand up to skyscrapers. A man of art who can match the purposeful deeds of Henry Ford. . . . Send no saint. Send an artist. Send a scientist. Send a Bolshevik. Send a man."[26]

Stevens brought to this milieu a more equivocal stance toward masculinist ideology, which allowed him to make his critique despite certain assumptions he may have shared with Gold and others. Thus the poem "Life on a Battleship," a withering critique of totalitarianism, begins as follows: "The rape of the bourgeoisie accomplished, the men / Returned on board *The Masculine*." In the poem the battleship's name becomes an emblem for authoritarian modes of behavior and thinking: "'*The Masculine*, much magnified, that cloud / On the sea, is both law and evidence in one.'" Throughout the poem the linkages between masculinity and an aggressive, even ruthless rationalism, an ersatz science, and unrestrained power and violence are continually implied when not asserted. Although the poem does not end with an image of woman, or of a feminized male, but rather ends with a chastened male wielding new, albeit diminished power ("Our fate is our own. The hand . . . must be the hand / Of a man, that seizes our strength"), the force of the critique as it pertained to the masculinist left must nonetheless be acknowledged.

In "Mr. Burnshaw and the Statue" Stevens directed a more formidable challenge to the masculinist left through his depiction of revolutionary, sensual, feminized passion. Despite his acceptance of a hypostatized equation between sensuality and femininity, he turned his own idealizations against the more dangerous ideals of an aggressive and misogynist masculinity, which at the time had great currency on the left. In the poem it is arguably Stevens's muses, acting as his alter ego or perhaps even as the alter egos of the revolutionaries, who become a radiant ring of women; in either role they clearly represent an alternative to the poem's dour proletarian critic, whose earlier praise for the statue's lack of "subterfuge" served to suppress the kind of joy that can be legitimized by art, and is described by the poem's aroused narrator:

> Dance, now, and with sharp voices cry, but cry
> Like damsels daubed and let your feet be bare
> To touch the grass. . . .
>
> . . . .

> Be maidens formed
> Of the most evasive hue of a lesser blue.
>
> . . .
>
> Let your golden hands weave fastly and be gay
> And let your braids bear brightening of crimson bands.
>
> . . .
>
> Speaking and strutting broadly, fair and bloomed,
> No longer of air but of the breathing earth,
> Impassioned seducers and seduced, the pale
> Pitched into swelling bodies, upward, drift
> In a storm blown into glittering shapes, and flames
> Wind-beaten into freshest, brightest fire.
>
> (*OP* 51–52)

This orgiastic dance is an altogether remarkable fantasy, not least for what it tells us about the "pulse pizzicato" of Stevens's own voyeuristic and objectifying imagination; but also because such a passage possessed genuine critical force given the virulent masculinity of the sexual politics Stevens was opposing. By presenting the doctrinaire left with its repressed, bodily other, Stevens affirmed the need for a new openness to personal and sensual modes of transformation. Whether one agrees that the poem deserves the kind of generous praise Adrienne Rich gave to "The Idea of Order at Key West"—that the poem's unconscious celebrates the life, power, and energy of the female principle[27]—it is nonetheless significant that in the face of the orthodox left's incessant demands for restraint and discipline, Stevens insisted on putting fantasy and the unconscious into play. Whatever we may think of the ideological content of the material manifested as a result, we must remember that the passage is in the imperative—its content is not presented primarily as an act of personal disclosure but rather as a direct challenge to his inhibited addressees to act out, and act on, the full range of their desires.

In other poems of "Owl's Clover" the importance of the unconscious becomes paramount, and I would like to explore this emphasis in terms of Stevens's implicit critique of Marxism's refusal to take the unconscious—indeed, psychology in general—into account. Sounding extraordinarily up-to-date, Stevens asked in "The Irrational Element in Poetry" (1936), "Does anyone suppose that the vast mass of people in this country was moved at the last election by rational considerations? Giving reason as much credit as the radio, there still remains the certainty that so great a movement was emotional and, if emotional, irrational"[28] (*OP* 225). Before 1968 few Marxists and certainly no doctrinaire Marxists would ever have asked such a question, much less supplied an answer. Marxism simply gave no priority to developing an understanding of the psychological dimensions of social change. Indeed, barely a year before Stevens began work on "Owl's Clover," Wilhelm Reich had been expelled from the German Communist Party for his attempt to unite Marx and Freud in an explanation of fascism's psychological appeal.[29] For his part Stevens devoted much of "Sombre Figuration" to an exploration of individual and social psychology, their relation to current historical developments, and the consequences of this relationship for art.

To begin examining these matters in "Sombre Figuration," it is necessary to be clear about the meaning of the confusing but very important image of the portent. Some critics have taken it to be a symbol of the Jungian collective unconscious, whereas others define it more vaguely as a deathlike presence. But at least in this case it seems wisest to begin with Stevens's own explanation, offered to Hi Simons in a letter written in 1940, which contained a gloss of the poem:

> When we were facing the great evil that is being enacted today merely as something foreboded, we were penetrated by its menace as by a sub-conscious portent. We felt it without being able to identify it. We could not identify what did not yet exist. . . . It was, after all, ourselves, all of us, all we had reason to expect from what we knew. The future must bear within it every past, not least the pasts that have been submerged in the sub-conscious, things in the experience of races. We fear because we remember.[30]

I take this explanation to be both a historical and a social-psychological one, the portent being a metaphor not for war and fascism, but rather for the subjective experience that these historical phenomena gave rise to, the feelings of fear and foreboding that had become widespread and constitutive enough in their own right to comprise what Raymond Williams might have called a "structure of feeling."

The poem is quite specific on this count:

> It is the form
> Of a generation that does not know itself,
> Still questioning if to crush the soaring stacks,
> The churches, like dalmatics stooped in prayer,
> And the people suddenly evil, waked, accused,
> Destroyed by a vengeful movement of the arms,
> A mass overtaken by the blackest sky, . . .
>
> (*OP* 68–69)

This passage indicates that Stevens, although certainly attuned to the historical moment, was not mainly interested in determining the "objective" historical reasons for the crisis. By focusing on the subjective response to war and fascism rather than on events themselves, the poem suggests that the doctrinaire definition of fascism was, at the very least, incomplete. This definition, codified in Dimitrov's famous formulation adopted by the Comintern in the summer of 1935 as Stevens labored over "Owl's Clover," was well publicized at the time: "Fascism in power," it asserted, "is the open, terrorist dictatorship of the most reactionary, the most chauvinistic, the most imperialistic elements of finance capital."[31] Stevens, rather than explain the rise of authoritarianism as the outcome of class struggle carried out under conditions of extreme economic crisis, was interested in the cultural and psychological dimensions of authoritarianism. He explored the connections between the social disaster at hand, sterile rationalism ("We have grown weary of the man that thinks" [*OP* 66]), and the unconscious, irrational, transformative impulses of "the man below," or the "subman" (*OP* 66). From these he sought to determine what the function of the statue, or art, might be under such trying circumstances.

Not unlike Adorno and Horkheimer in their analyses of fascism and "administered" societies, Stevens implicated insistently rational modes of thought in the social disaster, modes whose destructive power derived from the obsessive exclusion of the irrational. He maintains that "the man below," though "ogre"-like, is neither wholly destructive nor alien: "It is not that he was born in another land, / Powdered with primitive lights, and lives with us / In glimpses, on the edge or at the tip, / Playing a crackled reed, wind-stopped, in bleats" (*OP* 66–67). On the contrary, the unconscious is a powerful source for emancipatory change: "The man below / Imagines and it is true, as if he thought / By imagining, anti-logician, quick / With a logic of transforming certitudes" (*OP* 66). Moreover, not only is the unconscious integral to individual experience, it is also a constellation of moving shapes and sounds that helps to constitute the memory of the many we give the name history, as well as constitute individual memory, which operates synechdochically to bind the individual to a past and to a community. "He dwells below," Stevens explains,

> the man below, in less
> Than body and in less than mind, ogre,
> Inhabitant, in less than shape, of shapes
> That are dissembled in vague memory
> Yet still retain resemblances, remain
> Remembrances, a place of a field of lights,
> As a church is a bell and people are an eye,
> A cry, the pallor of a dress, a touch
>
> (*OP* 67)

The subman, or the unconscious, is intimately connected to the response to war and fascism ("The man below beholds the portent poised, / An image of his making" [*OP* 69]), thus the current crisis is not to be explained principally in terms of the bourgeoisie's assault on the masses. Any such historical explanation is dangerously incomplete without an understanding of how the subjective experience of social dislocation can deepen or ease that dislocation. If the latter is to occur, Stevens seems to suggest, it will be because committed attention to subjective experience, particularly to the protean, dynamic unconscious, lessens the claim that monolithic social structures have on the mind. The portent, therefore, is poised, "but poised as the mind through which a storm / Of other images blows, . . . " (*OP* 69). The mind does not experience a given crisis as immutable because it constantly generates alternative images that countervail the dominant structure of feeling.

But how are we to preserve the capacity for diverse imaginings, given the debilitating social developments that have narrowed the way we think about our social and personal possibilities, curtailing the old romantic vision of self-actualization and social transformation? Stevens compared the current situation unfavorably to the one Shelley faced in the wake of the failure of his revolution:

> images of time
> Like the time of the portent, images like leaves,
> Except that this is an image of black spring
> And those the leaves of autumn-afterwards,
> Leaves of the autumns in which the man below
> Lived as the man lives now, and hated, loved,
> As the man hates now, loves now, the self-same things.
>
> (*OP* 69–70)

Formerly change had run its course naturally, and imagination was exercised, not exorcised, by advocates of social change. For this reason Stevens portrays the future as ominous, and he ends "Owl's Clover" accordingly. Instead of insisting on hope, he sternly measures the final embodiment of the statue as monument to the ordinary against his embattled claims for the renovating imagination. In the poem's final section the statue looms as a *cordon sanitaire*—surrounded by black but itself sanitized white, stately, neatly proportioned, eminently sane and normal. It is the outward manifestation of an inward passion, or perhaps it gives rise to the passion roundly felt, to flee imagination and its seemingly illimitable, destabilizing uncertainties:

> Even imagination has an end,
> When the statue is not a thing imagined . . .
> Even the man below, the subverter, stops
> The flight of emblemata through his mind,
> Thoughts by descent. To flourish the great cloak we wear
> At night, to turn away from the abominable
> Farewells and, in the darkness, to feel again
> The reconciliation, the rapture of a time
> Without imagination, without past
> And without future, a present time, is that
> The passion, indifferent to the poet's hum,
> That we conceal?
>
> (*OP* 71)

In the face of the social crisis, art officially designated to alleviate misery and promote solidarity merely provides a sense of risk-free, palliative immediacy, which its desperate audience has come secretly to desire. Anything that threatens this equilibrium—memory, history, the unconscious, imagination—is consigned to night.

III

The basis for perhaps the most astute critique of Stevens's utopian politics was suggested by Irving Howe in a 1957 review of *Opus Posthumous*. "Stevens' insistence upon human possibility," he wrote, "can itself become mechanical, a ruthlessness in the demand for joy." But instead of demonstrating why this was so in light of the needs of the polity or of actual political movements, Howe retreated to the conventional and in my opinion false view of Stevens: "Perhaps the greatest weakness in his poems is a failure to extend the possibilities of self-renewal beyond solitariness or solitary engagements with the natural world and into the life of men living together." For whatever reasons—I suspect that the constriction of cold war politics played an important role—Howe turned his back on the riches in "Owl's Clover," calling the poem "unfortunate . . . an assault upon a subject which as a poet Stevens was not prepared to confront."[32] Today, however, our political and ideological inventory includes new possessions, whose sources are in the larger historical and intellectual currents cited at the beginning of the chapter. These new acquisitions enable us, I think, to take a more sympathetic look at Stevens's political poems. Although I have had little or nothing to say concerning such implied matters as Stevens's critique of the Dialectic, Marxism's domination of nature, cultural imperialism, and identity theory, I hope I have managed to make a contribution toward a change of attitude and approach to these poems. For the poems I have explored are not failures, not as political or nonpolitical poems. They are no more commonplace denunciations of communism than they are the self-absorbed utterances of an aesthete out of his element. Commentators who have emphasized the displeasing polemical, sometimes caustic tone of "Owl's Clover" have tended to encourage these views by drawing attention away from the substance of the poems. On the contrary, the author of "Owl's Clover" has always struck me as being intent on resisting his own occasionally impatient tone with an effort of persistence. He often seems, in other words, to check his stridency with a solicitude that constantly returns him to a careful consideration of the doctrinaire position. He takes due cognizance of the materialist position which argues that the self, and even the self that desires and imagines, is not prior to but is constituted by its relationship with others. Indeed "The Greenest Continent" is nothing if not an anti-imperialist attempt to show how the European and the African imaginations are culturally specific, historically constituted, and therefore fundamentally incompatible (this is not to gainsay Stevens's reliance on stereotypes in his analysis). At the same time, Stevens was never willing to *equate* self and what Marx called the ensemble of social relations. Although Stevens in the poetry of the 1930s rejected the radical division between the private world of the poet's imagination and the public world, he was not willing to subordinate the former to the latter by agreeing that the free development of the self could only be realized through the free development of all. The imagination is too unruly to be asked to wait so long. In "Sombre Figuration" Stevens wrote of "each man in his asylum maundering, / Policed by the hope of Christmas" (*OP* 68). At the end of "A Duck for Dinner," desire is called "that old assassin" (*OP* 66). If the social environment becomes hostile, desire rebels; the poetics of the self become destructive, no matter how ostensibly revolutionary the regime.

A. Walton Litz once called Stevens "a capitalist of the imagination"—a laissez-faire capitalist at that—because he is the partisan of "individuality, privacy, spontaneity."[33] True enough, Stevens's highly personal imagination makes cultural capital out of fugitive, fortuitous impulses. But I would argue, along with Stevens, that this need not be fatal to a leftist politics willing to settle for a mixed economy of the mind, in which poetic currency is valued for its depiction of desire that can meet the collective need. Stevens's challenge to the left in the 1930s has yet to be met: let us see, he proposed, "how easily the blown banners change to wings" (*CP* 508).

7

The Culture Wars of the 1940s: Literature, Popular Culture, and the Battle over a Usable Past

In this country there was a time when virtually all intellectual vitality was derived in one way or another from the Communist Party. If you were not somewhere within the party's wide orbit, then you were likely to be in the opposition, which meant that much of your thought and energy had to be devoted to maintaining yourself in opposition.

ROBERT WARSHOW, "*The Legacy of the 30's*"

The anonymous human being to whom [the makers of mass culture] bring their messages has at least the metaphysical advantage of being forced to deal with material things and real situations—tools, working conditions, personal passions. The fact that his experience has a body means that mass culture is to him like a mirror in an amusement park; whereas the "enlighteners," whose world is made up entirely of mental constructions, live inside the mirror.

HAROLD ROSENBERG, "*The Herd of Independent Minds*"

I

In 1938 Philip Rahv wrote an article for the social-democratic and anti-Communist labor weekly the *New Leader*, in which he condemned the continuing attraction of intellectuals to what he called "the Stalinist cultural front." By this he meant the Communist Party's Popular Front activities in the area of culture, and its success at winning over writers and critics. Rahv emphasized that this attraction had little to do with ideas and much to do with the seductive appeal of power itself. He maintained that intellectuals were attracted largely by the prospect of losing themselves in a powerful collective and that they even dreamed of wielding power within it. Rahv thought he saw evidence of this in the fact that what appealed to intellectuals most about Soviet society were not its cultural and intellectual achievements, but the "efficiency" of the regime. Assessing the situation in New York City in the late 1930s, he saw that the Communist Party was quite "efficiently" gaining real power of its own, and this frightened him a great deal. "During the last few years," he warned,

> the Stalinists and their friends, under multiform disguises, have managed to penetrate into the offices of publishing houses, the editorial staffs of magazines, and book review sections of conservative newspapers. The result has been that a kind of unofficial censorship is now menacing the intellectual freedom of those left-wing writers who are known to be opposed to the bureaucratic dictatorship in Moscow and to its representatives abroad. This G.P.U. of the mind is attempting—through intrigue, calumny, and overt as well as covert pressure—to direct cultural opinion in America into totalitarian channels.[1]

Rahv's vision of an ever widening and conspiratorial Stalinist presence stemmed from a fear shared by many anti-Stalinist leftists in the late 1930s that the Stalinists were gaining cultural power of major proportion, far outweighing their relatively small numbers and simpleminded notions. The perception of this threat grew over the years. In early 1946 George Orwell wrote to Philip Rahv inquiring into the likelihood of a favorable reception in the United States for his anti-Stalinist fable *Animal Farm*. Rahv addressed an extraordinary letter back to George Orwell in which he advised him not to publish in the United States because "public opinion here is almost solidly Stalinist, in the bourgeois as well as the liberal press."[2] Given the book's subsequent success, it is all too clear that Rahv's characterization of American culture was seriously flawed.

Yet it is not difficult to imagine how Rahv could have so badly misjudged the ideological climate in 1946 if we review some of the important political and cultural developments that had earlier affected his judgment—just before and during the war. Paramount in the minds of the *Partisan Review* critics was a profound sense that the capitalist world was teetering on the brink of collapse. According to the Trotskyist analysis, which many of these critics shared, the bourgeoisie was grasping at fascism in its desperate search for order and stability. German capitalism had already succumbed and in the remaining democracies there were ominous totalitarian trends slowly but surely working their way out into the open. As the United States geared for war, critics such as Philip Rahv, William Phillips, Dwight Macdonald, Clement Greenberg, and Lionel Trilling expected a qualitative change in American life. With no working-class movement to lead the fight against fascism, the war effort would swiftly degenerate into militarism, jingoism, coercion, and conformity. As Phillips and Rahv claimed in a major editorial statement in *Partisan Review*: "Our entry into the war under the slogan 'Stop Hitler!' would actually result in the immediate introduction of totalitarianism over here."[3]

Totalitarianism was making major inroads into American life according to the *Partisan Review* critics, chiefly by way of the Communist Party's Popular Front. To them the success with which the Party strategy was weaning intellectuals away from a position of critical independence and toward nationalism and reformism constituted a grave danger.[4] These changes encompassed a new and effective policy toward culture, as the Party, with characteristic zeal, sought power and respectability in broad areas of American life. To be sure, it did so by transforming its image into that of a loose association of progressives rather than the centralist organization it actually was. Not only did the outward organizational forms change as local union caucuses were turned into industry-wide organizations, and neighborhood groups turned into electoral groups—along with the change in and proliferation of front groups, the Party leadership made sure that the language, habits, and look of the Party members changed as well. Irving Howe and Lewis Coser described the changes dramatically:

> Here is a portrait of the old-style professional revolutionary as drawn by a Comintern "rep" to the American party: "The professional revolutionist . . . gives his whole life to the fight for the interests of his class. A professional revolutionist is ready to go wherever and whenever the Party decides to send him. . . . From these comrades the party demands everything . . . the matter of family associations and other personal problems are considered but are not decisive. If the class struggle demands it, he will leave his family for months, even years. The professional revolutionist cannot be demoralized; he is steeled, stable." The new party member, by contrast, was to be distinguished by the fact that he was not supposed to be distinguishable from anyone else: "To make our party a party of human beings who live and laugh just as everybody else does remains a task before us. Our shop papers often tell the Communist position on this or that important problem, but seldom do we find an article on the kick one gets from doing party work. We always play up the gloomy side of life with hardly any relief in sight before the social revolution; yet our party has picnics, outings, affairs, and such things."[5]

A combination of political reformism, organizational diversity, and cultural image-making won for the Party an influence and prestige it had not previously enjoyed. Within the labor movement the Party was winning unprecedented influence, and in both the political and cultural spheres it had become an important, though certainly not a dominant presence. By the summer of 1938, according to Howe and Coser,

> The Communist Party had become an important, if not yet a major, force in American political life. At the tenth CP convention, in 1938, the membership was announced as 75,000. In 1939 the party claimed to have reached 100,000, though there is internal evidence to suggest that the figure was exaggerated. It is possible, however, that between 80,000 and 90,000 people were in the party at one time or another during 1939. . . . There can be no doubt that the CP had taken some major steps toward becoming a "mass organization" that was now a powerful force in the CIO, the youth movement, the intellectual world, and in a few large cities.[6]

By 1946, when Rahv had written Orwell, the Party had experienced several potentially fatal crises and had still managed to expand its influence. These crises included the double reversal in policy in response to the Hitler-Stalin nonaggression pact of 1939 and its abrogation in 1941. This caused the departure of a significant number of prominent intellectuals, including Granville Hicks, Malcolm Cowley, and Archibald MacLeish, yet the Party continued to amass political gains and expand its influence.

The area of Popular Front penetration that most alarmed the *Partisan Review* critics was culture. By the early 1940s fellow travelers had gained strong footholds in Hollywood, in radio, on the stage, and in publishing. Numerous enthusiastic reviews of the latest progressive movie, best seller, or hit play appeared in *New Masses* during the period of Popular Front cultural ascendancy. A typical example was this composite review of the best films of 1939, in which the claim was made that "the long campaign of Hollywood's progressive film workers to bring reality to the screen was realized . . . with a handful of films . . . 'Juarez,' 'Confessions of a Nazi Spy,' 'Boy Slaves,' [and] 'Young Mr. Lincoln.'"[7] A review of the film *The Grapes of Wrath*, was entitled "The Great American Film," and subtitled "John Steinbeck's 'The Grapes of Wrath' comes to the screen as a motion picture masterpiece, unadulterated, uncensored, and triumphant. . . . A film to rally around."[8] This 20th Century Fox film was even called revolutionary. In the following issue of *New Masses*, Carl Sandburg's *Abraham Lincoln: The War Years* was described as "an epic biography of the Great Emancipator . . . no monument of bronze or stone inspires more reverence than Sandburg's biography."[9]

The Popular Front had a large stake in the emerging culture industry in the United States. But it is equally true that the *New Masses* was critical of a great many products of popular culture, though the *Partisan Review* critics ignored this criticism. Nevertheless, they were correct in claiming that aesthetic considerations did not amount to much for the *New Masses* critics. The latter judged art and writing on the basis of conformity to doctrine, a time-tested practice, only now the doctrine was reformism rather than sectarianism. The sole standard remained whether a work adhered to the "needs of the people," the difference being that now these political needs were defined more broadly. That the people's aesthetic needs might also be reassessed with equal care was never taken seriously as an important critical enterprise. Thus in a film review entitled "Two Brilliant Medical Pictures," both *The Fight for Life* and *Dr. Erlich's Magic Bullet* were liberally praised for their progressive stand on medicine, but nothing was mentioned about the artistry of the films other than to say the acting excelled. More important than aesthetic considerations was the need to observe that "Erlich" was about the doctor who found the cure for syphilis, "a disease which has reigned because the bourgeois moral code is unwilling to recognize sex as a function of life."[10] In a review of Orson Welles's *Citizen Kane*, one of the most formally compelling American films ever made, not a word was mentioned about the film's quality; the more important subject was William Randolph Hearst's attempts to have the film suppressed. Thus the reviewer implored that

> The case of "Citizen Kane" must not be judged by the Jew-baiting, Red-baiting studio vigilantes but by those who carry the weight of the little golden calf labeled Box

> Office; not by a bellowing old tyrant but by those ultimately responsible for having made movies a *mass* entertainment. Theirs, as always, will be the final verdict.[11]

The *Partisan Review* critics considered this sort of criticism to be both platitudinous and pandering in its eagerness to score political points and in its failure to engage seriously with the issues of artistic merit and considered standards. They spoke of this as criticism cum publicity, in which intellectual rigor and honesty were sacrificed in order to saturate the audience with the language of social uplift. Rather than being the vigorous, implacable criticism of capitalism it should have been, the criticism of the *New Masses* represented little more than a Chamber of Commerce appeal for civic-mindedness.

The *Partisan Review* critics believed that such spurious criticism had spawned a host of movies, books, and plays in which complex social and psychological problems were reduced to the theme of the people versus the corrupt boss or politician. From Howard Fast's historical novels to Frank Capra's transparent populist dramas, from sentimental WPA theater productions to Thomas Hart Benton's nostalgic paintings, Popular Frontism in the arts was "on the march" and well on its way to conquering the two largest establishment media centers: Hollywood and New York.

Not only were the interests of the masses being betrayed in the latest blockbuster movies, but major mass publications such as *Collier's* and *Life* were joining the established Popular Front magazines, the *Nation* and *New Republic*, in praising the Soviet Union. Henry Luce devoted the entire March 29, 1943, issue of *Life* to the subject of Soviet-American cooperation, where it was stated that the Russians were "one hell of a people." Even the *Reader's Digest* chimed in by publishing a condensed version of *Mission to Moscow*, U.S. Ambassador to Russia Joseph Davies's notorious whitewash of Stalinism (later made into a movie starring Walter Huston). It was in response to this combination of cultural and political success that *Partisan Review* mounted its attack against popular culture.

II

Throughout the late 1930s and early 1940s *Partisan Review* critics kept up a steady barrage of attacks on Popular Front products. In her regular column "Theater Chronicle," Mary McCarthy spared few barbs as she assaulted what she considered to be the sentimental populist plays of Clifford Odets, Ben Hecht, Maxwell Anderson, Marc Blitzstein, and others. In the book review column there were regular demolition jobs of the latest Popular Front effort at serious literature (generically known as "middlebrow"), efforts that inspired fellow-traveling critics to the heights of approbation. But *Partisan Review*'s legacy in the area of popular culture rests less in any single review than in the few influential statements that brought together the assumptions behind the reviews. Here I would call attention to Dwight Macdonald's early work on Soviet film, his later, more systematic "A Theory of Popular Culture" (1944), and Clement Greenberg's well-known essay "Avant-Garde and Kitsch" (1939).

Macdonald's first sustained study of popular culture was a two-part history of Soviet film entitled "The Soviet Cinema, 1930–1938," published in *Partisan Review* in 1938. He began with an effusive appreciation of the extraordinary breakthroughs in form and of the subtle rendering of political content achieved by the Soviet cinema of the 1920s, and singled out Eisenstein, Pudovkin, and Dozhenko for praise. Yet according to Macdonald, at the moment Soviet cinema was on the verge of its greatest triumph—the conquest of the new sound technology in the late 1920s—a precipitous decline in the quality of filmmaking set in. Largely as a result of pressure from the first Five-Year Plan of 1928, the avant-garde found it increasingly difficult to produce vital, experimental work. Part of the problem, according to Macdonald, was that the new themes of industrialization and collectivization were simply harder to dramatize than was the older and broader theme of the revolution itself, which offered opportunities for more dramatic and catalytic subject matter. But the major problem was that Stalinist coercion and censorship in the arts made it impossible by the mid-1930s to create anything but highly stylized and insipid encomiums to Stalin and his policies—otherwise known as socialist realism. Here Macdonald was careful to distinguish his critique of the Stalinist destruction of Soviet film from the liberal point of view that detached art from politics altogether. In "Soviet Society and Its Cinema" (1939), an elaboration of his earlier effort, he made it clear that the issue was not the politicization of film. In fact, Soviet film was never more politicized, never more "propagandistic," and intimately connected to the state, than it was during its most productive period beginning in 1924. As he observed, "it is not the mere *fact* of political control, but the *direction* of this control that has been damaging."[12]

One of the problems with this direction, aside from the fact that it was dictatorially and violently imposed, was that it borrowed the same rationale for its impoverished quality then being used by Hollywood: it was "what the masses wanted." But Macdonald maintained that Stalin had actually borrowed more than a rationale. Worse, he had borrowed the Hollywood film and made it the official Soviet film.

In "A Theory of Popular Culture" Macdonald turned to the phenomenon of the mass production of culture by Hollywood and other media centers. For close to a century, he claimed, Western culture had actually been two cultures: high culture, or the avant-garde, and popular, or mass, culture. Mass culture was described as culture "manufactured wholesale for the Masses."[13] In the twentieth century it had very much gained the upper hand with the creation of a mass, literate, and leisured audience, new media and genres, and new opportunities for capitalist profit-taking. Mass culture imposed itself on a subservient public, and tailored its formulas to fit its own idea of what would sell. Whereas during the precapitalist and early capitalist epochs folk art was an authentic expression of the people, mass culture was produced from above by technicians of art in the hire of the ruling class.

Mass culture, according to Macdonald, had become so virulent a force in modern society that it threatened the very life of high culture. It did so by debasing traditional forms and reducing elite audiences, themselves more often than not seduced by the appeal of midcult, which was really kitsch dressed up in the finery of high forms. Mass culture was thus well on its way toward undermining the avant-garde through absorption. A sort of Gresham's Law of culture was at work, according to Macdonald, where "the bad drives out the good, and the worst drives out the bad."[14]

For its part the avant-garde, by which Macdonald meant modernism in general (he did not single out individual writers), had always been distinguished above all by what he said was its "refusal to compete." Under late capitalism this meant that it was losing its tiny market among the cultured elite. In the face of encroaching mediocrity, the avant-garde was woefully unable to defend itself. However precarious the older bohemian and expatriate communities had been, they were at least sustained by a tiny minority of cultured bourgeoisie. But now the only thing the ruling class could afford to cultivate was power, not criticism, the avant-garde's specialty. It could not tolerate what it had once solicited. This was especially true for what Macdonald called "this deeply reactionary period," in which "men cling to the evils they know rather than risk possible greater ones by pressing forward."[15] For Macdonald this decline was inaugurated not by anything specifically American but by "the rise of fascism, and the revelation, in the Moscow Trials, of the real nature of the new society in Russia."[16] America, in other words, was the scene of European totalitarianism in embryo. "As in politics," Macdonald observed, "everything and everybody are being integrated—'coordinated' the Nazis call it—in the official culture structure."[17] Indicative of these trends were the examples of the talented writers who had been absorbed by the Luce organization since 1930 (actually at least one of them, James Agee, rather enjoyed working for Luce), the fact that Edmund Wilson's talents were being wasted by his taking over Clifton Fadiman's column at the *New Yorker*, and the hiring of avant-garde writers by the *New York Times* Sunday Book Section.[18]

It should be pointed out that Macdonald insisted on the distinction between capitalism and totalitarianism. "Capitalism," he maintained, "perverts art or makes its practice more difficult, but totalitarianism simply liquidates it." Whereas under capitalism the artist can take advantage of "conflicting forces" and survive in the available "crannies," in Russia "there are no crannies, no contradictions, no conflicting forces."[19] Nevertheless, a persistent motif in Macdonald's work on popular culture has always been that significant totalitarian trends were advancing on American society. Since he was often reticent about what it was that provided *resistance* to these trends, he never made it clear whether he considered the differences between the two social forms to have been differences in degree or kind. This was confirmed by the fact that neither he nor the other New York intellectuals were ever much concerned with exploring the "conflicting forces" and "crannies" of American society to which he alluded, especially as they were manifested in oppressed groups and marginalized cultures. For the New Yorkers the masses were not only victims, but passive victims of the culture industry. Portrayed in such unqualifiedly negative terms, it is hard to see how they could have had any role in the kind of social transformation Macdonald implied was necessary. Thus even though Macdonald made gestures toward distinguishing between capitalism and totalitarianism, his comments on the invasion of mass culture, and in particular his underlying attitude toward the masses, all emphasized his fear of totalitarianism rather than the capitalism that actually existed in the United States. Indeed in the article just cited, Macdonald was characteristically evasive about this crucial issue. In asking why it was that the dominant classes in America fed the masses dreams and romance whereas "their peers in Russia" gave them historical pageants "*although both have the same end in view*" (emphasis added), Macdonald was fudging a central issue when he declined to say what that same end was.

Much later, in 1960, we read in "Masscult and Midcult," that the end was "to transform the individual into a mass man." This is "the tendency of modern industrial society, whether in the USA or the USSR. . . . The masses are in historical time what a crowd is in space: a large quantity of people unable to express their human qualities because they are related to each other neither as individuals nor as members of a community."[20] In fact, continued Macdonald, the masses are not related to each other at all in any meaningful sense; more important is their relationship "to some impersonal, abstract, crystallizing factor," which unfortunately remained unidentified. Again, Macdonald failed to provide an analysis of the details of the American social structure or of the experiences of ordinary Americans, especially those experiences in which they managed to avoid, resist, or positively make use of the formidable power of popular culture. Macdonald defined mass culture as an instrument of social domination and exploitation. He considered mass culture to be crucial to the ongoing transformation of society in which the integrity of the individual—his or her individuality, imaginative life, and critical faculties—was under constant and increasing attack. Given this monolithic set of conditions from which no escape seemed possible, what could radicals do on behalf of the masses? Plainly very little, except assess the nefarious effects of mass culture on their impoverished lives.

One of the reasons for socialism's sterility, Macdonald claimed, was its neglect of these kinds of cultural factors. In this area, he said, "Hitler had more imagination." Radicals must understand heroes such as Superman, Tarzan, and the Lone Ranger; they must become familiar with the psychology behind their appeal, and begin taking the results seriously in working out their own politics. But this did not necessarily indicate a new solicitude for mass culture—for any positive effects, or for the possibility of an ambivalent, critical, or creative response by the masses. The criterion that radicals were to use to judge mass culture was nothing short of Trotsky's vision of "a potential[ly] new *human* culture."[21] Here again there was a conspicuous silence from Macdonald, for the standards he alluded to were as mute as was the classless society they were alleged to arise from. Ultimately Macdonald, like many of his fellow New Yorkers, found a source for the utopian values of this idealized society—not in the best portions of the lives of the masses but in literary modernism.

In assessing Macdonald's theory, it seems clear that he made too severe a dichotomy between the high and the low, the elite and the mass, modernism and other schools. Inevitably this resulted in a tendency to idealize modernism and demonize popular culture. As Macdonald wrote in the final version of "Masscult and Midcult": "Two cultures have developed in this country and . . . it is to the national interest to keep them separate. The conservatives are right when they say there has never been a broadly democratic culture on a high level."[22] By wishing to cordon off high culture from the onslaughts of the culture industry, Macdonald essentially wrote off the working class, which was in his view hopelessly mired in kitsch. Thus he ended up reinforcing the kinds of social and cultural distinctions, including the class distinctions that underwrite high culture for a cultivated elite, that his earlier Trotskyist politics meant to destroy.

III

In his well-known essay "Avant-Garde and Kitsch," Clement Greenberg defended much the same position as Macdonald. He and Macdonald shared similar politics during the late 1930s and early 1940s when the nation was mired in war. Both were Trotskyists, and both believed, as had Lenin during World War I, that the working class was faced with a war of interimperialist rivalry. What was necessary was to turn the imperialist war into a civil war, only after which America could successfully combat the fascist menace, thought to be an internal as well as an external threat.[23] For Greenberg the rampant commodification of culture under late capitalism had created an increasingly ubiquitous mass culture (kitsch) that fed off of high culture (the avant-garde) and in the process eroded and debased it. He described kitsch in unqualifiedly negative terms (in much the same way as Macdonald had described Hollywood and socialist realist films): as mechanical, formulaic, vicarious, and fake. If the avant-garde was experimental, bold, and dynamic, kitsch was derivative, cowardly, and static. And if the avant-garde righteously withdrew from the marketplace (we recall Macdonald's observation that the avant-garde could be characterized by its "refusal to compete"), the creators of kitsch fell over one another in the scramble for profits. Kitsch was thus the absolute, polar opposite of the avant-garde, the latter loftily described as "culture with its infinity of aspects, its luxuriance, its large comprehension."[24]

According to Greenberg, at its advent in the mid-nineteenth century, the avant-garde had flourished alongside the new "historical criticism" and "scientific revolutionary thought"—in other words alongside Marxism.[25] Emboldened by politics to pursue its critique of the bourgeoisie, the avant-garde nevertheless detached itself from the welter of political and revolutionary movements, as well as from the capitalist marketplace. In its detachment, in its renunciation of the world of "common, extraverted experience," it fixed its gaze inward upon its own artistic processes and its own medium—whence the origin of the "abstract," a major concern of Greenberg's. With abstraction the avant-garde had inaugurated its characteristic activity—"the imitation of imitat*ing*." This new introversion did not lead to experimentation for its own sake; rather it was a way of finding "a path along which it would be possible to keep culture *moving* in the midst of ideological confusion and violence."[26]

But the bohemians' retirement from the public sphere was not as complete as they had imagined, for always there remained what Greenberg termed, borrowing Marx's phrase, "an umbilical cord of gold" attaching the avant-garde to a small but influential elite among the ruling class who provided material support. What was potentially fatal to the avant-garde in the contemporary crisis, according to Greenberg, was the fact that this elite was shrinking dramatically due to the seductive influences of mass culture. Under these circumstances, when the very existence of the avant-garde hung in the balance, it fell to socialism to provide for "the preservation of whatever living culture we have right now."[27] Whereas Macdonald had at least intimated that socialism could provide a vision of and material basis for a new culture, Greenberg could only manage a ritualistic reference to the "inevitability" of socialism. Surrounded by sentences of doom and gloom, such optimism lacked conviction.

How then did Greenberg's and Macdonald's politics and social situation shape their criticism? The first thing to be said is that their political analyses and the ideological pressures therein caused them to place their socialism in uncompromising opposition not only to reformism, but to American life in general. Their socialism was oddly nullified by the fact that all of its values were placed in direct opposition to American life. Instead of offering a framework that would lead to judgments about what was good and bad about popular culture, Greenberg and Macdonald tended to offer otherworldly ideals that exposed the limits of the real but were not situated in the real. Within the visible, bourgeois world, each fallen process or object of culture had its counterpart in the rarefied precincts of the avant-garde, an avant-garde that functioned as a kind of avatar of the future classless society.

Here, unfortunately, dialectic was reduced to Manichaeanism. Greenberg and Macdonald could not extricate their analysis of popular culture from the reigning ideological antinomies of their time. In viewing every new cultural development as a step toward totalitarianism, they underestimated the adaptability and diversity of American life. They did not take into account the diverse modes of production, reception, and criticism of popular forms. On the production end they ignored several variables: the autonomy sometimes rendered artists and technicians by prudent owners or managers, the opportunities for solidarity and criticism afforded by collective artistic projects, the chance to imbue escapism with elements of social critique and utopian longing, and finally the potential for novel technical modes developing new art forms that might parallel and in some cases instigate achievements within the avant-garde.

As for the reception of mass culture, Greenberg and Macdonald simply assumed what they needed to prove—that is, that the masses were indeed being "warped," to use Macdonald's word, by the culture industry. To be persuasive such a conclusion would need to be based upon a careful, empirical study of the cultural experiences of the masses. As Richard Altick observed long ago in his introduction to *The English Common Reader*, any interpretation of the popular taste of an age must be based on a thorough scrutiny of the social composition, education, and general character of the group whose taste is undergoing judgment.[28] This was precisely what was lacking in Greenberg's and Macdonald's analyses. They simply were not receptive to the possibility of genuine critical judgment among the producers and especially the consumers of mass culture in the face of adverse institutional pressures against discriminating tastes. They failed to take into account, namely, what George Orwell called the masses' "mental rebellions" and "momentary wish[es] that things were otherwise," which even so modest an art as Donald McGill's comic, obscene postcards were able to take advantage of.[29] Orwell of course enthusiastically searched out such examples of resistance among ordinary British citizens, and he considered these examples to be parts of a working-class culture that spawned nonconformity if not socialism. What raised Orwell's ire was the ironic fact that many socialists refused to respond to or even to know about these small but daily examples of resistance. Thus the *Partisan Review* critics seemed transfixed only by those examples in which it seemed the mass mind was marching according to the bourgeoisie's orders. The only two sites of freedom in their analyses were both rigidly circumscribed: genuine folk art had been all but destroyed by capitalist hegemony, and the avant-garde was hemmed in by the surrounding philistinism. Perhaps their vision of realms of artistic freedom and pure value, surrounded by the encroaching catastrophe of capitalism, was but the logical response to their own confinement within the small social space accorded them by capitalism and which they accepted. As their connections to ordinary Americans diminished, they could only speculate that others were even more confined than they. This, perhaps more than anything else, was what caused them to try to impose the allegedly uncontaminated values of autonomous cultural spheres on the rest of the captive world.

IV

We have seen how, when faced with the Popular Front's resurgence of patriotism and literary traditionalism, the *Partisan Review* critics looked to the modernist avant-garde of the past generation for examples of independent, critical thought. Given their views of European versus American intellectual traditions, it is not surprising that they took an unusually large number of their examples from European literature.[30] If we return for a moment to the earlier essay "Literature in a Political Decade," we see that as early as 1937 Phillips and Rahv were making some very negative and very consequential judgments about American literature. Declaring the need to "enquire into the basic failures of American writing," they asked why American literature "has remained so primitive despite all the aesthetic and social movements that have been injected into it?" According to the editors, American literature had suffered its own national strain of Eliot's dissociation of sensibility by adopting the extremes of pure action and pure thought—hence the crude dualities that obtained between "the rigors of the act and the subtle delights of relaxation, between the tyranny of purpose and the self-indulgence of artistic myths."[31] The chief consequence of this persistent bifurcation of vital elements and the privileging of action over thought had been that "the progress of the American mind still has not been defined in a work of art." This having been the failure of American literature "for decades, . . . the failures of the left cannot be attributed to its Marxist elements but rather to the pragmatic patterns and lack of consciousness that dominate the national heritage."

Phillips and Rahv listed the factors most responsible for the imperfect state of American letters: endemic pragmatism, populism, regionalism, and false materialism. Each of these in turn reflected a larger "anti-intellectual bias which constantly draws literature below urban levels into the sheer 'idiocy of the village.'" As the editors pungently put it,

> Describe only those units of American life whose intellectual experience hardly exceeds their behavior; avoid those problems which harass the mind of mankind but have not yet stirred the consciousness of the average American—this is the Monroe Doctrine in American Literature.[32]

The way out of this "domestic stasis," which the editors labeled, echoing Trotsky, this "futile attempt to create a literature in one country," was to transform a political strategy into a literary one. Just as the American class struggle of the past decade had been "Europeanized" through the "native adaptations of general principles" common to all bourgeois societies, so literature must "subject the native reality to the full consciousness of Western man." Like the Russian Slavophile movement, American provincialism must be defeated by a Bolshevik trend that will "Westernize America." "The international program that will liberate American society from its economic fetters will have its necessary counterpart in literature."[33] More than twenty years after Pound's and Eliot's initial attempts to internationalize American literature, Phillips and Rahv were claiming that the task still needed doing, albeit differently. Because of continued fierce resistance among the "provincials," and because of the reactionary politics that Pound and Eliot subsequently attached to their versions of internationalism, the job presented itself as one that the left might complete. This was the initial political rationale for the very influential reassessment of the American canon undertaken during the 1940s by Phillips and especially by Rahv, who influenced Trilling and Matthiessen, who in turn influenced Richard Chase, Marius Bewley, Leslie Fiedler, Leo Marx, and others.[34] Here is yet another instance of the New York intellectuals' celebrated cosmopolitanism starting out as internationalism, or at least a particularly Eurocentric version of internationalism.

Although *Partisan Review*'s rejection of the realist/naturalist canon involved powerful literary insights and arguments, its impetus and shaping dynamic was always the magazine's ideological battle against Stalinism; Irving Howe has said as much about the left's concomittant relation to the Emersonian tradition. In a 1989 interview with William Cain, Howe remarked,

> It needs to be said . . . that we had a certain bias . . . with regard to American literature. We associated American literature with a narrow cultural chauvinism. Van Wyck Brooks, by the 1940s in his inferior phase, and others too were celebrating American writing as against what they looked upon as decadent European writing. And we accepted a little too uncritically this kind of counterposition. If you were concerned with socialist ideas, if you felt you were living in a moment of historical apocalypse—say, in 1939 or 1946—and were trying desperately to define what Stalin-

> ism was and what new kind of society had emerged that had traduced your socialist hopes, then Emerson's Concord could seem extremely thin, even genteel.[35]

As Howe points out, a crucial episode in the New York intellectuals' contribution to reshaping the American canon was the acrimonious dispute with Van Wyck Brooks during the early 1940s. I will characterize this battle shortly; first, however, it is necessary once again to focus on an earlier episode: Rahv's all-out retrospective attack on the proletarian literary movement. It was here that the "apocalyptic" ideological battle with Stalinism over the American canon was initially joined, establishing the contours of the later conflict over literary nationalism and Popular Front reformism.

When he wrote 'Proletarian Literature: A Political Autopsy" for Robert Penn Warren's and Cleanth Brooks's *Southern Review* in 1939, Rahv, and *Partisan Review*, had been two years out of the proletarian literary movement. The article reflected Rahv's deep concern, which he shared with many anti-Stalinist leftists, that the Communist Party had been consistently winning over prominent authors to antifascism and reformism and was once again subordinating literature to political exigencies, just as it had done earlier with the proletarian literary movement. Because most of the writers associated with the proletarian movement and later with the Popular Front were considered to be part of the realist-naturalist tradition rather than the growing tradition of experimental modernism,[36] the polemical confrontation between the two political camps contributed formidably to the reification of the American literary past into separate and opposing traditions.

As I have shown, for Rahv the proletarian literary movement had been monolithic and aggressively controlled from the top by the Party. Distinguishing the guiding principles of the movement from Marxism, Rahv claimed these principles were uniform and extremely constricting, reducible to the formula "the writer should ally himself with the working class and recognize the class struggle as the central fact of modern life."[37] To insure conformity, the Party directed its apparatus of national and international organizations, publications, and political commissars "to supervise the public relations of the new literary movement and to minister to its doctrinal health."[38] Although Rahv mentioned other factors that shaped the movement—the Depression, the accompanying emergence of the common man as literary hero during the 1930s, the prior "exhaustion" of literary modes during the 1920s—he contended the movement was essentially an artificial creation of a Party culture and not a class or national culture. "It is clear," he concluded in a famous remark, "that proletarian literature is the literature of a party disguised as the literature of a class."[39]

Rahv was remarkably incisive concerning some of the ill effects of the movement's politicization of literature and culture. For example, he raised the pertinent points that the movement's prescriptions ignored questions of form or craft and lacked aesthetic principle and direction. He also pointed to a weakness that continues to plague those who would politicize literature and criticism today, namely the absence of "defensible frontiers" between art and politics: the two were indissolubly merged. Put another way, with regard to its treatment of politics the movement drew no distinction between what Rahv called "the politics of writing in a *generic and normative sense*" and the politics of a specific writer in a specific historical period.

Rahv's insights, however, were accompanied by certain distortions, which contributed to the subsequent polarization of American literary history. First of all, he overstated the extent to which the Party exerted control over the movement. Here Daniel Aaron remains our best guide, having observed long ago that, in spite of meddling and manipulation, in the end the Party simply did not care enough about specifically literary matters to pay close attention to every significant development within the proletarian movement.[40] The very example of the early *Partisan Review* and Phillips's and Rahv's own revisionism within the boundaries of the proletarian movement is evidence enough that the Party allowed for a degree of debate and diversity, albeit within a narrow range of positions. But even if we were to agree with Rahv that the Party attempted to exert nearly total control over the movement, we need not concur that the Party was successful at achieving anything close to unanimous compliance. Again as Aaron has pointed out, even in the mid-1930s writers on the left did not form a cohesive and malleable group. Rahv took note of this fact, but only in passing, and only to suggest that whatever real quality writers were able to muster was imported into the movement from other realms of experience. "The better writers," granted Rahv, "such as Josephine Herbst, Grace Lumpkin, Robert Cantwell, and Kenneth Fearing avoided [the movement's] fantasies by sticking to what they knew."[41] "What they knew" according to Rahv, necessarily had its source wholly outside the Party's milieu. Overcompensating for the widespread attitude of condescension toward the bourgeois cultural tradition, or worse the ultra-left repudiation of all things bourgeois, Rahv simply inverted the terms of the orthodox dualism. For Rahv, and here he was following Trotsky's views as articulated in *Literature and Revolution*, bourgeois culture, having had centuries to develop and gain hegemony, was the only viable culture in existence. The working class, lacking both the material means and the self-consciousness that are prerequisites of cultural creation, has functioned and would continue to function for the foreseeable future as a "cultural consumer, liv[ing] on the leavings of the bourgeoisie." The conditions of existence of the working class "allow it to produce certain limited and minor cultural forms, such as urban folklore, language variations, etc., but it is powerless to intervene in science, philosophy, art, and literature."[42]

One need not subscribe to any of the doctrinaire definitions of proletarian literature to realize that Rahv seriously underestimated the contributions of the writers in or near the proletarian literary movement, as well as the extent to which their writing was enhanced through contact with working-class life and culture. This is not to suggest that bourgeois culture was irrelevant to their work—far from it. But it is to suggest that the best writers and artists of the 1930s *combined* bourgeois and working-class cultural forms, the latter especially providing rich sources for American literature, most notably for African American literature. Sterling Brown, for example, introduced black folklore, dialect, and speech patterns into poems whose debt to Wordsworth, Sandburg, Robinson, and even Housman are readily apparent. Zora Neale Hurston did the same, in the process reconfiguring both the traditional American novel of self-discovery and the cautious Afro-American narrative of uplift. Richard Wright, who served his literary apprenticeship in the John Reed Club, rewrote the working-class novel by introducing the theme of race, by including unprecedented descriptions of urban despair, anger, and violence, and by incorporating unresolved endings notorious for their political ambiguity. Kenneth Fearing integrated jazz, blues, and slang into his raucous, Whitmanesque poems. Nathanael West saturated his highly experimental narratives with a wide range of popular forms. And Tillie Olsen, Meridel LeSueur, and Tess Slesinger, among others, quietly transformed proletarian and masculine formulas so as to reflect the unique experiences of poor and working women. The list, of course, goes on, the point being that many writers in or near the proletarian movement or later associated with the Popular Front heroically dissented from the orthodoxies of dissent at the same time that they benefited from their association with the proletarian and Popular Front movements.[43] Phillips and Rahv simply failed to fulfill the promise they had made in the editorial statement that introduced the revived *Partisan Review* in 1937, in which they committed themselves "to distinguish[ing] between the tendencies of [the Communist Party] and the work of writers associated with it."[44]

To return to Rahv, it is difficult to resist the conclusion that the explosive ideological environment of the late 1930s prevented him from judging the proletarian movement in measured fashion. In his haste to discredit the Communist Party and all that was associated with it, he lost an opportunity to broaden the culture of anti-Stalinist socialism so as to include works that did not easily fit into a modernist framework. This would have made for a more genuine internationalism because it would have made room for a wider range of traditions and cultures. As salutary as the New York intellectuals' and New Critics' canonization of modernism may have been, it came at the high price of hypostatizing dramatic literary divisions and excluding works written out of other experiences and traditions that did not meet a specifically modernist understanding of ironic difficulty. The irony, of course, is that the anti-Stalinist canonization of modernism perpetuated a dualism that Stalinism itself had a hand in creating.

V

In critiquing Rahv's retrospective assessment of proletarian literature, I do not want to suggest that he or the *Partisan Review* circle was solely or even predominantly responsible for our legacy of a bifurcated literary history. Rahv wrote his influential piece at the time of the Popular Front's growing success. Popular Front critics and their allies were hard at work attacking modernism, and their intemperate wartime campaign against what came to be known as "coterie" literature was responsible for further exacerbating an already polarized situation. What was particularly distressing to the anti-Stalinist left about the Communist Party's growing success at gaining allies in the Popular Front was that success had been won despite what it regarded as political opportunism and intellectual irresponsibility. In his important article "The Anatomy of the Popular Front," Sidney Hook charged, in classically Leninist fashion, that the Popular Front strategy was mired in reformist illusions about persuading capital to transform itself. The day-to-day struggle had become paramount, supplanting the long-term goals of the movement. Dubious maneuverings, unprincipled alliances, small "victories"—these became the order of the day as political wheeling and dealing replaced traditional political work. Hook believed such nearsightedness and naiveté were the outcome of a politics that had chosen to ignore ideas and theory. His analysis thus paralleled Lenin's critique of economism, for Lenin had also singled out for reproach the neglect of theory and long-range goals by his rightist opponents.[45] "Ideas" and "mind," of course, became values absolutely central to *Partisan Review*'s defense of modernism. The alleged betrayal of these values became the basis for the magazine's campaign against literary nationalism.

A major task of the cultural wing of the Popular Front was to rewrite the past so that the Party's nationalist slogan "Communism is twentieth century Americanism" would ring true. As Irving Howe has pointed out, this demanded that a "whole new school of historiography" be founded whose purpose was to uncover a tradition of "progressive" leaders, artists, and causes that prefigured and legitimated the Popular Front movement. Proceeding anagogically in much the same way that the early Christians reinterpreted the Old Testament as a prefigurement of the New, Party and fellow-traveling critics reinterpreted the American past, finding progressive political and literary precursors in many historical figures. In literary history V. L. Parrington, Granville Hicks, and Bernard Smith had already provided a running start with, respectively, *Main Currents in American Thought* (1930), *The Great Tradition* (1933), and *Forces in American Criticism* (1939). In these revisionist accounts of American literature, only the "progressive" American writers counted—for example, Franklin, Samuel and John Adams, Paine, Freneau, Cooper, Bryant, Stowe, Whitman, Twain, London, and Dreiser. "Reactionary" writers were either ignored (Dickinson, Stein, Eliot, Pound) or scorned (Mather, Edwards, Poe, Melville, Hawthorne, Henry Adams, James, Wharton, Cather, Cummings). Although the "progressive" literary historians, as Trilling acknowledged in a famous passage on Parrington's "saving salt," possessed an admirable grasp of ordinary material reality ("Parrington knew what so many literary historians do not know, that emotions and ideas are the sparks that fly when the mind meets difficultly"), they were hemmed in by their own social determinism and could not see how artistry, form, and sensibility might produce something other than ultimate political opinions and attitudes.

The organization most responsible for spreading Popular Front values in the cultural field was the League of American Writers (LAW), an outgrowth of the Writers' Congress of 1935. Although repeated policy reversals ultimately did in the organization by the end of World War II, several of its most prominent members, namely Van Wyck Brooks and Archibald MacLeish, continued to advocate a strongly "progressive" and nationalist literary perspective. In doing so they attacked the modernist avant-garde of the early century, which of course

Partisan Review strongly defended. For the *Partisan Review* critics, these literary judgments not only reinforced the values of the Popular Front, they also fed a growing propaganda machine that churned out patriotic banalities asking the public to make ultimate sacrifices without making any significant political demands.

Van Wyck Brooks had always had his problems with Marxism, even as he labored at maintaining working relations with Marxists. A lifelong socialist, in general he tended to keep his literary and cultural criticism relatively detached from explicit politics, the exceptions being precisely the pieces I will examine shortly. During the 1910s he and Randolph Bourne had written their pioneering essays calling for the creation of an independent, critical, and politicized intelligentsia capable of supplying America with the discrimination and skill needed to inaugurate a thorough cultural transformation of society. Such a transformation would need to eliminate the paralyzing effects of the Puritan tradition on the American mind. In part this would be accomplished by doing away with Puritanism's strict division of culture into highbrow and lowbrow.

But this was before World War I. During the 1920s, although he never repudiated this earlier perspective, Brooks was discouraged from insisting on immediate social change by the antiapocalyptic mood of that decade. His criticism became less ambitious in a partisan sense, although it always remained a source of controversy. His two major works of the decade were *The Ordeal of Mark Twain* (1920, rev. ed. 1933), a psychological study, and *The Pilgrimage of Henry James* (1925), in which he condemned expatriation and judged James's later novels to be inferior because of the author's distance from life in his native land. *The Life of Emerson* (1932), also written during the 1920s, represented Emerson as the great unifier of art and life, and of what Brooks termed, respectively, "Highbrow" and "Lowbrow." But it was the unusual success of his next effort, *The Flowering of New England, 1815–1865* (1936), that placed him once again at the center of the critical scene. The book went through forty-one printings in its first sixteen months, and became a Book-of-the-Month-Club selection. The critics were effusive in their praise. Blackmur said of the book's upbeat nationalism that it proved "Where we [the nation] are aware of a lack, we can feel . . . that the cure is within us."[46] Carl Van Doren hazarded that if future volumes in the projected series were written "with the same knowledge, range, insight, precision, and grace," the series would be "one of the best literary histories in any language."

As James Hoopes has observed, Americans in the mid-1930s were more than ready for an inspiring book. But the *Partisan Review* critics were not ready, certainly not after Brooks associated himself with LAW and its ethos of national uplift. It was time, thought the *Partisan Review* editors, to have a look at the decline of Brooks's career. F. W. Dupee's "The Americanism of Van Wyck Brooks" (1939) was a piece not wholly unsympathetic to Brooks insofar as Dupee mourned the loss of the more historically minded and radical younger critic. Dupee credited Brooks with having had the perspicacity to understand how valuable an autonomous, critical intelligentsia might be for social change in general and for a labor movement vitally in need of ideas and culture. Brooks had once known how damaging the cleavage in American life was between a culture for the elite and another for the masses; his earlier recommendations in *America's Coming of Age* (1915) that a heterogeneous and complex culture was needed to replace one of invidious contrasts were highly salutary. But for Dupee, Brooks's views became muddled over the years: the "materialist" in him "was always coming into conflict with the 'organic' visionary, the social historian with the psychologist."[47] Brooks was ironically accustomed to thinking of matter and spirit as polar opposites, as were his New England ancestors, and he was unable to approach cultural phenomena from the point of view of its connection to its material surroundings. In *The Flowering of New England* Dupee saw the socialist and historian being supplanted by the impressionist. Brooks's lack of clarity on culture's connectedness to material reality caused his criticism to veer back and forth between the extremes of free will and determinism such that at one point the writer is monadic, at another a mere slave to his times. The result was that the critic was sometimes omniscient and sometimes merely a mouthpiece for conventional views. When Brooks "had finished trying to reconcile politics and literature, mysticism and science, he was left with an ideology as diffuse as that of an Emerson or a Whitman."[48] In a time as ideologically predatory as the late 1930s, he would be forced

> to pay in the unprofitable coin of centrism and compromise the price of his synthetic ambitions. . . . Endeavoring to embrace the Whole, [he] ended by losing touch with its parts. . . . In time he was to seem almost the type of that Liberal critic whom T. S. Eliot from one angle and Mencken from another, were to assail with so much effectiveness.

In the face of the major problems of the epoch, Brooks had chosen to preoccupy himself with "ancestral attitudes" that permitted "the visionary to consume the skeptic." Alas, wrote Dupee, "the powerful critic of the United States has turned into a zealous curator of its antiquities."[49]

Dupee's attack on Brooks was an early skirmish in a much wider battle that pitted the New York intellectuals against the growing trend of literary nationalism. This battle heated up significantly in 1940 when several well-known intellectuals—Waldo Frank, Lewis Mumford, and Archibald MacLeish—joined Brooks in making public their contempt for writers who insisted on remaining neutral in America's impending war against fascism. The greatest stir was created by MacLeish's statement, "The Irresponsibles." Appearing in the May 8, 1940, issue of the *Nation* just a few weeks after the Nazi invasion of Belgium and the Netherlands, the article attacked American writers for ignoring their moral and political responsibilities by refusing to support America's democratic institutions and antifascist policies. For *Partisan Review* it was an attack that represented the reverse of views held by Benda, Bourne, and formerly by Brooks, which emphasized the importance of intellectuals maintaining a critical distance from the centers of power. MacLeish, then working for the State Department, appealed to intellectuals to fight the menace of fascism with "the weapons of ideas and words." At stake for MacLeish were not practical matters—"The man whose only care is for his belly and his roof . . . can safely be indifferent to these troubles"—but "a condition of men's minds . . . the community of Western culture."[50]

MacLeish attributed the "inactivity" (but not "indifference") of American intellectuals to the demise of the unalienated, engaged "man of letters." The man of letters "was a man of learning whose learning was employed not for its own sake in a kind of academic narcissism but for the sake of decent living in his time."[51] Today, noted MacLeish, a Milton or a Voltaire would need to overcome the pseudoscientific notions of objectivity and a cult of specialization that prevents members of the two dominant "castes," the writers . . . and the scholars, from accepting responsibility for "the common culture" and its defense. He predicted that in the future the present "pure" devotion to culture will yield "little profit or much praise." What would loom much larger would be

> the one question asked of our generation: "Why did the scholars and writers . . . witnesses as they were to the destruction of writing and scholarship in great areas of Europe and to the exile and imprisonment and murder of men whose crime was scholarship and writing . . . fail to oppose those [destructive] forces while they could—while there was still time and still place to oppose them with the arms of scholarship and writing?"[52]

Van Wyck Brooks was among many to respond enthusiastically to MacLeish's broadside. In a letter to the editors of the *Nation* he stated that MacLeish's article was indeed inspiring and would "turn the tide in American literature." It would "rouse our thinkers and set our intellectual world in action."[53] Some months later Brooks pressed MacLeish's attack even further in an address entitled "On Literature Today":

> [I]t seemed as if the most powerful writers, from James Joyce to Hemingway, from Eugene O'Neill and Theodore Dreiser to Eliot of *The Waste Land* were bent on proving that life is a dark little pocket. . . . What did the novelists say if not that nothing good exists, that only the ugly is real, the perverted, the distorted? Faulkner and Dos Passos seemed to delight in kicking their world to pieces, as if civilization were all a pretence and everything noble a humbug.[54]

Less than a year later, at a Columbia University conference, Brooks delivered his polemical coup de grace, "Primary Literature and Coterie Literature." Although this was not as direct an appeal for political engagement as was MacLeish's "The Irresponsibles," the two pieces shared important assumptions about modern literature. Brooks maintained that much of the literature of the twentieth century was "coterie literature," by which he meant literature that has lost all faith in human progress, human nature, and human goodness. Conversely, primary literature "somehow follows the biological grain. . . . It favors what psychologists call 'the life-drive'; it is a force of regeneration that in some way conduces to race-survival on what might be further called the best possible terms. . . . A great man writing is one who bespeaks the collective life of the people, of his group, of his nation, of all mankind."[55]

In the course of his address Brooks had occasion to attack most of the major modernist writers. Eliot, who received more than his share of abuse, had succeeded merely in substituting the secondary for the primary with his so-called "transvaluation of English literature." Pound had "chucked out" Pindar and Thucydides, whom he had the arrogance to call a mere journalist, to make room for the likes of Corbière and Rimbaud ("a neurasthenic wretch").[56] Joyce was described as "the sick Irish Jesuit . . . who had done more than Eliot to destroy tradition." Mallarmé "wrote verses for ladies' fans," et cetera. What seemed most objectionable to Brooks was that the moderns "have resigned life and made a virtue of it." Whereas the primary writer responded with "enthusiasm" and was possessed with "matter," the modern writer was antisocial and placed "construction" before matter: "Here was the heart of [the] case against these coterie-writers, who doubt the value of life and of literature also."[57]

In making his critique Brooks defended a canon that was entirely distinct from the one the *Partisan Review* critics defended. For Brooks the "primary" writers were Homer, Socrates, Erasmus, Rabelais, Goethe, Hugo, Emerson, Whitman, Whittier (as a "folk-poet" and "poet of freedom"), Dickens, Tolstoy, Dostoevsky, Ibsen, and Mann; the "primary" critics were Saint-Beuve, Taine, Renan, and Arnold. His secondary writers were *Partisan Review*'s primary writers: Joyce, Pound, Eliot, Valéry, Hemingway, and Farrell; and his secondary critics were *Partisan Review*'s primary critics: Eliot, Pound, Richards, Winters, Tate, and Ransom.

The *Partisan Review* critics referred to the position held by Brooks and MacLeish as the Brooks-MacLeish Thesis, and they set out post hoc to destroy it. The opening salvo was fired by Morton Zabel in a two-part article, "The Poet On Capitol Hill," the title indicating just how seriously he regarded MacLeish's position as spokesman for the established powers. Zabel began his piece by acknowledging MacLeish's phenomenal success at winning over the public and in the process winning power for himself as well:

> During the past eighteen months Americans have witnessed an event that constitutes an unmistakable milestone in their cultural history. A poet has become a national leader—not a poet of mere household or academic honor like Whittier, Longfellow, or Bryant . . . but a poet of recognized standing and a leader of official rank. Mr. Archibald MacLeish has been made Librarian of Congress. He has been saluted by the President of the United States and receives the nation-wide acclaim of the press and public.[58]

Mentioning in passing that MacLeish's speeches were in the process of being printed by the U.S. Government Printing Office, Zabel identified what he considered to be most distinctive and most ominous about them:

> These writings [have not] followed the tradition of conventional ceremonial amenity. They are a poetic call to arms, a rallying-cry for the American way of life, an indictment not only of the powers of darkness that threaten Democracy . . . but of Mr. MacLeish's own fellow-writers and scholars, "the irresponsibles." . . . An indictment of writers and scholars equal in severity to this has seldom been heard in public places in modern times, in America or elsewhere. Perhaps the maledictions of Hitler and Goebbels alone have surpassed it within living memory.[59]

Although he linked MacLeish to the Nazis, Zabel wisely chose not to pursue this line of argument. This was left to Dwight Macdonald, which he accomplished in a piece I shall examine shortly. Zabel was content to confine his critique to a discussion of MacLeish's careerism: "[MacLeish] impress[es] on his fellow-citizens the fact that a Milton not only should be living in this hour but by miraculous good fortune *is*." Who is the contemporary Milton? MacLeish, of course. Yet Zabel pointed out that MacLeish's career revealed a record in many ways quite at odds with his latest impersonation. He quoted from the poetry to demonstrate MacLeish's former defense of artistic detachment, and from earlier political writings that showed leanings that were clearly isolationist, even pacifist. He cited certain of MacLeish's literary judgments as examples of inconsistency, and adduced respected critical opinion about the salient weaknesses of MacLeish's verse. What all of this pointed to, according to Zabel, was a spiritual condition in MacLeish that made him eminently susceptible to the appeal of the collective, irrespective of the content of that appeal. "MacLeish," he observed, "is obviously the type of man (the type is legion in our age) whose mortal dread is the fear of becoming isolated from his props and stays in society."[60] For Zabel the important issue was less that the threat of fascism was real than the utter conventionality of this particular claim that it was real. Given MacLeish's unreliability, presumption, and need for self-aggrandizement, Zabel implied that here was a writer eminently susceptible to the blandishments of fascism even as he ostensibly fought against them.

In "Kulturbolschewismus Is Here" (1941), Dwight Macdonald further raised the stakes in the dispute. The Brooks-MacLeish Thesis, he claimed, was nothing more or less than Stalinism. "Where have we heard all this before?" he queried,

> Where have we seen these false dichotomies: "form" vs. "content," "pessimism" vs. "optimism," "intellect" vs. "life," "destruction" vs. "construction," "esthete" vs. "humanity"? Where have we known this confusion of social and literary values, this terrible *hatred* of all that is most living in modern culture? Where have we observed these methods of smearing an opponent, these amalgams of disparate tendencies, this reduction of men's motives to vanity and pure love of evil? Not in the spirit of abuse but as a sober historical description, I say these are the specific cultural values of Stalinism and the specific methods of the Moscow Trials.[61]

Far from being an idiosyncratic expression of Stalinism, the Brooks-MacLeish Thesis was in fact Stalinism become official. It was "totalitarian" because it rejected the spirit of free inquiry and criticism that had spawned the great creative works of the modern period. In Macdonald's view these modernist works were about the business of dissent—of rendering "the overmastering reality of our age: the decomposition of the bourgeois synthesis."[62] He granted that, in so rendering bourgeois decline, the approach of the modernist writers to the values of bourgeois society was "negative, cynical, and destructive," but he maintained that in a moribund age "it is only by rejecting the specific and immediate values of society that the writer can preserve those general and eternal human values with which Brooks is concerned."[63] Macdonald closed his polemic with the observation that modernism and its defenders, victorious in their cultural battles of the 1920s, would be forced once again to fight those old battles. "In a reactionary period," he sadly observed, "[modernism] has come to represent again relatively the same threat to official society as it did in the early decades of the century."[64] He predicted that, as the Soviet and Anglo-American war efforts converged, there would be a further confluence between American culture and the cultural values represented by the Moscow Trials, and thus an even greater need to fight for the values embodied by the modernist masters.

Taking up its duties as defender of modernism with characteristic urgency, *Partisan Review* organized a symposium on the Brooks-MacLeish Thesis for its next number.[65] An invitation to participate sent to Glenway Wescott by Nancy Macdonald on behalf of the editors revealed just how seriously the editors regarded the threat: "This is a life-and-death matter, we feel, involving the very survival of our culture. We can think of no difference between the writers today great enough to prevent them speaking out together against the Brooks-MacLeish Thesis."[66] By and large the participants in the symposium shared Macdonald's distress at Brooks's simplifications, although not always with the same fervor. Several contributors endorsed the view that Brooks's pronouncements had a decidedly totalitarian drift to them. James Farrell made comparisons to the Moscow Trials, Allen Tate alluded to the fact that Goebbels had also attacked modern art, and T. S. Eliot, in a belated letter to the editors, remarked that Brooks's use of terms like "life-drive" and "race-survival" was "depressingly reminiscent of a certain political version of biology."[67] Several of the contributors touched on important weaknesses in Macdonald's analysis. Lionel Trilling stated flatly that he agreed with Macdonald's literary judgments but did not share his feelings about the political implications of Brooks's and MacLeish's attitudes. He did not believe their views would be widely accepted by the public, or that socialism provided any meaningful counterposition. Allen Tate questioned the judiciousness of suggesting that modern writers were in any real sense "anti-capitalist": "the most we can say is that they have written out of a vision of life in our times; or out of a vision of the evils of life which are common to all times." And both Tate and Ransom in different ways raised the possibility that modern literature be viewed tragically—as flawed in some of the very ways Brooks suggested, but flawed "due to a scruple of conscience holding the writers sternly at home in their actual contemporary world." That writers were unable to "lapse back into the inadequate forms and immature affirmations of the last century," commented Ransom, was not their fault. "What Mr. Brooks should indict is not the writers, but the age which bore them."[68]

To these reservations we may add several more. First, history has shown that Zabel, Macdonald, and Dupee all overestimated the influence of Stalinism and the degree to which Brooks and MacLeish spoke for the Popular Front. All three ignored the fact that Brooks and MacLeish resigned from LAW after the Hitler-Stalin nonaggression pact. They also ignored the fact that the Party and its fellow travelers were attacking the two critics as warmongers at the same time that *Partisan Review* was claiming they were in effect Stalinists. The objection may be made, of course, that the Party's hostility was due solely to the latest policy reversal emanating out of Moscow and was thus nugatory—the real similarity between the Party and the two critics was their shared assumptions about literature. But it is questionable whether intellectual irresponsibility alone, no matter how crude, can amount to a "totalitarian" trend. Such an argument trivializes the absolute control over civil society and state terror that have been the basis for our understanding of totalitarianism. When the term is applied to an overzealous liberal like MacLeish and a left-liberal who called himself a socialist like Brooks, the charge seems spurious indeed. As it turned out, Brooks and MacLeish were not enormously successful in their strident appeals for civic-mindedness, despite the latter's prestigious position as Librarian of Congress. But even had they articulated ideas widely disseminated by an aroused ruling class, there was no all-encompassing mechanism or movement with which to enforce them. It was certainly true that during the war there was a great deal more public sympathy for the Soviet Union than ever before—that through the press, publishing, the movies, labor unions, and low-level government offices the Party acquired unprecedented influence. This did create, in the words of Irving Howe and Lewis Coser, a widespread "atrophy of moral sensibility among American intellectuals" willing to praise a bloody and brutally repressive regime. Still, there was little actual justification for the belief that the totalitarian danger had reached serious proportions in the United States. There was a single-mindedness about warnings of the growing Stalinist threat that tended to blunt the New Yorkers' critique of powerful centrist and conservative ideologies—namely those of corporate liberalism and McCarthyism—which in the long run proved to be much more powerful and harmful to the body politic than Stalinism. Ironically, less than ten years after the debates I have just summarized, the Popular Front artists and critics and not the avant-garde or its defenders were subject to the repressive powers of the state. It would be the avant-garde—or "Cultural Bolshevism," the name the Nazis, and then an ironic Macdonald, had given to cultural forms of dissidence and experiment—that would be absorbed and divested of its original adversarial impact by bourgeois culture, through appropriation by Madison Avenue and by museums, art galleries, and the universities.

In *Partisan Review*'s critique of literary nationalism, we see some of the limitations and inconsistencies that have caused many to question the New Yorkers' radicalism. There was the failure to encourage ties with other agents of social change, the failure to explore the variety of means of dissent by which writers and intellectuals might make their discontent felt, and an unwillingness to imagine ways in which modernism might be made to speak to ordinary Americans by taking into account their specific experiences and contributions ("native" modernists such as Crane, Williams, Zukofsky, and Marianne Moore were neglected). In their concern that authentic, individualized experience was being swallowed up by the pressure to conform, they ceded the public sphere and all notions of cooperation, community, and collectivity to the opposition. This was fully consistent with their literary tastes: the literary battlefield upon which they maneuvered discouraged them from fortifying their defense of modernism with an equally sensitive treatment of works from other traditions that effectively rendered everyday American life. This was a corollary of their politics. Many New Yorkers were obsessed with foreign affairs and the Soviet threat, and failed to seriously address problems of immediate concern to Americans in the postwar period: race relations, poverty, women's inequality, the discontent of the youth, pollution, militarism, the national security state, and corporate greed. Here the magazine's chronic focus on Europe was not helpful. The endless debates over the nature of fascism, Stalinism, and the meaning of the coming war belied a singular refusal to explore the concrete American situation. It was as if to do so would have involved these critics ipso facto in a rightist, nationalist deviation.

As I will make clearer, when the magazine did turn to the American situation, it tended to make European culture and politics the measure of all things. When the debate over whether and how to support the American war effort arose after Pearl Harbor, the terms of the debate exposed the inadequacy of *Partisan Review*'s Eurocentric radicalism. Thus when Clement Greenberg and Dwight Macdonald demanded that the so-called imperialist war against Germany be turned into a civil war and that the working class needed to exert leadership of the war effort if there was to be any hope of victory, their distance from the realities of American life was more than apparent.[69] Later, when history revealed the inadequacy of their initial response to the war, they overcompensated by unnecessarily moderating their criticism of the West in order to demonstrate its superiority over the East. This meant that their critique of American society, though at times incisive, tended to avoid the realities of class, gender, and race in postwar society. *Partisan Review*'s hard-won independence, its intellectual integrity, and its admirable defense of modernism against political dogmatism and philistinism notwithstanding, the magazine failed to identify with the very constituency leftists must identify with: ordinary intelligent people. In these respects their political and cultural project was both a triumph and a failure, both a breakthrough and a continuation of old patterns.

Ironically, in some respects *Partisan Review*'s critique mirrored certain features of Stalinist discourse. First of all, the rhetorical pyrotechnics with which the disputes were carried out obscured an important similarity having to do with a common approach to certain kinds of modernism. That is, *Partisan Review*'s strong defense of the avant-garde was nonetheless a partial one. It left out the more formally experimental and politically objectionable writers on much the same grounds that the Popular Front critics attacked modernism as a whole. Indeed an important source for both views was *Axel's Castle*, Edmund Wilson's 1931 study of modernism (what Wilson called "symbolism"), a volume whose importance as the primer of modernism to an entire generation can hardly be overestimated. This profoundly ambivalent book both damned the avant-garde and praised it. It damned it for its obscurity, its solipsistic subjectivity, and its retreat from the public sphere; it praised it for effecting a revolution in modes of perception and formal technique. Likewise Brooks and MacLeish made passing reference to modernism's extraordinary technical proficiency and also condemned modernism for the reasons we have seen. And thus the *Partisan Review* critics dismissed or ignored dada and surrealism, and, as I have indicated, the work of Pound, Wyndham Lewis, Djuna Barnes, Louis Zukofsky, and Joyce (his later work only) for reasons of political undesirability of "aestheticism." The difference between the Popular Front critics and the *Partisan Review* critics was finally one of degree and not kind.

Another similarity was brought out by Allen Tate in his response to Macdonald's attack on Brooks and MacLeish. Tate raised the possibility, obliquely to be sure, that the impulse to smear one's opponent was not Brooks's or MacLeish's alone, but was all-too-powerfully at work in Macdonald's piece as well. To generalize from Tate's observation, throughout its history *Partisan Review* was not above stooping to the lurid rhetoric of its political opponents, a fact that ought to cause us to question whether the magazine ever fully extricated itself from the factionalism it denounced in its opening editorial statement in 1937. I am reminded of something Robert Warshow wrote in commenting on Lionel Trilling's novel, *The Middle of the Journey*. Having discussed the acuity of Trilling's political insights and their psychological reverberations in the novel, Warshow noted that in the author's incessant critique of Stalinism he had paradoxically reproduced its main features. Like the Stalinists themselves, he wrote, Trilling "can respond to the complexity of experience only with a revision of doctrine. . . . The problem remains: how shall we regain the use of our experience in the world of mass culture."[70] What Warshow was objecting to was Trilling's failure to "imaginatively assimilate" the experience of Stalinism and render it nonprogramatically as the experience of fully realized individuals in a specific cultural situation. Trilling's was a failure to particularize. In a larger sense this was the problem with *Partisan Review*'s assessment of literary nationalism. The burden of the magazine's attempt to link literary nationalism to the most retrograde and powerful sectors of society, the burden of transforming Stalinism into a totalizing psychological and sociological phenomenon, was too much to bear. The result was that the particularities of American life and literature were blurred because they made so little difference to the apocalyptic ideological battles of the 1930s. In the end *Partisan Review*'s dialectical criticism—its attempt to unify and move beyond the classic dualisms of American life and literature (as identified by James, Brooks, Eliot, and Wilson)—succumbed to the polarizing pressures of ideological struggle during the 1940s. The dissociation of the American sensibility, including the American *radical* sensibility, would live on, finding new expressions in 1960s and post-1960s experience.

VI

Having characterized the rancorous literary and political disputes that conditioned the search for a "usable past" in the late 1930s and early 1940s, we now confront the task of examining the consequences for subsequent generations of critics. And without doubt, the critical corpus we inherit today contains innumerable battle scars. Carefully concealed for two decades and more during a time when literature and criticism were said to be above the fray, these ideologically induced injuries have finally attracted the attention of a generation of scholars cum coroners who have by now examined every wound with great care. Three such autopsies have been performed on Rahv, Trilling, and Matthiessen, and it is worth inquiring into the accuracy of the findings.

From the late 1930s to the 1950s Rahv and Trilling defined the cultural situation as one dominated by the narrow materialism, ideological opportunism, and unconsidered praxis of Stalinism and its critical allies. The influence of Stalinism was thought to have perpetuated a long-standing dissociation of sensibility within American culture. Updating the insights of James, the early Brooks, and of course Eliot, Rahv and Trilling wished to bring about an amalgamation of what they believed had long been mutually exclusive aspects of American life: "experience, loosely defined as action, praxis, interaction with obdurate material reality; and "mind," meaning considered thought and refined feeling, an openness to a full interior life. But Rahv's and Trilling's attempts to unify the fractured American sensibility were compromised by their inability to withstand the formidable ideological pressure to declare themselves in opposition to the "Redskins," Rahv's term for those writers who fetishized raw experience and were thought to be in command of the literary battlefield by virtue of their alliance with the Stalinists and the Popular Front. Both critics perpetuated America's split personality by giving priority to "mind," and subordinating "experience" in their version of a new American synthesis. Thus during the course of Rahv's career he had little if anything positive to say about such writers as Crane, Norris, London, Dreiser, Lewis, Anderson, Wolfe, Steinbeck, Farrell, Wright, Hurston—all "Redskins" and therefore at least potentially "crass materialists," "greedy consumers of experience," and "sentimentalists." To be fair to Rahv, in "Paleface and Redskin" he presented himself more reasonably than I have suggested—he did acknowledge that many of these writers were capable of "genuine and . . . even of admirable accomplishments." Unfortunately, he never said what these accomplishments were.

As for the American "Palefaces," whom Rahv averred were responsible for most of the "rigors and charms" of American literature, they too were inferior to their European counterparts, though not as markedly inferior as the "Redskins." As we have seen, James represented for Rahv the single example of an American writer who had managed to transcend the constricting dichotomies of American life. But it is worth reiterating that in the course of defending James against his many "Redskin" detractors, Rahv's enthusiasm had its limits. James may have been a "Founding Father" and he may even have authored a "Declaration of Rights," but for all that he was only the most compelling of the "Palefaces" and inferior to European masters such as Balzac, Flaubert, Goethe, Dostoevsky, Tolstoy, Proust, and Mann:

> While the idea that one should "live" one's life came to James as a revelation, to the contemporary European writers this idea had long been a thoroughly assimilated and natural assumption. Experience served them as the concrete medium for the testing and creating of values, whereas in James' work it stands for something distilled or selected from the total process of living.[71]

In all of American literature, in fact, only *The Scarlet Letter, Moby Dick*, "some poems" of Emily Dickinson, and "not a few incomparable narratives by Henry

James" were thought by Rahv to have boldly confronted reality with a balanced sensibility, thereby escaping the limits of "Paleface" writing. Emerson, Whitman, Poe, and most of Hawthorne and Melville (not to mention Douglass, Garrison, or Stowe) simply did not measure up. As much as these judgments are still to be counted as part of the salutary break from both the genteel and naturalist traditions in order to form a new modernist canon, the "internationalism" they were said to embody failed even to embrace the literature of one country. It is worth observing that when F. O. Matthiessen's *American Renaissance* appeared in 1941, it was not only the Popular Front and progressive critics whom Matthiessen rejected, but also those *Partisan Review* critics whose American canon was rigid and exclusive in another way. In both Rahv and Trilling, the inferiority of American literature was asserted so often and so forcefully that, were it not for their critical acumen with regard to other traditions (Trilling is arguably best known for his essays on British writers; Rahv's reputation as critic rightfully remains anchored in his trenchant treatment of European, especially Russian, writers), they could easily have seemed merely to perpetuate the disdain for American literature possessed by earlier generations of Anglophiles.

Unlike so much contemporary criticism, Trilling's dualistic analysis of the American tradition was not an academic exercise in which a theoretical synthesis could substitute for a public one. But neither was it part of a milieu that encouraged balance by way of a modulated response to one's adversaries (despite Trilling's own eloquent appeal for tolerance in an essay like "Elements That Are Wanted"). "Reality in America," for all its energetic, lucid articulation of the desire to overcome the antitheses of the American tradition, nonetheless bore their marks in its unqualified praise of James and nearly unrelenting dispraise of Dreiser. As an engaged, thoroughly reactive critic, Trilling admirably exposed the poor arguments of the Popular Front and progressive critics, whose grounds for derogating James and hailing Dreiser were weak both critically and politically. But Trilling responded too directly to the critical climate: in order to "straighten out the stick," to use Lenin's phrase, Trilling "bent it the other way," rather than use the current ideological disputes as the beginning basis for, and not the defining feature of, his examination of specific textual problems.

Part of the problem was that Trilling very rarely engaged in close reading: his treatment of Dreiser in "Reality in America" consisted of a great deal more quotation from Dreiser's critics than from Dreiser himself. Obviously such metacritical commentary can be extraordinarily revealing, as indeed it was in this essay, but when it becomes the basis for a major revaluation of Dreiser's work, the hazards become all too apparent. A critic can be too immersed in his own times. Trilling's reassessment of Dreiser was the product of an inadequate method leading to faulty logic. Because Dreiser did not live up to the exaggerated claims of his advocates, his work therefore was claimed to possess little of significant merit. The overindulgence with Dreiser displayed by Trilling's adversaries was met with Trilling's refusal to indulge Dreiser at all. This can be said to have been true with regard to his treatment of the realist and naturalist canon generally, which attracted precious little of Trilling's attention during his distinguished career.

Trilling, like Rahv, was interested in the intersection of consciousness and being, intellect and circumstance, willfulness and social necessity. But both critics, in order to compensate for the grave distortions of reductive Marxists and overzealous liberals, belittled literature's capacity to reflect and deflect social life and thereby provide knowledge of society. Neither critic paid sufficient attention to life lived in the aggregate under the pressure of powerful ideological and economic forms. As many critics have pointed out, the strong impulse in Trilling's criticism to downplay canonical literature's relation to this dimension of social life, and even at times to use this relation as a means of judging literary value (the less obvious and intimate the relation the better the literature), were consistent with the tendency of the postwar ideological consensus to depoliticize literature and to establish its own dominance—partly by declaring everything else to be ideological. But academic leftists must be careful to avoid the temptation to define Trilling solely on the basis of those features of his criticism that purportedly give credence to the views of conservatives and neoconservatives. This was Trilling's error in his one-sided assessment of Dreiser, in which he was unwilling to credit Dreiser for anything really—not his mastery of the plain style, or his unparalleled appreciation for the enticing glitter, seductive power, and actual workings of capitalist corruption. Nonetheless, as I have tried to show, Trilling's influential diminishment of Dreiser and the naturalist canon generally, although lacking the kind of "modulation" Trilling famously praised in others, did speak to some of the deficiencies of the American left at the time.[72]

VII

Led by critical schools that have given unprecedented attention to historical and ideological considerations shaped by gender, race, and class, contemporary criticism has embarked on a long-overdue revaluation of the American canon and the canon makers of the postwar period, the latter including F. O. Matthiessen, Lionel Trilling, Philip Rahv, Richard Chase, Leslie Fiedler, Perry Miller, and others. The revaluation has been extremely critical, partly because these postwar critics, in uneasy alliance with the New Critics, acquired for themselves substantial power as they defined the American canon and established the basis of literary study in the postwar period. In general, contemporary critics have found their work wanting in terms of its commitment to pluralism, diversity, egalitarianism, and an understanding of material reality. They have been amply criticized for excluding writers of color, women writers, and popular writers from their chosen canon. They have also been charged with perpetuating false dichotomies in their understanding of the American past (by employing, as we have seen, such antitheses as Paleface/Redskin and experience/mind), and of giving the impression that their privileged value, which often went under the name of "complexity," was incompatible with a dissident mentality and was instead compatible with established modes of authority in the postwar period. I consider all of what I shall have to say in this section consistent with many of the charges that have been leveled at these critics, whose race, gender, and historical moment all contributed to their limitations as critics. But I think they must be balanced by considerations of equal importance.

To be more specific, the concern for measuring commitment to diversity, egalitarianism, and material and social reality has led to valuable insights into the limits of the postwar reconfiguration of the "usable" past. But the almost exclusive concern for these matters has blinded many critics to a signal *political* achievement of these postwar critics, which was to offer the left and liberalism a language with which to discuss the impingement of a rich personal life, alive to aesthetic and moral questions, upon the community and its politics. What I am suggesting is that, despite its achievements, the contemporary academic left has not been attentive enough to the politics of the self, at least those politics of the self in which individuals exist as a source of creative behavior that allows them to resist oppression and manipulation, that provides them solace or inspiration, or enables them simply to endure. Concomitantly, left academics have not been attentive enough to those portions of their own tradition characterized by ambivalence toward the liberal values of open inquiry, debate, and the integrity of the individual. In the recent criticism of canonizers such as Matthiessen and Trilling, there has been too much emphasis on their work as expressions of cold war ideology and not enough emphasis upon their still very much unrealized agenda concerning the cultural limits of American progressive politics and the need to establish, in Alfred Kazin's phrase, an ideal of citizenship in some new and imperative moral order.

Put another way, contemporary critics have allowed their legitimate but inordinate concern about conservative methods, histories, and canons to obscure their obligation to address other and equally important needs—needs having to do with the left's ambivalent relation to myths pertaining to American exceptionalism and individualism that are shared by a majority of Americans.[73] In a very real sense, despite its superior sophistication the current academic left, with its sometimes disproportionate materialism and hypersensitivity to ideological culpability, has repeated some of the errors of the doctrinaire left of the 1930s—errors that Matthiessen and others wished to correct. There are, I think, compelling reasons why we must make haste to read the political hieroglyphics embedded in critical works whose politics, as flawed and obscure by turn as they may seem, nonetheless contain a crucial element lacking in contemporary left politics—namely, the endeavor to expand the meaning of politics so as to include the complexities of individual experience. In their best work, dissident canonizers like Matthiessen and Trilling offered a series of suggestions about how the encounter with certain literary texts (one may disagree with their choices) might afford readers the chance to think and feel more freely and more subtly than either commercial culture or the culture of dissent might allow.

As a case in point, I should like to examine the recent revaluation of F. O. Matthiessen's criticism. Matthiessen, of course, was not officially a member of the *Partisan Review* circle: he taught at Harvard, lived in Boston, and remained a fellow traveler until his untimely death in 1950. His political differences with the New York intellectuals were significant, and indeed his politics were the subject of a harsh attack by Irving Howe in *Partisan Review* entitled "The Sentimental

Fellow-Traveling of F. O. Matthiessen" (1948). Nonetheless, intellectually and critically Matthiessen shared a great deal with the New Yorkers, a fact reflected by his fairly consistent presence in *Partisan Review* throughout the 1940s.[74] What Matthiessen had in common with the New Yorkers in spite of his politics was an understanding of the weaknesses of the progressive community's intellectual and cultural life, and a belief that literature and literary criticism might provide an antidote to habits of mind and method that were preventing the left from developing a democratic and complex dissident culture. As was the case for Trilling, Rahv, and others, Matthiessen's criticism must be understood as an extension of his politics and an appeal for others to broaden their politics. It is only when contemporary critics adopt a utilitarian definition of politics—of the kind Matthiessen and the New Yorkers wished to critique—that they conclude that these older critics produced work that was incompatible with dissident politics, both their own and ours.[75]

A few examples will suffice. In her introduction to the influential *Ideology and Classic American Literature*, Myra Jehlen points to the political and ideological limitations of F. O. Matthiessen's *American Renaissance* and the criticism it spawned, but she does not attempt to see how that crucial text contained strategies for resisting what had become on the left habitually narrow conceptions of culture, community, and self. Here is Jehlen:

> F. O. Matthiessen, whose *American Renaissance* ushered in the forties and a new classical canon, shared the politically left attitudes of many of his predecessors. But unlike them he did not deal directly with political issues. Instead he recast these issues as artistic problems within a separate literary frame of reference. Writers as well as critics, Cowley and Wilson had also distinguished between literature and history. To Matthiessen, however, different meant apart: The language of literature was not only distinct, it was also self-sufficiently self-contained. If he located individual works and traditions in their historical context, he then read them as transcendent visions.[76]

This passage contains an important ambiguity, for in spite of the fact that Jehlen acknowledges that Matthiessen in his criticism has "recast" his politics, she abandons her metaphor by claiming these politics actually dissolve when they come into contact with the "separate," unalloyed language of literature. But separate for whom? Separate, certainly, for doctrinaire critics at the time, for whom the bourgeois antinomies of material and cultural life, and of the public and private spheres, were reproduced in a binary base-superstructure analytical model and a philistine attitude toward culture. Jehlen distinguishes Matthiessen from these predecessors, but she also distinguishes him, too hastily I think, from subtler critics like Wilson and Cowley. She seems unwilling to explore the possibility that Matthiessen, like Wilson and Cowley, may have been dealing with political issues indirectly and consistently in his encounter with literature during a time of crisis for the American left. Perhaps circumspection in the midst of stridency is not apolitical at all, but rather a tacit appeal for a refreshed and more measured, but no less dissident, political discourse.

In *F. O. Matthiessen and the Politics of Criticism*, admirable both for its scope and balance, William Cain comes closest to providing the kind of context I think necessary for a full understanding of Matthiessen and other canonizers of the 1940s and 1950s. Cain deepens our understanding of Matthiessen's limitations as a critic who, despite his socialist politics, was insufficiently responsive in his criticism to matters of ideology, race, class, and gender. Yet Cain is willing to complicate his judgment by considering the political and ideological situation out of which Matthiessen's criticism arose, particularly his masterpiece *American Renaissance*. "Matthiessen," explains Cain, quoting David Hollinger,

> sought in his own way to assemble a secular scripture that would, first, compensate for the deprivations and cruelties of the Depression in America, and, second, shore up men and women where, as was becoming increasingly clear by the late 1930s, doctrinaire Communism, if not socialism, had failed them. *American Renaissance* is, among other things, an extraordinary "turn" (or, better, "return") to enlightening and consolatory texts that will enable readers to make sense of, and to live beyond, such bad "investments of spiritual capital" as "the Agrarian program for the South" and, even more, "the revolutionary program for the industrial proletariat."[77]

Cain accurately but all-too-briefly elaborates on what the implications of Matthiessen's focus upon the cause of "democracy"—the stated theme of *American Renaissance*, might have meant to an audience of progressives in the early 1940s:

> Democracy means freedom from narrow and inevitably disappointing sects, factions, orthodoxies. It signals a resistance to the cult of ideology; it is the vital element, the breath of "open air," that prevents ideology from deadening and dehumanizing us. Democracy is organic, leading to an art that expresses (and gives evocative form to) the aspirations of the common man, whereas ideology is alien and artificial, mechanically imposed from the outside.[78]

Cain's insights into the specific political implications of Matthiessen's text are reinforced by a clear understanding of the ambivalent relationship that both text and author had toward the Popular Front. Qualifying Jonathan Arac's discussion of *American Renaissance* as a Popular Front text,[79] Cain points out that Matthiessen "spoke not only to and for the Popular Front, but to many who disputed it as well."[80] Citing in particular *Partisan Review* and two organizations spawned by members of its circle (the Committee for Cultural Freedom, organized by Sidney Hook with John Dewey as chair, and the League for Cultural Freedom and Socialism, organized largely through the efforts of Dwight Macdonald, discussed in chapter 3), Cain suggests that Matthiessen shared at least some of the anti-Stalinist left's disdain for the astonishing naiveté and political chicanery represented by the Popular Front's support of the Moscow Trials, its reversal of policy in 1939 with the signing of the nonaggression pact by Germany and the Soviet Union, its double reverse in 1941 with Germany's abrogation of the treaty, and its support of the Soviet Union's subsequent invasion of Finland.

Here, however, matters get more complex. Although Cain is, I believe, correct to argue that *American Renaissance* does not mirror the values of the Popular Front, and although he is absolutely right to speak of Matthiessen in conjunction with the *Partisan Review* circle, he mistakenly bases this conjunction on a shared political—or, more specifically, a shared programmatic—disagreement with Popular Front strategy. Yet to my knowledge there is no evidence to suggest Matthiessen ever publicly dissented from or even wavered from official Popular Front strategy, much as I would like to believe the contrary. Certainly Cain himself produces no evidence. It would be more accurate, I think, to say that Matthiessen's reservations concerning the Popular Front were not over policy but over culture and values. Cain begins to identify these issues in the passage quoted already in which he discusses democracy as an antidote to what Orwell called, in a slightly different context, "those smelly little orthodoxies."

We cannot do full justice to the political and democratic impulse behind Matthiessen's criticism if we regard that impulse as essentially compensatory, as intended mainly to provide solace in a time of pessimism and defeat. We must press Cain's claim that it was truly a "vital element" by saying what was and what remains politically alive in it. We must understand how a critical text that consistently falls short of what we now expect politicized criticism to be could be politically useful nonetheless.

To make a beginning in this effort, I would recall the remarkable passage from W. E. B. DuBois's "Criteria of Negro Art," quoted by Cain, in which DuBois stated that the political or "propagandistic" function of his art had always been to insure the right of individual African Americans, in conjunction with the larger community, to expand opportunities to *feel* more deeply and pleasurably. "All art is propaganda and ever must be despite the wailing of the purists," affirmed DuBois. "I stand in utter shamelessness and say that whatever art I have for writing has been used always for propaganda *for gaining the right for black folk to love and enjoy*" (emphasis added).[81] What is so extraordinary about this passage is DuBois's insouciant fusing of normally incommensurate elements: ideology and feeling, propaganda and love, the personal and the political. It is interesting that Cain recommends DuBois over Matthiessen as a critic who successfully merged his criticism with his politics, and he alludes to this passage as an example of DuBois's ability to link concerns about artistic, political, and social matters.

Now DuBois may arguably be the superior critic—I do not wish to argue the point one way or another. I would, however, want to point to the similarities between them, at least as reflected in the preceding passage from DuBois. This passage ought not, I think, be praised only because DuBois is willing to accent his commitment to social justice by using the word propaganda. It ought to be praised explicitly because he is willing to distinguish himself from most propagandists to show how his propaganda is only the means to an end—an end that, to be sure, is ideally shaped by the community, but more important, must inevitably dwell within and transact with the heightened sensibility of free, vital, and empowered individuals.

Similarly, Matthiessen's politics can only be understood in terms of his long-range goal of creating individual citizens with the capacity for vital thought and feeling, including love. In this respect his kinship with DuBois is strikingly borne out by a statement from his synoptic lecture, "The Responsibilities of the Critic," in which he observed that "one of the worst symptoms of sterility in our present culture is that of 'intellectuals without love.'" "No incapacity," he continued, "could be less fruitful in the presence of the arts."[82] According to Matthiessen, this statement was meant to be a direct challenge to the "grimly thin-lipped disciples of a more rigorous analysis" who shaped the progressive culture and criticism of the time.[83]

As is well known, *American Renaissance* has as its stated theme "the devotion to the possibilities of democracy" in Emerson, Thoreau, Hawthorne, Melville, and Whitman. But what can be left of the value of democracy once we subtract the exclusiveness of Matthiessen's canon, his neglect of slavery, abolitionism, and other pertinent social factors, and his apparently apolitical formalism?[84] The way toward an answer, I believe, is to stress the fact that Matthiessen read not as a socialist but as a Christian socialist and a democratic socialist, and this meant he had different priorities than most leftists of his day (and ours). Matthiessen's socialism, we recall, was never wholly defined by his very real commitment to a rationally planned, egalitarian productive mode, nor was it quite the equivalent of his active commitment, as we have seen, to an extraordinary range of reform movements from unionization to cooperation with the Soviet Union, from the fight against Jim Crow to the Wallace campaign. Above all, Matthiessen's socialism represented a moral commitment to the possibilities of unencumbered social expression for the multiple self—multiple in the Whitmanian sense of being imbued with often contradictory emanations of intellect, passion, spirit, and sympathy. In order to be all these things—unencumbered, social, and duly complex—such individual expression required cultivation by democratic, plural institutions and practices. Only democracy encouraged the articulation of diverse human resources, just as only democracy benefited by that articulation. As for guaranteed rights, specific institutional arrangements, and economic democracy, these were the necessary causes of democracy, but they were not sufficient causes: they must lead to cultural practices that appeal to all constituencies and encourage the full participation of the whole personality. Such comprehensive participation—comprehensive because politically and personally multitudinous—was what Matthiessen meant to encourage with *American Renaissance*:

> Emerson, Hawthorne, Thoreau, Whitman, and Melville all wrote literature for democracy. . . . They felt that it was incumbent upon their generation to give fulfillment to the potentialities freed by the Revolution, to provide a culture commensurate with America's political opportunity. . . .

Criticism and scholarship ideally accomplished a similar democratic mission:

> "If, as I hold," [Louis] Sullivan wrote, "true scholarship is of the highest usefulness because it implies the possession and application of the highest type of thought, imagination, and sympathy, [the scholar's] works must so reflect his scholarship as to prove that it has drawn him toward his people, not away from them; that his scholarship has been used as a means toward attaining their end, hence his. That his scholarship has been applied for the good and the enlightenment of all the people, not for the pampering of a class, that he is a citizen, not a lackey, a true exponent of democracy. . . . In a democracy there can be but one fundamental test of citizenship, namely: Are you using such gifts as you possess for or against the people?[85]

American Renaissance presents many examples of Matthiessen applying his own formidable gifts in the interests of democracy. Even when we confine ourselves to the first section of the study, that devoted to Emerson, we find ample evidence of an elusive but thoroughly coherent political subtext, now somewhat obscured by the intervening half-century. The political logic that powered this subtext was one in which Matthiessen continually but indirectly subjected the major philosophical and ideological presuppositions of progressive culture to insights derived from his encounter with his five major writers. In the section on Emerson I count at least three interrelated areas of concern in which Matthiessen tried to recuperate Emerson for a progressive culture convinced of his irrelevance or, worse, danger. These areas included his discussion of materialism and idealism, his discussion of the self and subjective life, and his discussion of the forms and function of literature.

By mooring his introductory remarks on Emerson in philosophical issues familiar to the dialecticians among his readers, Matthiessen immediately suggested Emerson's relevance to ongoing debates between materialists and idealists. The problem for the contemporary reader of Emerson, Matthiessen observed, was that, although Emerson stated things in opposites, he did not appear to be a dialectical thinker—that is, he rarely offered an understanding of the dynamic interrelation among opposites to the point of synthesis. Despite his reverence for Plato's divine reconciliations, Emerson despaired of ever finding a way to unify the modern "double consciousness," best articulated by Kant's distinction between the Reason and the Understanding. Matthiessen acknowledged that such despair often led to metaphysical leaps into such ethereal realms as that inhabited by the Over-Soul, and he did not wish to contradict the predominant opinion that here Emerson was "generally unreadable."

His case for Emerson was based on the more mundane, practical manifestations of his idealism. It was built on numerous examples in which the untrammeled desires and pleasures of private life are seen as giving sustenance to individuals who do not as a result turn their backs on practical, material life. Matthiessen placed Emerson's practical idealism in the tradition of late eighteenth- and early nineteenth-century revolutions and independence movements. Like the English romantics, he was part of the rebellion against the formulas of eighteenth-century rationalism "in the name of the fuller resources of man." He often rejected accepted categories, most notably the kinds of dichotomies that Blake had dismantled with so much spirit and Coleridge attempted to reconcile with his notion of the organic. At his best Emerson enfolded opposing perspectives into his own, thereby disarming opponents and limiting his own claims. He believed the individual "was the world," yet also believed he needed to explain to the world why this was so. In spite of the impression that he was self-indulgent, Matthiessen pointed out he was in reality "occupied with consciousness, not self-consciousness. He wanted to study the laws of the mind, . . . but he always felt a repugnance to self-centered introversion."[86] Here we can readily decipher the meaning of Matthiessen's political hieroglyphics as he attempted to sweep away the common belief during the 1930s that Emerson's was a particularly retrograde example of bourgeois individualist ideology. Rather, in an effort to graft onto the reigning materialism aspects of Emerson's practical idealism, Matthiessen quietly offered his reader examples of an individualism that was social in nature, spiritual life that bonded the self to others, and refined sensibility that could be possessed by all. "What saved Emerson from the extremes of rugged Emersonianism," observed Matthiessen, "was the presence not merely of egoism but also of a universal breadth in his doctrine that all souls are equal."[87] Matthiessen reminded his reader that it was because Emerson's idealism was grounded in the everyday, and because he "restor[ed] to the common man . . . all the rights of art and culture," that Dewey regarded him as "the philosopher of democracy."[88] The idealist written off by so many materialists on the left was in actuality more dialectical than they—not in terms of offering a vision of an apocalyptic synthesis, but in terms of describing those valued moments in life that offer evanescent yet proleptic experiences of union.

One of Matthiessen's unique contributions to the understanding of Emerson was his exploration of how these themes shaped the form and the function of his work. Matthiessen stressed the importance of oratory and eloquence to Emerson as communicative arts that were essentially social and democratic in nature. Oratory, the art on which he based his prose, was the chief means of communication and education among citizens and as such had a long and revolutionary tradition in America. Eloquence, which Emerson distinguished from sober, formal speech, was what enabled the orator, in Matthiessen's words, "to speak most directly and most deeply to men, breaking down their reserves, tugging them through the barriers of themselves, bringing to articulation their own confused thoughts, flooding them with the sudden surprise that the moment of their life was so rich."[89] Eloquence breaks through "mere intellectualizing" to glimpse "the deep subconscious forces that remain buried in men unless quickened to life."[90] These were the foundations of Emerson's art. His triumphs, according to Matthiessen, were to be found in his lectures and essays rather than in his poems, where detail, harsh experience, and the "implication of counterstatements" were often abandoned.

If in his poetry Emerson was "incapable of being restrained within the hard actuality of experience,"[91] in the more concrete transactions between citizens generated by his other forms he was able to employ the symbol, as Carlyle put it, in order to blend the Infinite with the Finite where the Infinite stands visible and attainable.[92] For Matthiessen there was something distinctly salutary about the sustaining, forward-looking function of the symbol. Far from being the literary manifestation of a defeatist, postrevolutionary romanticism and the site of privatization and depoliticization—allegory's reactionary counterpart—the symbol was in fact capable of generating a kind of intimate utopianism through its sudden conjunction of material existence and spiritual longing. Although severely circumscribed compared with other, more apocalyptic literary effects preferred by most of Matthiessen's politically engaged contemporaries (who remained committed to literary realism and its attendant shocks), the symbol was preferred by Matthiessen because of his skepticism about apocalyptic change. Although an advocate of wide-ranging social transformation, he was wary of those espousing a direct assault on institutions, individuals, and sensibilities.[93] Given the wide discrepancy "between the world of fact and world that man thinks," he preferred a route to radicalism whose starting point—and constant point of reference—was the direct, personal experience of individuals. In this kind of political journey the ideal is neither won by frontal attack nor defined by future goals: it exists in the process of change itself, in the quality of difficulty and fulfillment realized by individuals committed to the intensities of transformation both of the personal and public kind. Matthiessen valued Emerson's symbol because, when sufficiently grounded in material life, it introduced the reader to new modes of perception, including spiritual modes, which could as easily lead to empathy as to the empyrean. Of course, Matthiessen strongly implied that such a process would be greatly facilitated by readers who were able to gain contact with a community of dissidents expansive enough to welcome such unconventional and indirect forms of politics. Matthiessen seems finally to have been hoping for a political movement made up of citizens who resembled the reluctant but inspirited left-wing individualist of earlier times.

VIII

My reading of this small but symptomatic portion of *American Renaissance* is meant to add another perspective to the ongoing discussion of Matthiessen and of the larger issue of ideology's relation to the classic American canon. I have tried to show how Matthiessen and others, despite their limitations, undertook a critical project that ought to be seen as vital to the future of the left in this country, given a certain idea of what political radicalism might be. If for many critics it has been impossible to conclude that an essay like Trilling's "Reality in America" did a significant service on behalf of radicalism in the United States, or *American Renaissance* in any meaningful way expressed Matthiessen's socialist values, I would suggest that their conception of radicalism and socialism is simply too narrow. Although they are sensitive to the subtleties of cold war ideology, many nevertheless overemphasize its determinant power. Although they indefatigably lay bare postwar systems of power, many leave off exploring the power (or lack thereof) of dissident movements to resist or mitigate that power. Although militant in their embrace of an abstract materialism, many surrender what Marx called "subjectivity" and "the *active* side" to idealism. In short, little room remains in their schemas for individual experience that is not determined (or "overdetermined") by material factors. The only variable seems to be whether these factors serve to stun individuals into submission or, in rare cases, jolt them into frenzied and ultimately self-consuming opposition. There is precious little concern for the vital, intermediary range of creative responses entertained by the critics I have been discussing. These responses could be harnessed by a more capacious culture of dissent than the one we have. It is hard to avoid the conclusion that unless radicals begin seriously to come to terms with the critique leveled against an attenuated leftism by the critics I have mentioned, American radicalism will be in for continued hard times.

THE LIMITS AND USES OF CRITICISM

8

The "Dark Ladies" of New York

One can't be angry with one's time without damage to oneself.

ROBERT MUSIL

I

Histories of formations of intellectuals, artists, or activists traditionally have been heavily institutional, dwelling on the one hand on spurned affiliations, hierarchies, practices, and ideas, and on the other on forms created in opposition by recognized leaders, namely journals, manifestos, exhibitions, communities, salons, and the like.[1] According to Raymond Williams, an early advocate of the term, "formation" was first employed precisely to offset the undue attention given to institutional life in favor of a renewed concern for the less formal and more fluid practices of movements and tendencies. The term therefore suggests the need for "analytical procedures radically different" from those developed to study institutions.[2] But it is worth pointing out that Williams did not take into account the matter of gender as a significant impetus for new procedures. Thus in his expanded treatment of formations, in *Culture* (1981), he presented an ambitious taxonomy of the many organizational forms adopted by artists and intellectuals engaged in different types of collective action. These included guilds, unions, academies, exhibitions, professional societies, movements, schools, fractions, the avant-garde, as well as other permutations. But nothing was said about how gender affects these categories, how the experiences of women might alter our understanding of the dynamics of these groups. Can women be adequately represented by focusing on these particular organizational forms? Do these forms give us the best way to understand the particular kinds of relationships that intellectual, artistic, or activist women have with one another? Do they offer the best framework within which to explore the modes of women's participation within male-dominated groups? For feminist scholars, the answer to all of these questions has been a resounding no. Inquiry into the contributions of individual women as well as groups of women has shown that new terms and categories are needed to render their experiences in full.

When it comes to understanding the New York intellectuals, one of the major American intellectual formations of the twentieth century, the questions pertaining to gender are extremely relevant. This was a formation that at one time or another included an impressive number of formidable women writers and intellectuals: Hannah Arendt, Tess Slesinger, Mary McCarthy, Diana Trilling, Eleanor Clark, Elizabeth Hardwick, Midge Decter, Ann Birstein, and Susan Sontag. Indeed, simply to call attention to gender and claim these women as New York intellectuals raises the controversial matter of who belongs and who does not, who is included and who excluded, who is inside and who outside. In many cases it is difficult enough to determine membership in such groups when such things are decided on the basis of more or less formal institutional affiliations. Here, the criterion can be which publication one writes for (I believe it was Norman Birnbaum who offered a working definition of the New York intellectual as anyone who read and wrote for *Partisan Review*), membership on editorial boards, participation in a common creative or critical discourse, displaying a characteristic style, joining certain organizations, or signing manifestos or other public statements. But formations are always changing in response to historical circumstances and to personal and professional conflicts; thus it is not easy to establish boundaries even given these guidelines. How *many* contributions and of what kind qualify an individual to be a member? Must the individual maintain a close geographical proximity with the center of the group? And what is considered *close*? In the case of the New Yorkers, their center in the mid-1930s was lower and mid-Manhattan. By the 1940s the center had expanded to include the Upper West Side. By the 1950s, when a good number of the New York intellectuals took academic positions in and around New York and places north to Boston, it became still less clear just how determinate geography was in ascertaining membership in a group whose name, after all, was also its location.

When dealing with women, the fact of membership becomes even more difficult to ascertain, for several reasons. First, women involved with formations are nearly always underrepresented at the upper levels; hence their names are not often found attached to mastheads, editorials, or manifestos. They are not, ordinarily, seen as public spokespersons. Second, most women, especially heterosexual women, no matter how serious they are about their artistic or intellectual work, have been obliged to spend a disproportionate amount of time fulfilling their responsibilities to families and friends. This has caused them to spend significantly more time within the sphere of home and neighborhood than have men, who generally have convened in public and professional spaces where they have sustained their organizations. Third, relationships among women are often formed backstage, so to speak, away from the spotlight of public alliances and battles. This is of course especially true of lesbians, whose lives and culture scholars are just beginning to explore. There are layers of mutual influence, sympathy, and animosity among women that the historian must search out. As yet, we have no accepted vocabulary for these kinds of connections beyond alluding to "traditions" of female experience, literary and otherwise, or to "communities" of women. Categories such as support group, affinity group, or even subculture hardly do justice to the actual scope of their experiences.

Some of the conceptual and methodological problems I have identified can be illustrated by briefly summarizing the treatment afforded to women by the major accounts of the New York intellectuals. I should add the obvious point that this study, by being conscious of the problem, has not overcome it. I am well aware, as the reader will be, that many of the patterns I am about to describe are etched quite deeply into this book.

In *Writers on the Left* (1961—recently reissued by Columbia University Press), the classic account of the American left from the 1910s to the 1930s, Daniel Aaron included subchapters on Max Eastman, Floyd Dell, John Reed, Randolph Bourne, Joseph Freeman, Mike Gold, H. L. Mencken, Theodore Dreiser, Edmund Wilson, William Z. Foster, Archibald MacLeish, and Kenneth Burke. These were followed by complete chapters on Eastman, Freeman, V. F. Calverton, Malcolm Cowley, John Dos Passos, and Granville Hicks. There were no chapters or subchapters devoted to women, and no woman was given sustained, substantive treatment. Those women mentioned more than twice were Mabel Dodge Luhan, Emma Goldman, Josephine Herbst, Mother Jones, Meridel Le Sueur, Edna St. Vincent Millay, Dorothy Parker, Gertrude Stein, Anna Louise Strong, Genevieve Taggard, Mary Heaton Vorse, and Ella Winter.

James Gilbert's *Writers and Partisans* (1968), despite its many examples of clarity and insight, showed roughly the same level of interest in women on the left as did *Writers on the Left*. Gilbert presented sustained treatment of two women—Mabel Dodge Luhan and Mary McCarthy. Only two other women—Gertrude Stein and Genevieve Taggard—were mentioned more than twice.

The title of Alexander Bloom's book speaks for itself: *Prodigal Sons: The New York Intellectuals and Their World* (1986) (as does the book's first chapter: "Young Men from the Provinces"). The men either directly discussed or cited at least several times included Lionel Abel, William Barrett, Arnold Beichman, Daniel Bell, Saul Bellow, Elliot Cohen, Morris Cohen, Malcolm Cowley, John Dewey, F. W. Dupee, Max Eastman, Jason Epstein, Leslie Fiedler, Nathan Glazer, Clement Greenberg, Michael Harrington, Granville Hicks, Alger Hiss, Richard Hofstadter, Sidney Hook, Irving Howe, Alfred Kazin, Irving Kristol, Melvin Lasky, Seymour Martin Lipset, Joseph McCarthy, Dwight Macdonald, Archibald MacLeish, Norman Mailer, F. O. Matthiessen, Daniel Patrick Moynihan, Reinhold Niebuhr, J. Robert Oppenheimer, William Phillips, Norman Podhoretz, Richard Poirier, Philip Rahv, Harold Rosenberg, Meyer Schapiro, Arthur Schlesinger Jr., Delmore Schwartz, Allen Tate, Lionel Trilling, Leon Trotsky, Robert Warshow, and Edmund Wilson. The list of women either directly discussed or cited at least several times is much shorter: it included Hannah Arendt, Midge Decter, Lillian Hellman, Mary McCarthy, and Diana Trilling.

Alan Wald's *The New York Intellectuals* (1987) included in-depth discussions of more women than any of these earlier accounts: serious treatment was accorded Hannah Arendt, Eleanor Clark, Midge Decter, Mary McCarthy, Tess Slesinger, Susan Sontag, and Diana Trilling; moreover, there was material on forgotten women such as Anita Brenner, Margaret De Silver, Elinor Rice, and Adelaide Walker. All but one of Wald's subchapters devoted to individuals, however, were devoted to men (the exception was a joint portrait of Norman Podhoretz and his wife Midge Decter). These included sections on Elliot Cohen, Lionel Trilling, Herbert Solow, Sidney Hook, James Rorty, Charles Rumford Walker, James T. Farrell, F. W. Dupee, Edmund Wilson, Max Eastman, Leon Trotsky, James P. Cannon, Max Schachtman, James Burnham, Dwight Macdonald, Meyer Schapiro, Irving Howe, Harvey Swados, and Irving Kristol.

The breakthrough in terms of gender and the New York intellectuals—actually in terms of gender and the political left during the 1930s as a whole—is the work of one of Wald's former students, Paula Rabinowitz. Her *Labor and Desire* (1991) focuses on the fiction of women who were part of male-dominated political movements yet who dissented in their work from the movement's strictures regarding gender and literary form. Rabinowitz points out that all of these women challenged the separation between "history" and "class struggle" on the one hand, and gender and domestic space on the other. She argues persuasively that the work of novelists such as Fielding Burke, Lauren Gilfillan, Josephine Herbst, Meridel LeSueur, Tillie Olsen, and Clara Weatherwax make up a subgenre, which she labels "revolutionary fiction." Such fiction, according to Rabinowitz, is distinctive because it explores the impact of gender on traditional working-class paradigms. What is germane to the discussion of the New York intellectuals is that she includes the fiction of Tess Slesinger and Mary McCarthy in this subgenre, despite the fact that they depicted women intellectuals rather than working-class women. Other writers who produced work that might be considered in this context were Eleanor Clark, Elizabeth Hardwick, Susan Sontag, Marge Piercy (whom Rabinowitz includes), Ann Birstein, and Grace Paley.[3]

If Rabinowitz's study is the first of its kind that has not relied primarily on the visible, public life of the American left, this hardly suggests that her predecessors willfully misrepresented the past. In all cases they have done nothing worse than write "history from below" from the top, so to speak, by attending to the documents and individuals that have been bequeathed them. The extent to which this legacy has been male-dominated is not hard to ascertain. The *Partisan Review: Fifty-Year Cumulative Index* (1934–1983) tells a portion of the story. Of the roughly 750 "Articles, Illustrations, and Commentaries" that have appeared in the magazine during the years stipulated, approximately 90 (12 percent) were written by women.[4] The figures are slightly higher when it comes to fiction and poetry: of the 228 pieces of fiction published, 40 (17.5 percent) were written by women; and of the 514 poetic entries, 101 (20 percent) were written by women. When it comes to interviews, a category that includes those individuals the magazine's editors deemed most interesting and important, only 4 of the magazine's 54 interviewees have been women (Luciana Castellina, Ivy Compton-Burnett, Iris Murdoch, and Muriel Spark).

If we look at the statistics for the *New York Review of Books*, one of several *Partisan Review* offspring, the numbers reflect the same basic pattern, only concentrated. The *New York Review of Books* has published twenty-one times a year since February 1963, and each issue has averaged thirteen contributions. Thus the approximate number of contributions has been 8,100. For the period February 1963 to December 1973, there were 251 appearances by contributors who were female; for the period January 1974 to December 1982, there were 229; for the period from January 1983 to July 1992, there were 244. This results in a total of 727 contributions by women (8.9 percent).

Because such publications are primary sources for literary historians, their neglect of women writers has encouraged historians to neglect them as well. This, however, is only part of the problem. Fearful and hostile attitudes have played a role as well, especially when directed at those few women who do, against the odds, manage to win for themselves a degree of recognition. Perhaps more accurate than the word recognition is notoriety, for what has historically greeted the few recognized women of letters in America, from Margaret Fuller to Mary McCarthy to Susan Sontag, has often been fear and reproach. Elaine Showalter has convincingly made this point by referring to Norman Podhoretz's remark in *Making It* (1967) to the effect that the only way a young woman could make it was by becoming the "Dark Lady," that mysterious and fearsome woman who could tantalize an entire generation of (male) intellectuals by virtue of her superior wit, style, wisdom, command, and combativeness.[5] For Podhoretz, as his diction indicates, each recent generation selects for itself only one Dark Lady. For Podhoretz's generation, that Dark Lady was Susan Sontag; for the generation of his fathers, the Dark Lady was Mary McCarthy.[6]

Certainly McCarthy and Sontag could be sharply critical and polemical when their opinions or the occasion warranted. But in this they were no different in kind from contemporary men of letters. One need only read Edmund Wilson on academics or Lionel Trilling on Dreiser to see bloodletting skills on display. But the difference in reputation is that the women were often identified with the aggressive aspects of their work, whereas the men were identified with the more balanced, circumspect aspects. In McCarthy's case the abuse was particularly intense; so much so that her biographer, Carol Gelderman, chose to begin *Mary McCarthy: A Life*, as follows:

> From the time she first appeared on the literary scene in the pages of the *New Republic* in the mid-1930s until the present, she has drawn the ire of commentators. John Aldridge has depicted her as strident and shrewish and her novels as "crammed with cerebration and bitchiness." Paul Schlueter has described "her approach to writing as reflective of the modern American bitch." Norman Mailer, less overtly hostile to McCarthy, nonetheless opened a review of one of her novels by portraying her as a virago: "Mary, our saint, our umpire, our lit arbiter, our broadsword, our Barrymore (Ethel), our Dame (dowager), our mistress (Head), our Joan of Arc, the only Joan of

Arc to travel up and down our raddled literary world, our poor damp kingdom, her sword breathing fire." After Alfred Kazin criticized McCarthy as harsh-tongued and snobbish in *Starting Out in the Thirties*—she set out to remind readers, he said, "of the classical learning they had despised, the social lapses they could no longer overlook"—he told Dwight Macdonald that dozens of people had written him, saying, "God Bless you Mr. K., for speaking out on THAT woman." The recurring sameness of the criticism prompted a Ph.D. candidate, preparing to write a dissertation on the critical reception of McCarthy's work, to ask if in a hundred years textbooks on American literature will still be using words like "knives, stilettos, switch-blades, cold, heartless, clever, cerebral, cutting, acid, or acidulous" in discussions of Mary McCarthy.[7]

Similarly, Susan Sontag endured virulent attacks in the early 1980s for her public denunciation of communism, which she referred to as "fascism with a human face." The attacks prompted Elizabeth Hardwick to comment,

> I think her being a woman, a learned one, a *femme savante*, had something to do with it. As an intellectual with very special gifts and attitudes, it was somehow felt that this made her a proper object for ridicule of a coarse kind. I believe the tone was different because she was seen as a very smart, intellectually ambitious woman."[8]

Another reason for the underappreciation of the women among the New York intellectuals concerns their relationship with feminism. It has come to pass that, despite the many commonalities contemporary feminism shares with the work of Mary McCarthy, Elizabeth Hardwick, and Susan Sontag, feminists by and large have remained indifferent or lukewarm toward that work. One cannot help but believe this is due to the fact that only Sontag has openly endorsed feminism or the women's movement, and that, more important, each woman in her own way has been critical. As far back as 1961, for example, Mary McCarthy eschewed the category of woman writer, or, as she archly put it, "a certain kind of woman writer who's a capital W, capital W."[9] In this category she placed Virginia Woolf, Katherine Mansfield, Elizabeth Bowen, and the recent work of Eudora Welty. She characterized the "Woman Writer" as a writer "interested in décor" and "drapery," preferring instead writers like Jane Austen and George Eliot, "who represent sense" over "sensibility." Later, McCarthy's attitude toward what she regarded as the official women's movement was hostile:

> As for Women's Lib, it bores me. Of course I believe in equal pay and equality before the law and so on, but this whole myth about how different the world would have been if it had been female-dominated, about how there would have been no wars—and Women's Lib extremists actually believe these things—seems a complete fantasy to me. I've never noticed that women were less warlike than men. And in marriage, or for that matter between a woman and her lover or between two lesbians or any other couple, an equal division of tasks is impossible—it's a judgment of Solomon. You really would have to slice the baby down the middle.[10]

Apropos of these attitudes, Carol Gelderman adds the following ironic comments:

> Of course there is no "woman" problem in the circles she moves in, nor have any of her husbands been anything but generous in supporting her in her work. McCarthy,

> understandably, is happy in the feminine sphere—marriage, domesticity, . . . pretty clothes, gardening, and, above all, cooking.[11]

McCarthy's close friend Elizabeth Hardwick has also expressed reservations about attitudes and ideas she has attributed to feminism and, in some cases, to herself. When questioned during an interview about aggressiveness in her own writing, Hardwick replied with ambivalence:

> And not only in my writing, alas. I don't like aggressiveness and I detest anger, a quality some feminists and some psychiatrists think one should cultivate in order to express the self. I was astonished by the number of obituaries of Lillian Hellman that spoke with reverence of her anger. I don't see anger as an emotion to be cultivated and, in any case, it is not in short supply.[12]

When asked whether she felt there are specific difficulties in being a woman writer, Hardwick replied "Woman writer? A bit of a crunch trying to get those two words together. . . . I guess I would say no special difficulty, just the usual difficulty of the arts."[13] On the other hand, Hardwick went on to make the remark, already quoted, about the double standard regarding female intellectuals like Susan Sontag. She also rescinded her earlier critique of de Beauvoir's *The Second Sex* for speciously seeking to aggrandize women by minimizing their physical inferiority to men (Hardwick had made the memorable remark, "Any woman who has ever had her arm twisted by a man recognizes a fact of nature as humbling as a cyclone to a frail tree branch. How can *anything* be more important than this?"). And of course Hardwick produced *Seduction and Betrayal: Women and Literature* (1974), a collection of essays exploring the experiences of women writers and literary heroines. These more or less explicit statements by Hardwick make it all the more plausible to argue, as I shall, that her work is as consistent with certain themes of contemporary feminism as it is inconsistent with others. But perhaps the single statement of hers that, although it is about Mary McCarthy, has application to all of the "Dark Ladies" of New York, is the following:

> A career of candor and dissent is not an easy one for a woman; the license is jarring and the dare often forbidding. Such a person needs more than confidence and indignation. A great measure of personal attractiveness and a high degree of romantic singularity are necessary to step free of the mundane, the governessy, the threat of earnestness and dryness. . . . With Mary McCarthy the purity of style and the liniment of her wit, her gay summoning of the funny facts of everyday life, soften the scandal of the action or the courage of the opinion.[14]

Wisely, and with a touch of irony and regret, Hardwick calls attention to the fact that an essential ingredient in McCarthy's success was her personal magnetism, a vital requirement in a culture dominated by heterosexual men. Hardwick also singles out independence, honesty, irony, and style as necessary though, unfortunately, not sufficient attributes for success. In exploring the work of Tess Slesinger, Mary McCarthy, Elizabeth Hardwick, and Susan Sontag, I shall focus on specific aspects of the output of these women, namely those that contribute most to our understanding of the dynamics and liabilities of American literary radicalism from the 1930s to the 1960s. This may strike the reader as an unpromising framework within which to discuss these women, since, as those familiar with them know, so much of their critical and imaginative work has fallen outside this particular rubric. Yet a fundamental implication of much of their work is that our normative understanding of political life has been too narrow. It has excluded experience that in reality not only shapes politics from without but informs it from within. These women dwell with the everyday, where contingency mixes with chance, determinism combines with will, convention merges with dissent, and ideals run into practicalities. They have explored the personal side of politics, in the process giving meaning *avant la lettre* to the familiar phrase "the personal is political." In doing so, however, they have not simply portrayed private life as so much ideological epiphenomena; rather they have turned life loose on the ideological, revealing the chinks in the armor of radical intellectuals whose personal lives are riven by lapses of character. Thus these "Dark Ladies" illuminate a great deal: marriage, divorce, sexuality, abortion, maternity, adultery–in short, the quality of personal interaction among progressive intellectuals and the texture of the culture they surround themselves with. It goes without saying that such an exploration challenges abstraction, dogmatism, and arrogance on the left. It also inquires into the manner in which the left, always concerned with organizing others, organizes itself.[15] In the sections that follow, I shall sketch out how each of these women has built a unique career devoted in part to elucidating these problems.

II

Tess Slesinger's life was short. Born in New York City on July 16, 1905, she was stricken by cancer at the age of thirty-nine and died in Los Angeles on February 21, 1945. She grew up with three older brothers in an acculturated, financially comfortable Jewish household. Unusual for the time, both parents pursued careers, her mother more successfully than her father. Anthony Slesinger arrived in America from Hungary as a child and he graduated from City College of New York in 1889. He went on to attend law school at Columbia University, but before receiving his degree married and went into his father-in-law's successful garment business. He remained there, modestly successful, the rest of his life. Tess's mother, Augusta Singer Slesinger, despite leaving school after the eighth grade, became active in the field of social welfare. She directed a guidance clinic for children, became executive secretary of the Jewish Big Sisters, helped to found the New School for Social Research, and at the end of her life completed a manuscript on psychoanalytic technique. The marriage lasted a full fifty years but not a day longer: the day after their fiftieth anniversary Slesinger's parents separated and the two never saw one another again![16]

Slesinger attended high school at the Ethical Culture Society School in New York, then went to Swarthmore College for two years, and finally the Columbia School of Journalism where she received her B.A. degree. She worked for several years as assistant fashion director at the *New York Herald Tribune*, and assistant literary editor at the *New York Post*, where she wrote book reviews. In 1928, at the age of twenty-two, she married Herbert Solow, the assistant editor of the *Menorah Journal.* This small but influential left-wing magazine of secularized Jewish culture, edited by the charismatic Elliot Cohen (who later founded *Commentary*), attracted a circle of soon-to-be influential Jewish intellectuals, which included Lionel Trilling, Diana (Rubin) Trilling, Clifton Fadiman, Felix Morrow, Anita Brenner, Henry Rosenthal, and Albert Halper. On the periphery were Sidney Hook, Max Eastman, and Lewis Mumford. Slesinger contributed book reviews, and in 1930 the *Menorah Journal* published her first short story. Other stories soon followed—in *American Mercury, Forum, Modern Quarterly, The New Yorker, Pagany, Scribner's, This Quarter,* and *Vanity Fair.* In 1932 she gained a degree of notoriety for her subtly ironic story "Missis Flinders," published in *Story Magazine,* in which the subject of abortion was broached before the broad American public. Slesinger expanded the story into a novel, *The Unpossessed* (1934), in which the politics and especially the sexual politics of the *Menorah Journal* circle and left intellectual circles like it were subjected to trenchant yet solicitous critique.

Slesinger (who retained her name) and Solow divorced in 1932. Shortly after the publication of her collection of short stories *Time: The Present* (1935; reprinted as *On Being Told That Her Second Husband Has Taken His First Lover, and Other Stories,* 1971 and 1990), she moved to Hollywood where she began a second career as scriptwriter. Enticed by an offer of a thousand dollars a week from MGM's Irving Thalberg, she embarked upon her first project, rewriting the script for *The Good Earth* (1937), most of which remained intact in the final version of the film. While working on the film she met, and soon married, the film's assistant producer Frank Davis. The two collaborated on writing scripts for numerous films, including *Dance Girl, Dance* (1940), *Remember the Day* (1941), *Are Husbands Necessary?* (1942), and *A Tree Grows in Brooklyn* (1945).

Always politically active, Slesinger supported the campaigns to free Tom Mooney and the Scottsboro Boys, worked on behalf of the Abraham Lincoln Brigade, joined the Hollywood Anti-Nazi League, and was an officer of the Motion Picture Guild, an organization that encouraged the production of progressive films. Janet Sharistanian reports she became disillusioned with the Communist Party—of which she was a fellow traveler, not a member—following the Moscow Trials (1935–1938) and especially the Nazi-Soviet pact of 1939. Her most enduring political work was her courageous public support for the embattled Screen Writers Guild, which was finally able to negotiate its first contract in 1941.

When she died, Slesinger was at work on her second novel, a depiction of Hollywood from the point of view of its overlooked rank-and-file artisans and professionals. Portions of the incomplete manuscript have been published under the title of "A Hollywood Gallery."[17]

Long before the feminist movement insisted on including personal life under the rubric of politics in order to remake leftist politics during the 1960s, Tess Slesinger was exploring with insight and acerbic wit the androcentrism, instrumentalism, and damaged personal lives of intellectual leftists during the 1930s. Not that she confined herself to writing about this group—she wrote abut servants, wealthy schoolgirls, department store clerks, society women, secretaries, black and white high school students, and bohemian types as well—but she has always been known as a satirist of the New York intelligentsia. This was partly because the radical men and occasionally the women of the period tended to write or influence the reviews, and to these reviewers what she had to say about *them* always seemed the more compelling.

Her chief work was *The Unpossessed*, a novel depicting the trials and tribulations of a small group of intellectual leftists, the sum of whose political activity is endless planning for a new leftist journal, which, not surprisingly, never sees the light of day. The six major characters—three men and three women—are divided into three couples, all of which are to varying degrees shaped by the needs of the men. Elizabeth Leonard is an independent, uninhibited artist who returns from the expatriate community in Europe and pursues a romantic relationship with her distant cousin, mentor, and leader of a circle of intellectual leftists, Bruno Leonard (who resembles Elliot Cohen). But the sterile Bruno, an English professor with no real connection to the working class, is unable to provide Elizabeth with anything beyond his wry wit—little emotion and certainly no commitment. Nor is he able to inspire or lead others due to his self-absorption and paralyzing ambivalence about every major question he faces, personal or political. Norah and Jeffrey Blake (who resembles Max Eastman) make up the more placid of the novel's two married couples, but only because the eternally nurturing Norah never fails to provide her narcissistic and philandering husband with comforting caresses. The other married couple, Margaret and Miles Flinders (who resemble Slesinger and her first husband Herbert Solow), endure a marriage shaped by the repressed Miles' brittle New England morality, his disappointments with his career, and his perceived political obligations. In the novel's powerful final chapter, "Missis Flinders," the moody Miles convinces Margaret to abort their child so that they can devote themselves more completely to a political movement that shows no regard for the personal, familial, and emotional needs of its participants.

Whereas most political novels of the 1930s center around public events such as strikes, meetings, foreclosures, or dispossessions, Slesinger's is a leftist novel of manners: sexuality (homosexual as well as heterosexual), the construction of gender, marriage, family relations, sensibility, the unconscious—these compose the real subject matter of a novel that insists on an unusually broad definition of politics. In treating these themes, Slesinger employs a movable point of view, taking the reader inside the minds of various characters in order to render their inner speech. Here evasion, rationalization, fantasy, and uncertainty are given full and often ironic expression. Her fiction of consciousness owes a particular debt to Dostoevsky and his treatment of alienated intellectuals. Her title is itself a play on Dostoevsky's *The Possessed*, indicating that her radicals are not so much possessed as depressed. One also finds in her work the imprint of Henry James, Katherine Mansfield, and Virginia Woolf with their intensive exploration of inner consciousness, especially where female sensibilities are concerned. There is, as well, the influence of James Joyce and his innovations in narrative point of view and stream of consciousness, and Dorothy Parker where her wit was cheerful rather than mordant. In terms of Slesinger's own influence, it is worth noting that the fiction of her younger and more celebrated contemporary Mary McCarthy had much in common with hers. McCarthy most certainly knew of Slesinger through Solow, who became McCarthy's friend soon after his divorce from Slesinger.[18] As we will see, McCarthy's witty novels, like Slesinger's, remain noteworthy for their bold explorations of female identity and sexual politics among the New York intelligentsia.

In her short stories Slesinger explores such provocative subjects as class exploitation, racism, liberal guilt and unacknowledged snobbery, marriage and divorce, and masculine and feminine identity. She does so through the prism of personal impression and interior monologue, techniques that deflate whatever and whomever is portentous, self-important, or ideologically rigid. In "On Being Told That Her Second Husband Has Taken His First Lover," for example, Slesinger depicts a vulnerable woman's complex response to her second husband's infidelity. Hardly a paragon of militant feminism, this twice-burned victim of masculine vanity struggles to suppress her emotions and perform the necessary roles that she believes will allow her to survive. "White On Black" is a rare example of a white author writing on black experience. The story charts the rise and fall of Paul and Elizabeth Wilson, black siblings who attend an exclusive and largely white private school on the West Side, whose early popularity gives way to isolation as the result of changing hormones and racial attitudes among whites. "Jobs in the Sky" dispenses with transitions and employs multiple narrative perspectives to lend a sense of confusion to the chaotic Christmas shopping season at a large department store whose employees are cynically exploited by the paternalistic owner and his managers. "The Friedmans' Annie" portrays a young and naive German housekeeper who must decide whether to marry her insensitive, unambitious working-class boyfriend or follow the advice of her employers who complacently assure her she can do better. Finally the highly experimental "A Life in the Day of a Writer" brilliantly satirizes a childish male writer with a writer's block who attributes his failures to his wife, all the while dreaming his work will be seminal in more ways than one.

By and large, Slesinger has been well served by her critics, most of whom have been appreciative of her talents; the problem has been that there have not been enough of them. Her first book, *The Unpossessed*, was enthusiastically reviewed. Horace Gregory commented in *Books* that Slesinger "is a writer of unquestionable ability."[19] Writing for the *New Outlook*, Robert Cantwell declared her book to be "one of the very best of recent American novels."[20] T. S. Matthews in *New Republic* described the "extraordinary promise" of the novel: "[I]n its conscious sentimentality, its conscious self-mockery, in its fundamental unwillingness to be content with either, this novel is authentically of our day."[21] J. D. Adams, writing in the *New York Times*, observed that "*The Unpossessed* is an imperfect book, it is both clever and wise, which in a writer as young as Miss Slesinger is a rare combination."[22] John Chamberlain, also writing for the *New York Times*, called the work "quite simply and dogmatically, the best novel of contemporary New York City that we have read."[23] Indeed, *The Unpossessed* enjoyed four printings within a month of its appearance, as well as a British edition. By the 1950s, however, Murray Kempton referred to the novel as "almost forgotten."[24]

When it was reissued by Avon in 1966 it was accompanied by an afterword by Lionel Trilling entitled "A Novel of the Thirties," an essay of genuine importance for our understanding of American radicalism and the tumultuous decade of the 1930s.[25] Although it did not propel Slesinger to the higher ranks, it did strengthen her reputation as a writer still worth reading. (It should be mentioned that Slesinger was one of the very few female or noncanonical writers to whom Trilling devoted a major essay.) Having known both Slesinger and Solow as a fellow member of the *Menorah Journal*, Trilling contributed some personal impressions and biographical data. His focus, however, was on the lack of a moral life in the political culture of the day. Slesinger's novel, Trilling claimed, was not true to the specific political circle it portrayed, but no matter: its enduring strength lay in its examination of what happens when "the conscious commitment to virtue" results in "an absoluteness or abstractness which has the effect of denying some free instinctual impulse that life must have."[26] Trilling argued that the novel was diminished by reading it as "that so often graceless thing, a novel of feminine protest." He interpreted the novel's gendered stereotypes (men are linked to intellect and spirit, and women to instinct and the body—Norah, for example, is said to embody a "mysterious private womanhood") as contributing to the novel's larger intention, which was to "set forth the dialectic between life and the desire to make life as good as it might be, between 'nature' and 'spirit.'"[27]

Turning away from the male critics' (Rahv, Kempton, Trilling) interest in whether Slesinger accurately portrayed the political left, Shirley Biagi and Janet Sharistanian have reoriented Slesinger criticism in several ways. First, they have given greater emphasis to such subjects as female sexuality, gender construction, and the power relations between men and women in Slesinger's fiction. Second, they have placed a new emphasis on Slesinger's short stories, showing her to be a more versatile and stylistically interesting writer than had been previously claimed. Third, they have taken seriously both her career as an accomplished scriptwriter and her political activism during her years in Hollywood. Finally, both scholars have provided important new biographical data that has shed light on Slesinger's fiction.

In "The Menorah Group Moves Left," Alan Wald has given a thorough account of the Menorah Group, and in *The New York Intellectuals* he has inquired into how the group is represented in *The Unpossessed*. He has pointed out that several of the chief characters in the novel are actually composites: Miles Flinders, for example, said to be based solely upon Solow, is Irish Catholic, whereas Solow was Jewish, an important difference. Wald has called attention to the novel's failure to capture the dynamic and sophisticated political character of the group, but has acknowledged that Slesinger's critique of the group's distance from ordinary Americans is accurate and central to the novel's overall meaning.

Paula Rabinowitz, as I have indicated, has provided yet a third approach to Slesinger, resisting the alternatives of placing her in a bourgeois feminine literary history or within the context of the male-dominated radicalism of the 1930s.

Rabinowitz has placed Slesinger in the neglected context of women writers of the 1930s who sought to "regender the revolutionary novel"—Tillie Olsen, Meridel LeSueur, Josephine Herbst, among others.

Most recently Philip Abbott, in "Are Three Generations of Radicals Enough?" has explored Slesinger's critique of dogmatism and male domination on the left in terms of the left's historical failure to construct organizations capable of attracting and sustaining a community of active citizens in a liberal society.

Over fifteen years ago Shirley Biagi ruefully observed that Tess Slesinger had been overlooked despite "the current flurry of rediscovery of gifted women writers." Today, although she has not been overlooked—both her novel and her collection of short stories remain in print, and scholars have contributed valuable assessments of her work—she has yet to be rediscovered by the public or, for that matter, by an academic culture enamored of politics but forgetful of the dangers of arrogant politics and therefore badly in need of her gifts.

III

So much of Mary McCarthy's life quite literally has been an open book that any biographical information rendered here will likely seem redundant. Not only did McCarthy produce three volumes of memoirs known for their frank revelations (*Memories of a Catholic Girlhood* [1957], *How I Grew* [1987], *Intellectual Memoirs* [1992]), but, as is well known, much of her fiction was based on personal experience. Nonetheless, a short sketch emphasizing her relationship with *Partisan Review* and the New York intellectual circle may be useful.

McCarthy was born in Seattle on June 21, 1912, the eldest of four children born to Tess and Roy McCarthy. Six years later both parents perished in the great flu epidemic of 1918, and the McCarthy children moved to Minneapolis where they spent several years under the sometimes cruel guardianship of an aunt and uncle. The children were dispersed by concerned family members when Mary was eleven, and she was able to return to Seattle to the affluent home of her maternal grandmother. She attended Sacred Heart Convent, depicted at length and with some sympathy in *Memories of a Catholic Girlhood* (1957) because its nuns imparted an appreciation for style and grace. She then attended the Annie Wright Seminary in Tacoma where she was encouraged to write, and this led to her acceptance by Vassar in 1929. Here she was "radicalized," as she put it, by reading John Dos Passos's *The 42nd Parallel* for an English class. Her interest in literature resulted in her joining with other Vassar students, notably Elizabeth Bishop, Eunice and Eleanor Clark, and Muriel Rukeyser, to bring out a short-lived literary review they called *Con Spirito*.

A week after graduation McCarthy married Harold Johnsrud, an aspiring actor (their wedding was the basis for the wedding that opens *The Group*), and the two moved to Manhattan where Mary began writing reviews for the *New Republic* and the *Nation*. There she soon gained a reputation for her independence of mind, which some construed as harshness (one commentator called a series attacking contemporary book reviewing the "The Saint Valentine's Massacre of reviewers and critics"). McCarthy also began writing on political subjects, including a piece on the Seattle labor strike in which she criticized Roosevelt Democrats and the American Federation of Labor, in her words, "from the left."[28] It was at this point in her career, 1936, when she became directly involved in politics, the occasion having been a party at the home of James T. Farrell, then a Trotskyist. During a conversation with Farrell about the Moscow Trials and the charges leveled against Leon Trotsky, McCarthy expressed the opinion that the former Bolshevik leader was entitled to a hearing and to the right of political asylum. Shortly thereafter McCarthy received a letter in the mail from the American Committee for the Defense of Leon Trotsky, with her name on the letterhead! Before she could protest, however, she was inundated with threatening appeals from Party members and fellow travelers to resign. The message was delivered by phone and by an open letter in the *New Masses* signed by fifty writers and intellectuals. In defiance of these coercive tactics McCarthy made it her business to remain on the committee, thus beginning her long career as an independent, anti-Stalinist intellectual of the left.

McCarthy had divorced Johnsrud in 1936, and through her subsequent relationship with Philip Rahv became associated with *Partisan Review*. She joined Rahv, Phillips, Macdonald, and Dupee in defending Trotsky at the Second American Writers' Congress, organized by the Communist Party. Also during the summer of 1937 she helped plan the revival of *Partisan Review*. When the magazine reappeared in December of that year she was on its editorial board, where she remained for three issues. In her reminiscences McCarthy has stressed her marginal position within the *Partisan Review* circle:

> [T]he magazine had accepted me, unwillingly, as an editor because I had a "minute" name and was the girl friend of one of the "boys," who had issued a ukase on my behalf. . . . I used to come down to the office on Saturdays . . . and listen to the men argue.
>
> And I remained, as the *Partian Review* boys said, absolutely bourgeois throughout. They always said to me very sternly, "You're really a throwback. You're really a twenties figure." . . . I was wounded. I was a sort of gay, goodtime girl from their point of view. And they were men of the Thirties. Very serious. That's why my position was so . . . lowly. I had been married to an actor and was supposed to know something about the theatre so I began writing a theatre column for them. Once a month, late at night, after the dishes were done, I would write my "Theatre Chronicle," hoping not to sound too bourgeois and give the Communists ammunition.[29]

With her marriage to Edmund Wilson a year later McCarthy moved to Connecticut and began writing fiction, but she remained close to the *Partisan Review* circle. In fact she became one of the magazine's mainstays for some thirty-five years, contributing well over thirty pieces through the 1960s. In 1942 McCarthy published *The Company She Keeps*, a novel based on the experiences of Margaret Sargent, the protagonist in all six stories McCarthy brought together to form the novel. Sargent in many respects resembled McCarthy, who based many of the novel's incidents on events of her own life. Like McCarthy, Sargent is an educated, attractive woman with progressive ideas who finds herself buffeted both inside and outside the movement by conventional values, especially where women, relationships with men, and sexuality are concerned. Above all, McCarthy explores the difficulties faced by Sargent, a woman who desires to become all the things suggested by the phrase she is fond of quoting—uttered by Chaucer's Criseyde—"I am myn owene woman, wel at ese." But the world and, no less, Sargent herself seem to conspire against power, independence, and self-possession.

McCarthy remained married to Wilson until 1945, after which she moved from Wellfleet, which had become their home, to take a teaching position at Bard College where Dupee chaired the English department. She stayed on for a year, after which she became involved in organizing nonaligned intellectuals interested in social change both in Europe and the United States. Known as the Europe-America Groups, they included Phillips and Rahv, William Barrett, Nicola Chiaromonte, Sidney Hook, Alfred Kazin, Dwight Macdonald, Delmore Schwartz, and Elizabeth Hardwick. Soon, however, the group factionalized over the issue of how much weight to give criticisms of the Soviet Union. McCarthy used her experiences as a basis for the satirical novel *The Oasis* (1949), which depicted the disagreements and foibles of middle-class radicals trying to construct a utopian community for themselves in the hills of Vermont.

McCarthy spent the next several years in Wellfleet and Portsmouth, Rhode Island. She published *The Groves of Academe* in 1952, a novel based in part on her experiences at Bard. She again became involved in political organizing, this time devoting nearly a year planning for a new magazine, *Critic*, intended to offer social and cultural criticism of the status quo, as well as to condemn those anti-Stalinists who refused to oppose McCarthyism.[30] Unable to secure financing for the venture, she left for Europe where she would live for much of the following decade. And a productive decade it was: she published her novel *A Charmed Life* in 1955, in 1956 she published *Venice Observed*, a historical account of the relationship between culture and commerce, *Memories of a Catholic Girlhood* in 1957, *The Stones of Florence* in 1959, *On the Contrary*, a collection of previously published reviews and essays in 1961, and finally in 1963 her only best seller, the novel that some claim helped inaugurate the women's movement, *The Group*.

The Group weaves together the stories of nine Vassar classmates, class of '33, as they live out their lives against the backdrop of the economic and political crises of the 1930s. For the most part financially secure, possessed of above-average social and intellectual skills, and certainly believers in progress, technology, and opportunity for women, most of these women nevertheless end up uncertain about themselves. Only one, Lakey, has escaped the strong influence of the men who invariably shape their lives. She survives the decade with her potential for strength, confidence, and grace intact (actually in her case some of these qualities were acquired—she had been aloof and cold in college), largely because of her many years in Europe and her lesbianism.

In 1966 McCarthy was asked by the editor of the *New York Review of Books*, Robert Silver, to travel to Vietnam to report on the growing war. She produced three articles emphasizing what she regarded as an immoral war steeped in brutality, obfuscation, and Orwellian language. She concluded by calling for withdrawal. But it was too early: relatively few were interested and the articles attracted little attention. Shortly thereafter McCarthy and her fourth husband, James West, bought a house in Castine, Maine, near Mary's close friend Elizabeth Hardwick and Hardwick's husband Robert Lowell. McCarthy devoted some of her time to the antiwar movement, attending meetings of draft resisters and giving talks. Although supportive of the movement, she was disappointed by "the mixture of total incoherence and dead clichés" she encountered.[31] In 1968, a request she had made of the North Vietnamese to visit Hanoi, by then heavily bombed, was granted. The result was *Hanoi*, a book that added to the familiar political discussions a passionate moral sensibility based in precise observation of what was seen and close attention to McCarthy's own responses as observer. McCarthy's Vietnam writing was collected in *The Seventeenth Degree* (1974).

McCarthy completed her next novel, *Birds of America*, in 1971. Not nearly as popular as *The Group*, it reflected her concern for the threatened environment and for what she saw as the diminished quality of everyday life owing to the interference of technology—prepackaged food, trifling labor-saving devices, and the like. In 1973 McCarthy covered the Watergate hearings for London's *Observer*, produced six articles, and added to these three essays from the *New York Review of Books* to form her book on Watergate *The Mask of State* (1974). With the death of her dear friend Hannah Arendt in 1975, McCarthy spent much of the next two years serving as executor of Arendt's estate and editor of the unfinished *Life of the Mind*. The latter needed extensive work, which McCarthy lovingly provided in bringing out the two-volume study in 1977. Two years later McCarthy published her most topical novel, *Cannibals and Missionaries*, whose plot involved the terrorist hijacking of a plane full of liberal American politicians and art collectors bound for Iran.

In 1980 McCarthy became involved in a celebrated public and legal controversy with Lillian Hellman, which was not put to rest until Hellman's death in 1984. Appearing on the Dick Cavett Show, McCarthy had replied to a query about overrated writers that the only overrated writer she could think of at the moment was Lillian Hellman, who in addition to being overrated, was "a bad writer, and a dishonest writer."[32] When asked what was dishonest about her, McCarthy replied "Everything. I said once in some interview that every word she writes is a lie, including 'and' and 'the.'" Hellman sued for libel, asking for $2.25 million. The suit, which had been proceeding through the courts, was dropped after Hellman's death. McCarthy could finally devote herself entirely to her writing. She published her only volume of pure criticism, *Ideas and the Novel*, in 1980, and collected her recent essays in *Occasional Prose* in 1985. The first volume of her intellectual autobiography, *How I Grew*, appeared in 1987, and turned out to be the last book published during her lifetime. She died in 1989 in New York City. The second volume of her intellectual autobiography, *Intellectual Memoirs: New York 1936–1938* (1992) was published posthumously.

Among McCarthy's many books that deal with the interaction between politics and personal life, *The Oasis* is perhaps the most direct. (Sadly, this "veritable little masterpiece" in the words of Hannah Arendt, remains out of print.) As is well known, two of the novella's main characters, leaders of factions within a New England utopian community, were modeled after prominent New York intellectuals: Macdougal Macdermott, head of the purists, was Dwight Macdonald in disguise; Will Taub, head of the realists, was Philip Rahv. Indeed the two factions, as Alan Wald has pointed out, resemble somewhat the two groups that formed within the *Partisan Review* circle over the issue of whether and how to support the Allied war effort during World War II.[33] More recently, Carol Gelderman and Carol Brightman have suggested that the two wings of the Europe-America Groups, with their differences on how to create a third alternative to capitalism and communism, provide the more relevant context.[34]

At any rate, in the novel these matters are transmuted into the less portentous but in many ways more revealing debates over whether the utopians should admit a businessman into their midst, and how the group should deal with an impoverished family who trespasses onto their property and picks their wild, ripe strawberries. Of course, like so much effective satire, the exquisite humor of the novel derives precisely from the incongruity between grandiose ideals on the one hand, and mundane circumstances on the other. In the end it is not the circumstances that seem trivial at all, but rather the leaders with their self-important pronouncements and endless machinations.

By preferring to explore the consequences of ideas rather than their content, and by insisting on inquiring into the motives, styles, and attitudes of those who hold them, McCarthy presents the reader with one of our most discerning and comprehensive examinations of the morals and manners of the American left, or at least an important portion thereof.

For McCarthy, morals and manners inevitably involve gender. Although never explicitly stated in the novel, the incessant discussions and debates that take place among the utopians constitute a masculine discourse, one that prevails until the story's heroine, Katy Norell, offers an alternative at the very end. Until then, women are obviously subordinated by the commune's male leaders. When Eleanor Macdermott, for example, interjects her opinion into the discussion about whether to admit the businessman Joe Lockman into the community (her rather vapid contribution is "And what is Utopia but the right to a human existence"), her husband responds enthusiastically: "'Eleanor's right!' Mac shouted, slapping his knee in delighted admiration. He applauded his wife thunderously, as if she were a team of acrobats, whenever she performed what to him was the extraordinary feat of arriving at a balanced opinion."[35] The wife of the other leader, Will Taub, is also regularly diminished by her husband. Taub exits the same discussion, his wife in tow, as follows: "'We'll go,' he abruptly announced, tapping his wife familiarly on the shoulder, as if to apprise her that a show was about to begin into which he had privately written a sardonic star part for himself. His wife, inured to surprises, merely raised her penciled eyebrows."[36] When the men consider expelling Joe Lockman for jokingly sneaking up on Taub with a gun and embarrassing him, the wives are not included (although Susan Hapgood, a wide-eyed young novelist who venerates Taub, is). When it is suggested that the next meeting be held in "the chieftain's living-room" (Taub's), his "male associates ruminatively nodded their large heads in concord" at the prospect of stacking the meeting against Macdermott, who usually succeeds in meeting at his home because of his children (apparently his wife attends the meetings). One of the men readily volunteers his wife's services: "[Danny] Furnas now spoke up to remind them that his wife was assistant to a pediatrician. . . . 'Helen will be glad to stay with the kids.'"[37]

But McCarthy goes beyond depicting the men's complicity in subordinating and objectifying women. Indeed, one gathers from McCarthy's fiction that such dynamics are so pervasive and commonplace in society that calling attention to them with any regularity leads to predictability, stridency, and boredom, all enemies of good writing. Nor is she given to depicting women who resist patriarchal power and are therefore to be taken as inspiring heroines. Most of her women, as critics have long noted, have internalized their inferior status, and are plagued by self-doubt, self-contempt, guilt, dependence, and competition with other women. Nonetheless, some of McCarthy's profoundly conflicted women manage to create for themselves a degree of independence and self-knowledge that is, for the most part, admirable.

In *The Oasis*, it is Katy Norell who is capable of bringing fresh insights to bear on the stale alternatives proffered by the men. To be sure, she is hardly a tower of strength. In one of the novel's more important incidents she is singed when she attempts to light a stove, which Joe Lockman has mistakenly filled with oil. Instead of taking full responsibility for the accident and thereby protecting from possible expulsion the man whom she and her husband are in effect sponsoring, she tries to minimize what remains his mistake: "'It was only an accident,' she said finally, in a feeble and unveracious voice. 'Somebody was careless and left the oil turned up without lighting the stove.'"[38] Her husband Preston is furious with her:

> Preston Norell's long fingers dropped from his wife's arm, and he pushed his way out in disgust. For those words *only* and *somebody*, he wished her in hell. . . .
>
> She was following him out now; he could hear her footsteps running behind him; she caught him just beyond the lawn, her face distorted with tears, which he could envision with perfect distinctness while keeping his eyes averted. "Forgive me," she cried. "Forgive me!" Plainly, she was not going to pretend, as she sometimes did, not to know how she had offended; the others were watching curiously, and he perceived, with a certain savage satisfaction, that she felt she must deflect him from whatever course he was planning, before their rupture was public. "Go in and get the breakfast," he said sternly, shaking his arm free. "Pull yourself together. You disgust me."[39]

Yet Katy Norell's greatest humiliation comes when she is forgiven. Despite the fact that following her husband's outburst she "had made a great resolution to resign herself to his condemnation . . . to take it, that is, seriously," when he suddenly decides to forgive her (wickedly portrayed by McCarthy as the result of being inspired on the porch as he gazes into the New England sublime), she immediately relents.

> She did not know what to say. "Do I really disgust you?" she whispered, looking nervously behind her in the direction of Eleanor Macdermott. "Yes!" he cried emphat-

> ically, but with a shout of laughter; the conspiratorial manner of the question struck him as splendidly farcical—in his very tenderest moments, he looked upon Katy as a comedian. He now relented and patted her sharply on the buttocks. Katy returned to her work, in some peculiar fashion well pleased; though she had broken her resolution and come off with nothing to show for it, it seemed to her undeniable that she had acted for the best.[40]

These painful passages are typical of McCarthy, who out of hatred of oppression and impatience with women exposes with relentless and morbid exactitude every grim wound inflicted or self-inflicted. As indicated earlier, this is perhaps especially true when the wounds result in destructive behavior. This is a concern and at times an obsession in McCarthy that can be taken as misogynistic, but which does after all describe some women. Moreover, because McCarthy's men are often more seriously flawed and certainly more dangerous than her women, one needs to take care when arguing that McCarthy's interest in womens' weaknesses can be understood as misogynistic. Finally, McCarthy sometimes suggests that weakness can become a source of strength.

In the case of Katy Norell, weakness provides the basis for strength of character and uncommon insight. It is Katy's voice, after all, sometimes alone and sometimes in close association with the narrator's, with which the novel ends. Her voice represents a third alternative to what Marguerite Duras has called "the theoretical rattle of men."[41] The stale debate among the men is over how to handle the poachers—by threatening force or pacifically. Unlike the voices of the men, Katy's projects understanding and composure (maybe too much composure due to her state of inebriation). At any rate hers is a voice that is by all appearances sincere. Her ideas seem to emerge directly out of her feelings, and thus there is no hint that they are manufactured by a haunted intellect. Able to silence Taub for the first time, she declares

> "You conceive the problem incorrectly. . . . If the problem is to get rid of the berry-pickers, it follows that force is the answer—to that extent, you are right. Ultimately, it will have to be resorted to, if they will not respond to moral coercion, which is simply force still withheld. But," she went on, growing more excited, "supposing there is no problem, but simply an event: the berry-pickers are in the meadow; the sun is in the sky. If you do not wish to eject them, there is no problem, there is only an occurrence."
>
> Taub shrugged; he did not understand what she was getting at.[42]

Katy goes on to claim as virtuous "the body and the body's objects," until, that is, they become the object of "a mental desire"—presumably any abstract, doctrinal, possessive, or power-enhancing desire. Lying on the grass, her hand entwined with her husband's,

> [s]he recognized, with a new equanimity, that her behavior would never suit her requirements, not to mention the requirements of others; and while she did not propose to sink, therefore, into iniquity or to institutionalize her frailties in the manner of the realist faction, still, seen in this unaccustomed light, the desire to *embody* virtue appeared a shallow and vulgar craving, the refracted error of a naive

> and acquisitive culture which imagined that there was nothing—beauty, honor, titles of nobility, charm, youth, happiness—which persistency could not secure.[43]

In the end she sees the colony itself, with its "cycles of recession and recovery," as "a kind of factory or business for the manufacture and export of morality." Although she envisions no alternative, it is hard to believe she will remain the cooperative, subordinate citizen she has been in the past.

How much of the "manufacture and export of morality" we are meant to associate with male domination remains unclear. Certainly McCarthy has not gone as far as contemporary feminists in their efforts to demarcate what is masculine and deficient about the dominant culture (and, for that matter, adversary cultures). Nor has she been as interested in defining what is unique and salutary about the experiences women share.

Nonetheless, McCarthy's critique in *The Oasis* runs deep, and its consequences spread to wherever gender touches on the culture of radicalism. It calls attention to the exclusion of women, to the enfeebling of women, to the unhappiness of women. It mocks habits of mind such as abstraction, idealism, rigidity, and unprovoked combativeness which, if they are not necessarily shown to originate in male prerogative, nonetheless enhance it.

So let us, finally, forget about swords, daggers and switchblades. In *The Oasis* Mary McCarthy performed precise surgery on the American left, and she found malignancies. Today the patient is barely alive, not because McCarthy's care amounted to malpractice, but because the left has since replaced her with more benign physicians who avoid surgery at all costs. Suffice it to say, there is a great need today for those with talent and courage to emulate McCarthy by wielding the life-saving knife.

IV

Elizabeth Hardwick, longtime resident of New York, was born in Lexington, Kentucky, on July 27, 1916, to a mother who bore nine children and a father in the heating and plumbing business. Hardwick attended the University of Kentucky where she earned a B.A. and M.A. in English, and then left the South for Columbia University's doctoral program in English at the age of twenty-three. She remained in the program for two years before she quit in 1941 to become a writer. Four years later she published her first novel, *The Ghostly Lover*, a loosely autobiographical portrait of a young woman's quest for autonomy from her traditional Southern roots in the intellectual life of New York. What Hardwick chooses to stress, interestingly, are not the young woman's romantic interests but rather her complex relationship with her absent mother.[44]

Very soon after publication, Philip Rahv asked her to write for *Partisan Review*. From the mid-1940s until 1960 she was a frequent contributor to the magazine and an active participant in the cultural and political efforts waged by the *Partisan Review* circle. Hardwick has stated, in fact, that she had begun reading *Partisan Review* as a student at the University of Kentucky, where she "had already been a

Communist and an ex-Communist."[45] In New York she joined Mary McCarthy, with whom she would continue a forty-year friendship, and twenty-nine other writers in signing a manifesto founding the Europe-America Groups, formed to support dissidents in Europe who were independent of both Soviet and American power.

Hardwick wrote several prizewinning short stories during the 1940s, and received a Guggenheim grant in 1947. Like McCarthy, she acquired an exaggerated reputation for writing disparaging reviews. In reading these reviews, one does not find a pattern of gratuitous complaint, purportedly induced by ambition or malice; on the contrary, one finds the usual number of negative-reviews and no more—most are about the business of applying high standards reasonably. A typical example is Hardwick's review of Eleanor Clark's first novel, *The Bitter Box*. No doubt upsetting some of her readers, she pungently remarked that the novel was an "answer to those people who pride themselves upon never having taken an interest in politics simply because, by their apathy, they were spared disillusionment."[46]

In 1948 Hardwick met Robert Lowell at the Yaddo artists' colony. They soon married, and remained married for over twenty years despite Lowell's recurrent mental illness, frequent hospitalizations, and occasional affairs. The first political project the two engaged in was an unfortunate one. The FBI had recently investigated Yaddo because it was alleged that Agnes Smedley, an infrequent visitor, was a Soviet spy. She was cleared, but Lowell soon began a campaign to have Yaddo's director, Elizabeth Ames, fired for her involvement in Smedley's political activities. Nothing came of Lowell's charges, which were certainly in some measure the result of his paranoia, of which Hardwick was as yet unaware. At any rate this episode and others like it during the McCarthy period, if they did not typify the anti-Communist left as a whole, certainly represented a dangerous propensity within this political tendency to sacrifice balance, civility, and, in some cases, rights in order to oppose the exaggerated threat of Stalinism. Certainly more appropriate was the intervention shortly after the Ames affair of Hardwick, Lowell, McCarthy, and Macdonald at the pro-Soviet Cultural and Scientific Conference for World Peace in the Waldorf-Astoria, in which each was unexpectedly given two minutes to address the failures of the Soviet Union.

Hardwick finished her second novel, *The Simple Truth*, in 1954. Ostensibly a murder story based on a sensational murder trial in Iowa City during the time Hardwick and Lowell stayed there in 1950, it is actually a study of the responses of two people swept up in the excitement. Both are middle-class—the man a would-be writer and the woman a faculty wife interested in psychoanalysis. In addition to asking how social class may have affected the crime and the community's response to it, the novel inquires into the role of liberal values in mainstream America. It considers whether and in what degree the silenced voice of the female murder victim is representative of women's voices in general.

Hardwick and Lowell bought a home in Boston in 1955 and stayed there until they moved back to Manhattan in 1961. There Hardwick gave birth to a daughter in 1957. While caring for her and easing her husband through several bouts of mental illness, she managed to produce essays throughout the 1950s, later collected in *A View of My Own* (1962). One year later, during a newspaper strike that shut down the *New York Times Book Review*, Hardwick joined with others, some of whom were associated with *Partisan Review*, to bring out the *New York Review of Books*. She and Lowell were among six on the board of directors,[47] and Hardwick served as the advisory editor.

Over the past three decades Hardwick has contributed heavily to the *NYRB*, having published an astounding eighty-odd pieces to date. Some of these essays were collected in *Seduction and Betrayal: Women and Literature* (1974) and *Bartleby in Manhattan* (1983). Early on Hardwick covered the civil rights movement for the *NYRB*, and she was responsible for the magazine's sending Mary McCarthy to Vietnam in 1967. She herself produced reviews of books, film, and theater, and in 1967 she won the George Nathan Award for drama criticism, the first woman to receive the award.

A number of her essays, directly or indirectly, addressed concerns raised by the emerging women's movement. Some of these found their way into *Seduction and Betrayal*, a collection singular for its steady look at the confining realities faced by women—fictional and real—in this and the last century. These essays combine practicality with a sense of injustice that expresses itself in subtle observation rather than broad generalization. In discussing how the Brontës shaped a realm of freedom out of confinement, or how Nora Helmer in *The Doll's House* achieved significant autonomy despite her domestic responsibilities, or how Dorothy Wordsworth and Zelda Fitzgerald were suppressed and used by their famous male relations yet in some measure benefited as well, Hardwick rarely if ever romanticizes, polemicizes, or overly condemns.

Her third novel, *Sleepless Nights* (1979), met with broad critical acclaim, culminating in a nomination for the National Book Critics Circle Award. Like her earlier novels it is partially autobiographical, but unlike them the first-person narrative is experimental—time, story, and syntax are each unconventionally manipulated to express the instability of the narrator, Elizabeth, especially where her acute and protean sensibility is concerned. Set in Lexington, New York, Amsterdam, and Maine, when the many fragments are added up the narrative spans nearly the whole of Hardwick's life. Joan Didion has identified Hardwick's chief concern: "In certain ways, the mysterious and somnambulistic 'difference' of being a woman has been, over 35 years, Elizabeth Hardwick's great subject."[48] Many of the essays of *Bartleby in Manhattan* develop the subject further, especially in light of the so-called sexual revolution of the 1960s, and the militance and anger of the feminist movement. Although Hardwick found little of value in militance and anger, and was at times bitterly critical of the culture of the 1960s, she had no doubts about the necessity for and the power of the women's movement. "The women's movement," she wrote in "Domestic Manners,"

> is in some respects a group like many others, organized against discrimination, economic and social inequities, legal impediments: against the structural defects of accu-

> mulated history. Perhaps it is that part of the movement the times will more or less accommodate in the interest of reality. The other challenges are more devastating to custom, uprooting as they do the large and the small, the evident and the hidden. The women's movement is above all a critique. And almost nothing, it turns out, will remain outside its relevance. It is the disorienting extension of the intrinsic meaning of women's liberation, much of it unexpected, that sets the movement apart. It is a psychic and social migration, leaving behind an altered landscape.[49]

The women's movement shall triumph, Hardwick implies, for good *and* ill, in pleasure and pain. But of course she addresses a movement often reluctant to acknowledge that progress has its repercussions, very often at the level of personal experience. Typical of the New York intellectuals, she is not content with grudging acknowledgments. More is needed. For Hardwick the proportion of good effects to ill effects finally depends more upon the close attention the movement is willing to pay the latter than it does upon the strenuous promotion of the former.

Currently Hardwick continues as advisory editor for the *New York Review of Books*. Aside from continuing her essay writing, most recently she has introduced and edited Mary McCarthy's unfinished, posthumously published *Intellectual Memoirs*.

In *Sleepless Nights* Hardwick revisits the old left through the reminiscences of her narrator, Elizabeth, who recalls several friends and acquaintances from the 1940s who have been shaped by their political sentiments and/or involvements. Elizabeth's longtime friend Alex, for example,

> could look back to the 1940s, to the anti-Stalinist radicals. How happy he had been then; in the old *Partisan Review* days, the night when Koestler first met and insulted Sidney Hook; when Sartre and the French discovered Russia long after "we" had felt the misery of the trials, the pacts, the Soviet camps.[50]

But politics of this kind, dissident and highly intellectualized, engages only part of the self, and woe to the individual who mistakes the portions engaged for the whole. When this happens, nothing escapes direct politicization and everything atrophies, finally even the disembodied political head. In Alex's case the handsome head, whatever else may be said about it, was still attractive twenty years after Elizabeth had been seduced by it. She remembers vividly: not the sex but the words, the manner, and the character. She remembers that all three were poor—worn out, merely polite, guilty. At the time, she confesses, she saw all of this as "agreeably intimidating." He was also charming and solicitous, convinced like the intriguing Casanova that giving pleasure to women was the source of the greatest spiritual joy. "Some reason to doubt the truth of that," Elizabeth mutters, observing further that "[h]aving more charm than money" likely had something to do with producing the feeling. It meant that love, "like churchgoing," was being "kept alive by respect for the community." Should lovemaking be "evangelical?" she wonders.

No part of Alex's life is experienced for its own sake, least of all the intimate parts. They too are shaped by political sentiments, although not the way one might expect. Alex, it turns out, long ago had spurned Elizabeth, a subtle, thoughtful nonconformist with strong political interests. He preferred, as he now avers, wealthy women, because "only those women with money can violate the laws of probability." Elizabeth notes a pattern:

> [W]hen I first came to New York, I observed that a number of intellectual men, radicals, had a way of finding rich women who loved them in the brave and risky way of Desdemona. A writer or painter or *philosophe* sailed into port and a well-to-do woman would call out, *Evviva Otello*![51]

The men could feel they were engendering commitment in these women, not merely sharing it. Elizabeth's analogy is telling:

> Sometimes the moneyed women with their artists and thinkers were like wives with their vigilant passion for the Soviet Union, the huge land mass that had long ago aroused in them the blood loyalty and tenderness felt for a first child. And what are a child's "few mistakes"?[52]

These sorts of lapses in private affairs notwithstanding, Alex and his impressionable companions would revel in conversations about "the inability of the ruling class to *imagine*, to *experience*."[53] Elizabeth looks back in time and takes stock: "Insurgencies, insolvencies," she sighs.

Here and elsewhere Hardwick writes about the political left as if it were an old thoroughfare overgrown with weeds. Once a direct route to the future bypassing the tangle of local roads, it now seems sadly deserted. Yet Hardwick, not unlike earlier American radicals such as Emerson or Fuller (to whom she once compared Mary McCarthy), has a passionate interest in the local, whether it be defined in terms of geography or individual experience. It is evident in her richly nuanced diction and syntax, and it is evident in the attention paid to the details of sensibility. When politics bypasses and abandons the local, human life shrinks, it migrates, or it dies. A former fellow traveler is described by Elizabeth:

> In the sixties we saw Marie in Italy, where she had been as a girl to meet the past without pain, to flower like a fuchsia or a gardenia in the midst of paintings and churches. But Marie had no interest at all in paintings, in the structure of churches, in the true difficulties of music and poetry. She did not know this, since she had been well brought up. A calm philistinism, courteous, and unpolemical, was deeply rooted in her spirit and gave the clue—or part of it—to her wincing confusion about the claims of all those oddly named writers, poets, and painters imprisoned or killed. This was not callousness; it was blankness, a blankness without meanness such as one finds sometimes in priests and nuns.[54]

Certain readers will be provoked by this, duly taking note of Elizabeth's elitism. Perhaps, they may observe, Marie has immersed herself in struggles for political change and hasn't had the time to cultivate herself. Or perhaps she has cultivated an interest in, or a talent for, the popular arts? Why has Elizabeth missed this? Is it not because she is above the fray and therefore has the time to immerse herself in the refinements of "serious" culture? But even were this the case, as it may very well be, it does not invalidate Elizabeth's observation concerning Marie's incapacities. Nor does it gainsay the tendency of portions of the left to hypostatize the discrepancy between the "high" and the "low," even as they decry it. Of course, a crucial point is that anyone reading Elizabeth's words will likely be part of the same educated, elite group she is a part of. The question then becomes, which attitudes and values expressed perpetuate, and which erode, the cultural hierarchy? The assumption held by Hardwick—and here she is typical of the larger New York intellectual group—is that a vital democracy will make it its business to spread excellence around, will hope to enliven its citizens to a degree that no cultural achievement will be lost on them. If this has meant the neglect, or worse, the denigration of popular cultural forms, we ought to condemn it as such. But in doing so, we would do well to remember that our view of culture is part of a binary in which the popular viewpoint has consistently disparaged "high" culture as "bourgeois" culture, thereby weakening the capacity of people to partake in it, to reshape it, and thereby to make their lives more pleasurable and meaningful.

V

One hardly knows what to make of the fact that a recent feature article on Susan Sontag in the *New York Times Magazine* quotes Norman Podhoretz's twenty-five-year-old claim that Sontag had replaced Mary McCarthy as the new "Dark Lady of American Letters." Presented as part of a biographical sketch, the fact tantalizes by virtue of its very neutrality. Are readers expected to see through the preposterousness of such a designation? Or does its presence signal without irony that Sontag has indeed been a very important intellectual for a very long time? Whatever the case may be, it is a salient fact that the *Times* article passes over the issue almost entirely, proffering instead an honorific quote from the man of letters Carlos Fuentes comparing Sontag to Erasmus. Of much greater concern to the *Times* is Sontag's "vast" library, to which the author devotes over ten paragraphs, including one on why it is not alphabetized, another on the living room books, and yet another on "'the art river'" that flows along the hallway.

Susan Sontag was born in New York City on January 16, 1933. Her earliest years were spent in Tucson, Arizona, often under the supervision of a housekeeper who cared for her and her younger sister while her alcoholic mother and her father, who owned a fur business in China, stayed abroad. Following the death of her father and her mother's remarriage, the family moved to Los Angeles where Sontag attended high school. At the age of fifteen she enrolled at Berkeley. She spent a year there before transferring to the University of Chicago where she received a B.A. in 1951. Next was Harvard and master's degrees in English (1954) and philosophy (1955). She did additional graduate work there (she was an ABD), at Oxford, and at the Sorbonne (1957–1958). Next began a series of teaching positions at the University of Connecticut, City College of New York, Sarah Lawrence, and Columbia, as well as a stint as editor at *Commentary*.

Sontag produced her first novel, *The Benefactor*, in 1963, and thereafter devoted herself entirely to writing. Modernist in form, the novel presaged the celebrated essays that were to follow shortly. The narrator, Hippolyte, is overwhelmed by his dreams, and struggles valiantly but fruitlessly to interpret them along psychoanalytic, religious, and ethical lines. He ends up literally enacting his dreams, a free, self-knowing hero able to merge the real and the fantastic. This desire for immediate yet self-conscious experience was more emphatically applied by Sontag to the aesthetic realm in the famous *Partisan Review* essays "Notes on Camp" (1964) and "On Style" (1965), the former making her a celebrity nearly overnight. (These essays and others, including five more from *Partisan Review*, were collected in *Against Interpretation and Other Essays* [1966].) Among other things these essays provoked many intellectuals, New Yorkers and others, to reconsider their views of a number of key matters: the relation between "high" culture and popular, commercial culture; the "new sensibility," or counterculture; the politics of form; and, finally, the very nature of radicalism, both cultural and political.

Sontag's second novel, *Death Kit* (1967), further explored the limits of interpretation and the irreducibility of subjective experience through a protagonist unsure he has committed a murder. Married to a woman who is blind and who knows things palpably and directly, the protagonist's agony over the alleged act and its attendant temporal, social, and moral abstractions gradually abates in favor of a mode of being whose vitality, like his wife's, depends upon living willfully and sensuously in the present. In *Styles of Radical Will* (1969), her second collection of essays, Sontag explored the work of European artists such as Cioran, Godard, and Bergman to show how, as expressions of will, their artistry undermined conventional meanings, subverted interpretive projects, and asserted the necessity of experiencing artistic works fully, not merely cerebrally. Some of the political consequences of this outlook were spelled out in "Trip to Hanoi," by far the most controversial essay of the collection. Interestingly, although Sontag was highly critical of portions of American society whose capacity for self-deceptive abstraction was cataclysmic for many Americans and certainly for the Vietnamese, nonetheless she was not willing to relinquish to the political right any claims she or other dissidents might legitimately make to an American identity rooted in a rebellious past. Sontag later made controversial visits to Cuba and China, where her sympathy for certain aspects of these societies and the problems they faced seemed by some to outweigh whatever she may have admired about life in the Western capitalist democracies.

After making four films in the late 1960s and early 1970s, Sontag published *On Photography* in 1977. The book pursued the matter of interpretation into the realm of seeing and "realist" representation. She argued that photographs force us to reconsider how we "see" works of art by undermining a number of our expectations. One expectation, that the artwork reveals its hidden value only to those expert at interpretation, is undermined by photography's tendency to make the world accessible through concrete images. Applying some of the thoughts of Walter Benjamin, Sontag maintained that the reproducibility of photography meant that the art object could not contain its own meaning but rather produced meaning according to context and audience. Also, the technical nature of photography tended to weaken the role of the creator and encourage the demystification of artistic processes usually thought to arise from the inspired genius of rare individuals. For all of these reasons, photography tended to subvert traditional elitist notions and was an art form particularly suited to the building of a democratic society.

Illness as Metaphor (1978), written after Sontag discovered she had breast cancer, examined the ideological nature of the discourse of disease—etiology, diagnosis, description, public perception, myth. Rejecting the common "metaphors" of disease that linked them to character types and specific values, Sontag argued these associations must be seen as historically determined interpretations whose sources had little to do with the actual illness involved and everything to do with the logic of capitalism and the interests of the medical profession.

Contemporary capitalism's appropriation and domestication of a once-subversive modernism was a major concern of the essays collected in *Under the Sign of Saturn* (1980). In some ways comprising an elegy for modernism's ingenious, energetic strategies of resistance recently co-opted by a commercial culture with infinite absorptive capacities, the essays show Sontag at something of an impasse in terms of the future, both the culture's and her own as critic. Indeed, after 1980 she turned to editing, bringing out *A Susan Sontag Reader* in 1981 (minus "Trip to Hanoi," as Alan Wald and others have noted) and *A Barthes Reader* (1982). She assumed the presidency of PEN in 1987, and in 1989 she published a second inquiry into the ideology of illness, *AIDS and Its Metaphors*. She has recently published her third novel, *The Volcano Lover*, a new venture into the genre of romance.

In her introduction to *A Susan Sontag Reader*, Elizabeth Hardwick gives us at least one good reason to consider Sontag a New York intellectual. Having identified her cosmopolitan "ocean-spanning curiosity," she quickly refines her meaning lest it be misunderstood:

> I do not wish to suggest too great a degree of the ambassadorial in her career, to make of a free and independent intelligence a bringer of news or even of *the new*. She is too much of a New Yorker for that, too much at home here where you cannot tell anyone anything.[55]

There are other reasons to make the association, as Sohnya Sayres does in *Susan Sontag: The Elegiac Modernist* before concluding that the relationship is finally a tenuous one. First of all there is the familiar myth of origins: Sontag, marooned at North Hollywood High in the San Fernando Valley, journeys beyond the Santa Monica Mountains to the literary corner of Hollywood and Highland (surely the second longest journey in the world after the one from Brooklyn to Manhattan) where she picks up a copy of *Partisan Review* at the newsstand, reads Trilling's essay "Art and Fortune," and resolves to move to New York to write for the illustrious magazine.

This, of course, she did do, beginning with the celebrated early essays mentioned already. She would go on to contribute over fifteen pieces to *Partisan Review*, and over twenty to the *New York Review of Books*. Like her New York predecessors, and unlike most academics today, she was very much a public intellectual. She was also very much a situational critic, meaning that her criticism was always responsive not only to texts but to the critical, cultural, and broad political needs of the moment as she understood them. Typical was the way she ended "Against Interpretation": "The aim of all commentary on art *now* should be to make works of art . . . more, rather than less, real to us" (emphasis added). Why? Because "[i]n a culture whose already classical dilemma is the hypertrophy of the intellect at the expense of energy and sensual capability, interpretation is the revenge of the intellect upon art."[56]

Her methodology was no less historical, even though her chief concerns have always been style and form, elements of art notoriously difficult to grasp in their full historical dimensions. In "On Style," she wrote,

> [T]he notion of style, generically considered, has a specific, historical meaning. It is not only that styles belong to a time and a place; and that our perception of the style of a given work of art is always charged with an awareness of the work's historicity, its place in a chronology. Further: the visibility of styles is itself a product of historical consciousness. . . . Still further: the very notion of "style" needs to be approached historically. Awareness of style as a problematic and isolable element in a work of art has emerged in the audience for art only at certain historical moments—as a front behind which other issues, ultimately ethical and political, are being debated.[57]

One of the main points in these two early essays was precisely that this historical sixth sense, as it were, should not lead to conceptual statements regarding the work's verisimilitude or moral or political correctness, but rather should enhance the other senses by enlarging one's capacity for pleasure. One is strongly reminded here of Lionel Trilling's insights in his essay "The Sense of the Past" regarding the historical sense as an aesthetic sense. What was needed, Sontag argued, was an intensive engagement of the whole sensibility, "something like an excitation, a phenomenon of commitment, judgment in a state of thralldom or captivation."[58] "For it is the sensibility," she states further on, "that nourishes our capacity for moral choice, and prompts our readiness to act."[59]

These are familiar emphases: the stress on the historicity of form, of style, of reading, we have encountered before, as we have the stress on sensibility. So too, Sontag's insights concerning the actual relation between form and content echo the prescriptions, if not the practice, of the earlier generation of New York intellectuals in their struggles against literary "leftism" thirty years earlier. One hardly need mention the most obvious similarity with the older generation: the championing of a subversive, often politically objectionable, European modernist art.[60]

For these reasons, I would argue that Sontag's criticism should be seen as a continuation of the New York intellectuals' project. But it has been a continuation into a new time and for a new generation, and thus every element of consistency is also an element of revision as well: in updating the *Partisan Review* project, she changed it. Thus the initial targets of her manifesto against interpretation were not just the legions of pedestrian critics in the academy and in the popular press, but also the New York intellectuals themselves, some of whom had allowed the heated battles with the Stalinists, with McCarthyism, with the anti-anti-Communists, and with the Beats and the emerging counterculture, to simplify their criticism. In some cases attention to ideological content had further diminished efforts to treat form carefully. In other cases a strident tone had replaced an urgent one. More to the point, the earlier battle against the Popular Front critics and against literary realism had not succeeded in effecting a balance in American culture between "experience" and "mind." The successful war against rigidly materialist versions of experience had resulted in the privileging of consciousness, and by the 1960s "mind" struck many as having become sterile and oppressive. In effect, Sontag was recalling the New York intellectuals to their earlier recognition of the deeply intertwined, undulatory relation between aesthetics and morality, aesthetics and politics. Ironically, this had been the very recognition that, at the literary level at least, had caused the New York intellectuals to break from doctrinaire criticism and establish their independence. By the mid-1960s it was time for a major reorientation.

Sontag's most obvious break with her predecessors was over the issue of "high" versus "low" culture. Bad enough that Sontag's capacious, generous, and democratic receptiveness took in an entire group of late modernists such as Artaud and Ionesco, whose radical experiments in form were thought to lack seriousness by other New York critics; worse from the point of view of many established critics was her interest in film, photography, pornography, science fiction, lesbianism, transvestism, and the marginal sensibility of camp. This radical pluralism on Sontag's part was often taken as faddishness at best; at worst it was considered an unprincipled, irresponsible defense of the new sensibility's hedonism, escapism, superficiality, and nihilism.[61]

In the polarized 1960s, qualified argument was a rarity. Sontag may have ventured into forbidden terrain in her search for originality and vitality, but lost on Irving Howe and other defenders of traditional modernism, as it were, was the fact that Sontag brought demonstrable standards with her. She encouraged her audience to develop an appreciation for new arts, new genres, new subgenres, and little-known subcultures and values—but she discriminated among the art works within each group. In a 1975 interview she expressed her undiminished belief in quality in response to a question about feminist criticism:

> It's not the appropriateness of feminist criticism which needs to be rethought, but its level—its demands for intellectual simplicity, advanced in the name of ethical solidarity. These demands have convinced many women that it is undemocratic to raise questions about "quality"—the quality of feminist discourse, if it is sufficiently militant, and the quality of works of art, if these are sufficiently warm-hearted and self-revealing. . . . One common denominator of New Left polemics was its zeal for pitting hierarchy against equality, theory against practice, intellect (cold) against feeling (warm). . . . What was denounced in the 1960s as bourgeois, repressive, and elitist was discovered to be phallocratic, too. That kind of second-hand militancy may appear to serve feminist goals in the short run. But it means surrender to callow notions of art and of thought and the encouragement of a genuinely repressive moralism.[62]

Sontag's response reminds us that she has been a critic of the left as well as a critic of traditional values; indeed, she has assumed a position as an independent leftist critic whose lineage goes back at least to the 1930s and the break with Stalinism. This was first made clear following her trip to China, when she made critical remarks about the impossibility of art in that repressive society. Since then, especially in her Town Hall speech of 1982, she has denounced communism explicitly. In that highly controversial speech she said, among other things, that the "utter villainy of the Communist system" was "a hard lesson to learn. And I am struck by how long it has taken us to learn it. I say we—and of course I include myself." To the justifiable perplexity and anger of many, she unfortunately added, "Communism is fascism—successful fascism, if you will," needlessly confusing the issue.[63] Be that as it may, elsewhere she has shown scorn for habits of mind found on the left that do the cause of social justice little service and, in fact, resemble the tactics of official Communist societies. When asked, for example, whether she considered Ingmar Bergman a "reactionary" artist, she replied that she was "extremely reluctant to attack anyone as a reactionary artist. That's the weapon of the repressive and ignorant officialdom in you-know-which countries, where 'reactionary' is also associated with a kind of pessimistic content or . . . with not providing 'positive images.'"[64] Clearly such remarks have continued relevance today in the battle to rid the left of dogmatism and "political correctness."

VI

In his minor classic *Politics and the Novel*, Irving Howe endorsed the Jamesian thesis concerning the thinness of American culture and its inability to sustain a rich novelistic tradition "Very few American writers," observed Howe,

> have tried to see politics as a distinctive mode of social existence, with values and manners of its own. . . . Those massive political institutions, parties, and movements which in the European novel occupy the space between the abstractions of ideology and the intimacies of personal life are barely present in America.

Americans, Howe claimed, were "bored or repelled" by politics, however tempted they may have been.

> The Americans failed, and they could not help but fail, to see political life as an autonomous field of action. . . . Personalizing everything, they could not quite do justice to the life of politics in its own right, certainly not the kind of justice done by the European novelists. Personalizing everything, they could brilliantly observe how social and individual experience melt into one another so that the deformations of the one soon become the deformations of the other.[65]

Howe went on to discuss whatever circumscribed political life there was to be found in Hawthorne's *Blithedale Romance*, Henry Adams's *Democracy*, and Henry James's *Bostonians*. He brought his account into the twentieth century with some very brief remarks on Dos Passos's *USA*, Trilling's *Middle of the Journey*, and Robert Penn Warren's *All the King's Men*. The chapter in which these novels were discussed was duly entitled "The Politics of Isolation."

Howe's understanding of politics gave priority to the often violent, sometimes cataclysmic confrontation between ideologies, movements, parties, and powerful individuals. Having written his study in the wake of totalitarian violence and during the depths of the cold war, this should not be surprising to us. But this concern with apocalyptic politics foreclosed other possibilities, as Howe himself has acknowledged in the preface to the 1992 edition, expressing regret that he did not pay attention to the novels of Eliot, Meredith, and Trollope, "which portray the political life of a settled society, that is, a society in which the usual interplay of group conflicts is regulated by democratic procedures."[66] Had he examined these writers, no doubt his vision of the political novel would have been enhanced by greater attention to the nuances of politics in ordinary life.

But there is another category of experience—the ordinariness of political life—that seems to elude even this approach. For Howe, even the novel about settled societies "still tests extreme situations, the drama of harsh and ultimate conflicts." But what of the common situations, the drama of harsh and seemingly *trivial* conflicts, the small but consequential encounters among the politically involved that allow us to take the measure of their ideals in light of their personal lives, instead of vice versa? It seems to me that the women discussed in this chapter have performed this political task exceedingly well. How ironic that the "Dark Ladies" have been so named because, to paraphrase Adorno, they have illuminated the crevices of the body politic. They have shown us life before and life after the climaxes: not the manifesto but the writing of it; not the political rally but the trip home. They observed from below the people who claimed to observe history from below. Not only do they deserve to be taken seriously as writers of political fiction, but they should be given credit for having deepened and enlarged the very tradition Howe has adumbrated.

9

"Their Negro Problem": The New York Intellectuals and African American Culture

All of us are bereft when criticism remains too polite or too fearful to notice a disrupting darkness before its eyes.

TONI MORRISON, *Playing in the Dark*

"Sort of stay undercover," Jules said. "You know."

JAMES BALDWIN, "Previous Condition"

I

Readers familiar with the subject will know my chapter title derives from Norman Podhoretz's "My Negro Problem—and Ours" (1963), perhaps the most controversial essay ever written by a member of the New York intellectual circle. With astonishing honesty, Podhoretz confessed in the essay to having held deeply ambivalent feelings toward African Americans arising from numerous unhappy encounters during his youth—this despite his forceful public advocacy of civil rights and his credentials at the time as a well-known intellectual of the left. What makes the piece especially interesting today is that it appeared at the very beginning of a period of increasing trouble for the African American–Jewish alliance, trouble that recently has threatened the very existence of the long-standing relationship. Of course, as Kenneth Clark pointed out long ago,[1] African American–Jewish relations have often been fraught with tension. But in the early phases of the civil rights movement the relationship had taken the form of a righteous alliance with African Americans and progressive whites, many of whom were

Jews, fighting and in some cases dying side by side in the bitter desegregation battles of the South. But for several well-known reasons, the tactics and the objectives of many blacks and Jews began to diverge rather dramatically in the mid-1960s. First, once legal equality seemed closer at hand with the passage of civil rights legislation in 1964 and 1965, portions of the integrationist movement went beyond demands for formal legal equality in favor of equal socioeconomic results and thus a more equitable redistribution of resources. The specific issue that crystallized debate around these two different objectives was affirmative action, which many Jews could not bring themselves to support. In part this was because affirmative action seemed to challenge the very universalist, color-blind principles responsible for their own successful integration into mainstream American life; in part it was because Jews themselves, overrepresented in certain professions, stood to lose job opportunities to blacks. A second factor was the growing challenge to integrationism itself within the African American community, at first manifested as a challenge to biracial organization within CORE and SNCC, and later coalescing into new and powerful nationalist, Pan-Africanist, and internationalist, anti-imperialist tendencies. These tendencies, especially the latter one, caused some African-American activists to repudiate Zionism and the state of Israel on behalf of both the Palestinian national liberation struggle and black South Africans struggling against an apartheid regime aided by Israel. Finally, the fact that the civil rights struggle moved from the South to the North exacerbated tensions between blacks and Jews. Sometimes, as in the case of the Ocean Hill–Brownsville School desegregation controversy of 1968, the existence of quiet racist attitudes among Jews and quiet anti-Semitism among blacks became a volatile mix as the largely Jewish teachers' union confronted the largely African American community demanding greater control of the schools.

Undoubtedly these familiar explanations of black-Jewish troubles go some distance toward identifying differences between the two communities, but they tend to suggest that changing attitudes among African Americans were largely responsible for the turn in black-Jewish relations. In this chapter I wish to add another perspective by exploring the New York (Jewish) intellectuals' attitudes toward, knowledge of, and representation of African-Americans both before and during the divergence of which I speak. I shall argue that despite important contributions toward a better understanding and closer relationship with the African American community, the New York (Jewish) intellectuals remained the New York (white) intellectuals when it came to their inability to see what Ralph Ellison has famously called the "concord of sensibilities" that comprise the whole range of African American experience. That is, although the New York intellectuals were not part of the process of suburbanization and white flight by which many American Jews assimilated and also changed some of their attitudes toward race in the postwar period, they nonetheless remained relatively remote from African American life. This was especially true as their own success as intellectuals encouraged them to superimpose an ethnic, assimilationist model on an American dilemma fraught with racial and class (not to mention gendered) contradictions. Such a perspective will, I hope, provide a context for understanding not only relations between African Americans and Jews, but also relations between African Americans and white progressives.

Surely the New York intellectuals have had an important impact on race relations over the past fifty years. As one of the nation's most influential and prominent intellectual groups, they have done a great deal toward shaping American attitudes concerning race and ethnicity. Its chief publications–*Partisan Review, Politics, Dissent, Commentary*, and the *New York Review of Books*, were instrumental in this regard. In the pages of these magazines Nathan Glazer, Daniel P. Moynihan, Dwight Macdonald, Irving Howe, Hannah Arendt, Sidney Hook, Norman Podhoretz, Michael Walzer, Michael Harrington, Ralph Ellison, Amiri Baraka (then LeRoi Jones), and James Baldwin all discussed the broad range of problems that continue to define race relations in the country. A close, albeit selective, look at what they had to say will tell us something about responsibilities only partially fulfilled, and about the tasks still facing white progressives, and whites in general, today.

II

One of the dim ironies in the annals of *Partisan Review* is that the editors of the old Party-affiliated magazine invoked the "struggle . . . against racial oppression" in their editorial statement of 1934, whereas the inaugural issue of the celebrated independent *Partisan Review* of December 1937 contained no such invocation in its editorial statement. By itself, of course, this fact meant little, since by repudiating any requirement that the magazine conform to any organization's political agenda, the 1937 statement contained no programmatic passages or political goals at all. The omission was, however, symptomatic of the distance from the political movement and working-class constituencies which the *Partisan Review* editors considered inevitable, given the circumstances surrounding their struggle for autonomy and critical integrity.

Those circumstances were of course shaped by the Communist Party, which exerted varying degrees of influence in nearly every significant mass movement of the day, especially in and around New York City. Whatever Phillips, Rahv, and others may have felt about the corrupt content and quality of that influence, it was widespread within movements for social change, never more so than in the case of the African American community of Harlem. The Party played a key role in Harlem throughout much of the decade: in the realm of culture alone, it influenced at one time or another a very high proportion of African American writers and artists, among them Richard Wright, Ralph Ellison, Paul Robeson, Langston Hughes, Sterling Brown, William Attaway, Chester Himes, Countee Cullen, Dorothy West, Arna Bontemps, John Oliver Killens, and Margaret Walker. In addition, prominent liberal and social democratic figures such as Adam Clayton Powell, Jr., A. Philip Randolph, W. E. B. DuBois, William Lloyd Imes, Juanita Jackson of the NAACP, Alain Locke, and Ralph Bunche were willing to cooperate with the Party on specific issues and campaigns.

Beyond influencing individuals, the Party was active within the larger community on a number of cultural fronts. It was deeply involved in promoting black theater through the Negro People's Theatre and Hughes's "Harlem Suitcase Theatre"; it fought discrimination and promoted the teaching of black history in the schools; it organized concerts, recording sessions, and nightclubs to popularize jazz and blues musicians; and it waged a long-standing campaign to end discrimination in professional sports.[2] As is well known, much of this activity was later condemned as opportunistic and undemocratic by many of those who participated. Wright's contribution to *The God That Failed* (1950) and Ellison's *Invisible Man* (1954) remain the classic statements here and, as would be expected, many of their insights were consistent with those of the New York intellectuals.

It would be absurd to condemn the New York intellectuals for not organizing similar campaigns on a scale comparable with those of the Party. But it is proper to ask how important the struggle for racial equality was to the New York intellectuals; what their own role was in relation to that struggle, stated or implied, direct or indirect; and how these views were manifested in the literary and cultural work of the circle. In making such inquiries I am susceptible to the fallacy of anachronistic thinking, by which the careless historian superimposes contemporary values upon the past and imperiously judges the past too harshly. This error is not uncommon, but I believe my mode of inquiry avoids it for two reasons. The first has been implied by my summary of the Communist Party's work in Harlem. The history of American radicalism has shown—and there have always been a few discerning individuals to point this out—that *any* group or individual seeking significant social change in the United States must deal squarely with racial injustice. The American past is littered with the remains of organizations and movements that have shunned multiracial unity based on a clear, public stance against racial oppression. This is not to suggest that such a stance is a sufficient cause of success or even that in the short term it has not served as an obstacle to success. (Some reasonable people believe that it in fact insures defeat in a racist society—indeed, some of the littered remains belong to radicals who were scrupulous about racial equality.) But the experiences of the populists, the Socialist Party, the new left, and even the Communist Party of the 1930s demonstrate the difficulties but also the necessity of developing multiracial unity based upon mutual awareness and equality. One can therefore reasonably expect the New York intellectuals, most of whom were at some point either announced independent socialists or social democrats, to somehow address this set of problems, however obliquely.

My second line of defense against the charge of anachronism is that some of the New York intellectuals' contemporaries offered them ways of addressing these problems. During the 1950s and 1960s, for example, both James Baldwin and, in his more decorous way, Ralph Ellison challenged the New York intellectuals, among others, to see African Americans and their cultural heritage with fresh and clear eyes. Beneath the immediate issues at hand—segregation in the South, legal and structural change, race and sexuality, black power, protest literature—there was always the plea and sometimes the demand to reconsider assumptions, to reassess attitudes and approaches, to embark on the task of educating oneself and the public so that ideas and actions might for the first time be grounded in a working knowledge of the actual range of African American experience. I believe this was Ellison's fundamental appeal in his essay "The World and the Jug," his two-part response to Irving Howe's "Black Boys and Native Sons." As some readers will recall, Howe's contention in that essay had been that as a protest novelist Richard Wright must be considered the father of all contemporary African American literature. This was so because by virtue of the historical experiences of the Negro, this literature has always been intrinsically dissentient in nature. Ellison's justly famous response is worth quoting in full:

> [Howe] seems never to have considered that the American Negro life (and here he is encouraged by certain Negro "spokesmen") is, for the Negro who must live, not only a burden (and not always that) but also a *discipline*—just as any human life which has endured so long is a discipline teaching its own insights into the human condition, its own strategies of survival. There is a fullness, even a richness here. . . . [3]

What is ultimately sought by Ellison is not a position, a strategy, a policy, or an ideological perspective, but simply an openness toward African American experience, an openness that at the very least consists in a willingness to absorb available influences, and ideally consists in something more—an alertness, a curiosity, an eager yet disciplined exploration of an admittedly little-known though rich culture, the understanding of which is a prerequisite to any solution to the race problem.

In this nation especially, as an expectation or as a basis for judgment, this is an exacting standard. But it is not an unrealizable one. There have been examples of a mature openness toward African American culture and experience, although they have more often than not occurred outside the literary sphere. (The Beats and their "white Negro" explicator Norman Mailer hardly qualify, as Ellison himself pointed out. Their eager embrace of African American "nihilism" was more reaction to highbrow white culture than receptivity to a new and complex one.)

It is within the realm of music, among certain white jazz, blues, and folk musicians and critics, that we find examples of the kind of energetic, solicitous, and critical engagement with aspects of African American culture that allowed for mutual regard and influence. I have in mind, for example, Woody Guthrie's regard for Huddie Ledbetter (Leadbelly), which occasionally lapsed into sentimentality but was grounded in a serious appreciation of the latter's artistic accomplishments.[4] Earlier, of course, a legion of white jazz musicians had eagerly absorbed many of the innovations of African American jazz and went on to make their own contributions: Bix Beiderbecke, Eddie Condon, Bud Freeman, Jimmy MacPartland, PeeWee Russell, Dave Tough, Nick LaRocca, and others. Although Amiri Baraka in *Blues People* rather begrudgingly maintained that the transgressions of these musicians did not reflect an increased understanding of African American life, he nonetheless acknowledged that their receptivity served to make African American culture an object of an "intelligent regard it had never enjoyed before."[5] The same, of course, can be said of important impresarios and critics of African American music like John Hammond, Alan Lomax, Moe Asch, Nat Hentoff, Martin Williams, and Leonard Feather.

Of how many white writers and critics can we make the same claim? With literature we are, admittedly, dealing with a different art form whose mode of production is individual rather than collective, and thus not likely to generate collaborative effort and the kind of heightened reciprocal awareness this can give rise to. Moreover, by most accounts the dominant African American contribution to music has not been matched by the contribution to literature, as substantial as the latter has come to be regarded. As to the first point, literature may not be a collaborative art, but criticism is—if not in the strictest sense, then certainly in the sense that it is collective: generated by exchanges of opinion, and often by circles and schools defined by common values and approaches. Surely there is nothing about criticism per se that mitigates against a full exploration of a given culture and its literature. Indeed the best criticism engages in just such an encounter by "collaborating" with what is insufficiently understood in order to produce new meanings and new judgments.

Conversely, despite the collective nature of jazz, collaboration among black and white musicians was never easy. Until the 1970s jazz bands often integrated themselves at their peril. If they were fortunate, they merely suffered enormous inconvenience and, for black musicians, great personal discomfort to the point of humiliation. Of course, it could be argued that the quality of African American music was what enticed white musicians—that without the distinctiveness of the music few whites would have ventured beyond their own cultural horizons. This implies that African American literature was not sufficiently lustrous to attract the attention of whites. But who today would make the claim that James Weldon Johnson, Jean Toomer, Nella Larson, Jessie Fauset, Claude McKay, Sterling Brown, and Zora Neale Hurston, all of whom published in the 1920s or the 1930s, produced work that was neglected because not sufficiently alluring? Was the Harlem Renaissance of the 1920s not worthy of the attention of critics in the 1930s and after? True enough, many critics of the 1930s displayed impatience with the aesthetic experiments of the modernist and expatriate 1920s, but the New York intellectuals, once they left the Party's orbit, were not among them. The New York intellectuals were known precisely for their interest in the experimental literature of the previous decade, but their interest in high modernism did not extend to the hybrid literature of the Harlam Renaissance in which modernist and vernacular traditions were joined. Blaming African American writers for their neglect at the hands of white critics is like blaming the stars for not shining on a polluted night. To be sure, the obfuscating dynamics of race have been an important factor in the white response to music as well as literature, as has elitism. It has, after all, taken mainstream music critics decades to accept jazz and blues (as well as folk and country) as music worthy of serious attention. But in that effort there were pioneer white critics who joined the African American experts to help lead the way. The same, unfortunately, cannot be said of the belated attention directed toward African American literature by portions of the critical mainstream. That effort had to wait for the social movements of the 1960s to leave their mark on a new generation of scholars and critics. When this generation looked to earlier generations for mentors—to the New York intellectuals among others, they found few to guide them. Those predecessors they found were without exception African American. This is not to suggest that the New York intellectuals contributed nothing to an understanding of African American authors, or to better race relations in the country; on the contrary some of their achievements, as I shall demonstrate, were impressive indeed. But in the final analysis they remained relative strangers to African American life. Thus they were in no position to help effect the sea change in the structure and consciousness of American life needed to secure full equality, which we continue to await today.

III

Several generalizations can be made about the New York intellectual response to matters of race and ethnicity. The first is that, like the response of nearly all whites, it was reactive. Before 1957, the year President Eisenhower sent federal troops to Little Rock to prevent interference with school integration, the New York intellectuals had relatively little to say about the emerging civil rights movement or about race in general. Second, once the New Yorkers began to take these matters seriously, they made significant contributions, mostly of a political nature, by which I mean that the vast majority of essays which appeared in the publications of the New York intellectuals dealt with tactics and especially policy. In this area the influence of certain New Yorkers—Norman Podhoretz, Irving Kristol, Nathan Glazer—was sometimes felt at the highest levels of power during the 1960s. As for cultural issues, and here is my third generalization, the New York intellectuals did not adequately address the consequences of the civil rights and black power movements for literature and culture, ironically their own areas of expertise. My claim here shall be that this failure to expose a largely white audience to a representative range of African American cultural expression, especially written expression, while not a failure unique to the New Yorkers (they did better than most whites), nonetheless provides a historical record of how much reciprocal cultural interaction was lacking during the civil rights era.

Let me now explore at greater length *Partisan Review*'s coverage of African American literature and the subject of race. It turns out that the failure to mention the struggle against discrimination in the 1937 editorial statement, to which I alluded earlier, signaled but a minor shift in emphasis. This was because the magazine, even while affiliated with the Communist Party, did not distinguish itself by its coverage of African American writers. There were no articles on African American writing or on the subject of race in the early *Partisan Review*. The first such article did not appear until 1940, over two years after the magazine's reappearance on a new basis. As for book reviews of black writers, there were but two during the 1930s: a 1934 review of Langston Hughes's *The Ways of White Folk*, by Leon Dennen, and a 1938 review, of Richard Wright's *Uncle Tom's Children*, by James Farrell. The former was a fairly typical product of Leninist dogma, despite its astute observation that Hughes's characters feel, all right, but seem to lack the ability to think (we recognize the familiar Eliotic distinctions here). Noting that the main character in the story "Cora Unashamed" is a primitive and therefore unrepresentative of the Negro masses, Dennen would substitute another, albeit more noble stereotype: "But is the story of the thinking and fighting Negroes any less dramatic? Is the story of Angelo Herndon any less dramatic? . . . What about the relation between Negro and white sharecroppers fighting hand in hand in the South?" he dutifully asks.[6] Happily, Farrell's review does not lapse into declamation. The novelist praises Wright for his brutal, violent depiction of "the bitter experience of the Negro in a white man's world," but he is not pleased with some of Wright's mannerisms, such as his habit of capitalizing letters for emphasis. In a paternal mode that hardly seems justified given Farrell's mere four-year seniority over Wright, he assures his readers that such mannerisms "can easily be dropped with a little more work and some self-conscious reflection on writing as a technique."[7]

The 1940s brought no change in the quality or quantity of coverage of African American culture. In 1940 a "slice-of-life" column, which appeared infrequently in the early years of the magazine, entitled "Cross Country," offered a nostalgic glimpse of Stephen Crane's birthplace, Newark, New Jersey. In the piece, whose peculiar tone unaccountably shifts from serious reportage, such as one might have found in *New Masses*, to the detached bemusement of *The New Yorker*, the author noted that one of the changes taking place was the "virtual *occupation* of [Weequahic Park] by negroes" (emphasis added).[8] In the same issue Richard Wright's *Native Son* was reviewed by David Daiches as part of an omnibus effort with five other novels—all in the space of three-and-a-half pages! Given Wright's complex treatment of the Communist Party in the final portions of the novel, not to mention the novel's sociological detail and psychological subtlety, one might have expected something more substantial from *Partisan Review* than what, in fact, amounted to but a single lengthy paragraph. Daiches, for his part, pronounced *Native Son* a powerful and important book, despite reservations about the novel's several lapses into melodrama.[9]

Four years later *Strange Fruit*, by the extraordinary white southerner Lillian Smith, was reviewed by Robert Gorham Davis, once again as part of a multiple review, in this case with three other novels. In the same year, 1944, James Agee contributed a "Variety" piece in which he elaborated on the theme of the corruption of folk (popular) art in contemporary culture. (This subject had been broached in an earlier article by Louise Bogan.) Agee focused on examples of what he regarded as the dilution of the more vigorous, creative jazz of the 1930s by commercial interests intent on gaining a wider white audience. He reviled the pseudo-folk jazz of Hazel Scott, whose jazz "one could probably pick up, by now, through a correspondence school." And he attacked several of Paul Robeson's efforts—among them *Othello* and *Carmen Jones*—for betraying both Robeson's talents and his rich musical inheritance.[10]

One final word concerning *Partisan Review*'s coverage of race during the war years. It is well known that during World War II the African American community and some of its allies worked tirelessly to end segregation in the armed forces. The irony of fighting Hitler and his ideology of racial superiority in the name of equality with an army that discriminated against its black troops escaped no one. Yet it did escape the pages of *Partisan Review*. As we have seen, this war was a war that, at least initially, the magazine was very skeptical of, but what is revealing is that few of the magazine's doubts seemed to stem from an awareness of the immense hypocrisy and injustice of maintaining segregated armed forces. The exclusive fear was that, inspired by the Party and the Popular Front, American intellectuals would lead the nation to war and "*to military dictatorship and to forms of intellectual repression far more violent than those evoked by the former war*" (emphasis in the original).[11] Only Clement Greenberg and Dwight Macdonald, in "10 Propositions on the War," mentioned the need to struggle for racial equality.[12] It was not until Dwight Macdonald left *Partisan Review* to establish *Politics* that segregation in the armed forces and Jim Crow at home became major issues for any of the New York intellectuals.

The years immediately following the war saw little change in *Partisan Review*'s habits of attention. In a 1945 review of Wright's *Black Boy*, the magazine's first review devoted exclusively to an African American text, Elizabeth Hardwick suggested that the plague of African American writers had been "the endless need to explain and to underline truths already implicit in his subjects." The disparate roles of writer and spokesman for an entire race have greatly affected even so great a talent as Wright, Hardwick observed, and are responsible for "the appalling naiveté of writers such as Zora Neale Hurston, Langston Hughes, and Countee Cullen."[13] The starkness of this precipitous judgment, made with elaboration and offered as if self-evident, reveals several important assumptions regarding the lack of sophistication and complexity in African American writing. These assumptions remained unexamined in the pages of *Partisan Review*, and more broadly by American criticism as a whole until the 1960s and after.

It would be seven more years before *Partisan Review* published a significant piece of African American writing. Ralph Ellison's "Invisible Man: Prologue to a Novel," appeared in the January–February 1952 issue, and several months later the novel was reviewed by Delmore Schwartz in yet another of the magazine's omnibus reviews, this time with four other novels. Schwartz's review was glowing to the point that he despaired of being able to do the novel justice. "Some books," he asserted, "have so much actuality in them that the reader and the critic is forced to say to himself: it is somehow dishonest or hypocritical or irreverent to speak of such books merely in the language of literary criticism."[14] One is tempted to conclude from this encomium that Schwartz has simply added another reason to that of inferior quality for refraining to speak of books by African Americans: the book might be too good![15] In Schwartz's case this cynical interpretation is probably unfair, for it seems he was being unusually modest here: he actually ended his review by suggesting that someone else should have reviewed the book! This was unheard of among New York intellectuals, for whom mandarin self-assuredness was de rigueur. That Schwartz suggested the someone else be William Faulkner does diminish his modesty somewhat, but I think my point stands nonetheless: I can think of no other example of such deference in the criticism of other New Yorkers.

Modesty, in any event, was not the reason the New York intellectuals kept silent about African American writing. If we look at the range of African American writing from the 1930s to the 1960s, we see that nearly all of it was ignored by *Partisan Review*, not to mention nearly every other white publication in the country.[16] Among the writers who produced noteworthy work during this period yet whose work was never reviewed were W. E. B. DuBois, Lorraine Hansberry, Margaret Walker, Arna Bontemps, Sterling Brown, Robert Hayden, John O. Killens, William Attaway, Chester Himes, Frank Yerby, Ann Petry, Willard Motley, Dorothy West, William Gardner Smith, Frank Marshall Davis, William Demby, John A. Williams, Owen Dodson, and J. Saunders Redding. *Partisan Review* was also silent when it came to important political and cultural developments that affected African Americans during this period: the early years of A. J. Muste's CORE, floggings and increased Ku Klux Klan activity, the persecution of W. E. B. DuBois, the early desegregation cases, Montgomery, Little Rock, and the struggles for independence in Africa and among developing nations. Indicative of *Partisan Review*'s silence on such matters is William Phillips's autobiographical history of the magazine, in which there are no entries in the index under civil rights, Montgomery, Martin Luther King, black power, or race. Wright, Ellison, and Baldwin are the only African American writers listed.

Partisan Review did make a major contribution, however, to the dialogue on race in the United States. In the early 1950s its editors had the foresight to open the magazine's pages to a young African American writer, a graduate of DeWitt Clinton High School in New York who had been groomed to become a preacher but chose instead to write. In May 1949 the magazine reprinted James Baldwin's now famous critique of reform-minded fiction "Everybody's Protest Novel." The piece had appeared that spring in the Parisian little magazine *Zero*, and William Phillips was so impressed with it he published it immediately despite a policy of publishing original work only. In this brief but controversial essay, Baldwin had underscored the limitations of two progressive novels cherished by American progressives, *Uncle Tom's Cabin* and *Native Son*, claiming that each had woefully simplified human character in an effort to render individuals as products of powerful and inhumane social forces. Two and a half years later the magazine published Baldwin's equally well known critique of Richard Wright's *Native Son*, "Many Thousands Gone." Thus *Partisan Review* joined the *New Leader* and *Commentary*, two other publications of the New York Jewish intellectual milieu, in bringing out Baldwin's earliest work. Although Baldwin's first published piece, "Maxim Gorki As Artist" had appeared in the *Nation* in 1947, subsequently he had a dozen reviews published in Sol Levitas's the *New Leader* and five pieces published in Elliot Cohen's and Robert Warshow's *Commentary* before *Partisan Review* went with its first Baldwin piece. Over the course of the next several years *Partisan Review* published "Paris Letter: A Question of Identity" (July–August 1954), "Faulkner and Desegregation" (Winter 1956), "Sonny's Blues" (Summer 1957), and "Letter from the South: Nobody Knows My Name" (Winter 1959). By comparison, after 1951 Baldwin would write one more piece for the *New Leader* and three more for *Commentary*. By the late 1950s, of course, he was able to earn much more by publishing uptown in larger, more lucrative magazines such as *Esquire*, the *New Yorker*, and *Playboy*. Thus, as would soon be the case for Amiri Baraka, whose first essay was published in *Partisan Review* in 1958, the New York intellectual community gave Baldwin his start and supported him early in his career. This support, incidentally, went beyond publishing early work: Philip Rahv recommended an early version of *Go Tell It on the Mountain* to Random House (unsuccessfully), and in 1956 Baldwin received a much needed $3,000 grant from *Partisan Review*, not by any means a wealthy magazine.

Why did this rather bizarre relationship form in the first place between, as Baldwin's biographer James Campbell has put it, "a young Negro with no formal education beyond the age of seventeen . . . and the nation's top intellectual magazines"?[17] Was it because of a new commitment on the part of predominantly white, Jewish, radical publications to address long-standing racial difficulties, perhaps in response to events then taking place in the South? And why did Baldwin choose to submit his work to these particular publications, which heretofore had exhibited no great interest in African American writers?

The answer, it would seem, is that Baldwin's interest in modernism combined with the New Yorkers' awakened sensitivity to race largely through the catalytic effect of Baldwin's talent. To be sure, by the standards of the day Baldwin's early work dealt with race but obliquely. He certainly had no patience for the literature of racial uplift, as evidenced by his scathing review of Shirley Graham's hagiographic account of Frederick Douglass, *There Once Was a Slave*.[18] On the other hand, his personal and subtly psychological perspective on race relations—when, indeed, he chose to treat racial themes—certainly did not endear him to the more politicized elements of the black community. This meant there was little sympathy and less room in the black press, beleaguered as it was, for Baldwin's highly subjective prose and unpredictable, intense sensibility. In his introduction to *The Price of a Ticket*, Baldwin wrote of his futile attempts to gain access to black publications: "I had been to two black newspapers before I met these people [Levitas, Warshow, Cohen, Rahv, and Randall Jarrell] and had simply been laughed out of the office."[19]

But Baldwin's search for support from fellow African Americans was far from exhaustive, and his sarcastic characterization was probably not entirely objective. Nonetheless, his precarious situation speaks to a mitigating circumstance, which must be registered when judging the New York intellectuals regarding race. For despite the limitations of the left-liberal white community when it came to race, it did make resources available to some independently minded African Americans when none were readily available within the African American community. Put another way, if, as Robert Bone has remarked, Richard Wright's briefly felt sense of belonging within the ranks of the Communist Party was "a commentary on the failure of the democratic left" to support him, then it is to the credit of this same democratic left that for a time it was able to provide James Baldwin with, if not a sense of belonging exactly, certainly a knowledge that others were paying attention and taking him seriously.

To hear Mary McCarthy tell it, Baldwin's appeal had little if anything to do with race. His reading, she once asserted, "was not coloured by colour—this was an unusual trait. He had what is called taste—quick, Olympian recognitions that were free of prejudice."[20] But Baldwin also knew and recorded a great deal about African American experience even in his earliest work, and although his judgment still strikes us as exceptional, it was never wholly detached from considerations of color. That McCarthy so readily equated discussions of race with bias demonstrates an important limitation in the thinking of the New York intellectuals at the time, who resembled the larger white population when it came to attitudes toward African Americans. By equating the very subject of race with controversy and bias, and by preferring therefore to avoid it, the tendency among whites has often been to hide their actual hypersensitivity to the subject and instead charge African Americans who insist on discussing it openly as being themselves hypersensitive and therefore unreliable.

In the case of the New York intellectuals, the difficulty was compounded because many of them had chosen to evade their own ethnic experiences as second-generation Jews. Much has already been written on the Jewishness, or rather the lack thereof, of the "New York (Jewish) intellectuals," as Ruth Wisse has rather ironically identified them.[21] Nearly all of it testifies to the fact that, in their universalist embrace of Marxism, modernism, and liberal values, the New Yorkers minimized their connections to ethnic and certainly religious Judaism. As Lionel Trilling put it: "I cannot discover anything in my professional intellectual life which I can significantly trace back to my Jewish birth and rearing. I do not think of myself as a 'Jewish writer.' I do not have it in mind to serve by my writing any Jewish purpose."[22] Other readers will perhaps have read one of the several memoirs written by New York intellectuals tracing their paths out of the ghetto and into the self-perceived universalism of modernist culture, memoirs such as William Phillips's *Partisan View*, Alfred Kazan's *Starting Out in the Thirties* and *New York Jew*, Irving Howe's *Margin of Hope*, Sidney Hook's *Out of Step*, Norman Podhoretz's *Making It*, or Lionel Abel's *Intellectual Follies*. Each of these memoirs in its own way confirms Norman Podhoretz's observation that "One of the longest journeys in the world is the journey from Brooklyn to Manhattan."[23] Indisputably, the New York intellectuals successfully completed the journey, but they retained aspects of their Jewishness despite assimilationist pressures to distill the earmarks of style and sensibility—hence the triumph of the Jewish American novel of the 1950s, ushered in by a supportive phalanx of New York critics. In these novels the expressions of a secular Jewish identity flourished under the rubric of an encompassing modernity.

It is here, in the tension between acculturation based on universalist values, and the stubborn residues of ethnic life, that we find a resemblance between the New York intellectuals and James Baldwin. In his double role as champion of a cosmopolitan modernism and reluctant inheritor of a ghetto world, Baldwin was in some ways the black analogue of the New York Jew. Both, for a time at least, found themselves in limbo between ethnic or racial particularism and the liberal, universalist values espoused by dissident portions of the dominant culture, and this perhaps accounted for their convergence in the early and mid-1950s. But their paths would soon diverge as the New York intellectuals (and nearly all American Jews) continued their unprecedented centripetal journey, whereas Baldwin would increasingly gravitate toward an African American community for whom such a journey centerward has always proved to be far more difficult.

If we are to assign a large cause for these divergent paths, we will want to follow Morris Dickstein in noting that whereas the efflorescence of Jewish American literature of the 1950s occurred as part of a skeptical ethos regarding the possibilities of political renewal, the upsurge of African American writing during the 1950s and 1960s was part of a broad social and political movement with strong ties to the everyday lives of the black masses.[24] Thus, unlike the New York intellectuals, Baldwin was strongly encouraged by the African American community to intervene directly in a series of volatile political and cultural disputes during the 1960s, revolving around the subjects of racism, violence, homophobia, black nationalism, and the Black Muslims. And although these interventions produced work of uneven artistic quality, the passion and depth of Baldwin's engagement during the 1960s surpassed, I believe, that of the New Yorkers, for whom the mass movements of the period were often viewed as incipiently totalitarian, whatever the ostensible justness of their causes was acknowledged to have been. This overwhelming suspicion, part cause and part product of the position they attained at the apex of the American intellectual world, further prevented them from instigating new connections with popular constituencies such as youth, labor, or minority groups—connections that might have prevented some of the polarized encounters between old and new lefts during the 1960s.[25]

Writing in *Partisan Review* in the late 1960s, Susan Sontag succinctly identified the problem: "I do not think white America is committed to granting equality to the American Negro. So committed are only a minority of generous and mostly educated, affluent white Americans, few of whom have had any prolonged social contact with Negroes."[26] Indeed, white critics who attempted to explore the complexities of the African American community in order to educate other whites, did so at some risk to their nerves, if not to their reputations as expert authorities on literature. Other whites sometimes regarded them as having eccentric interests, while blacks sometimes regarded them as well intended but incompetent. In 1958 *Partisan Review* ran portions of Stanley Edgar Hyman's pioneering paper (pioneering, that is, for a white critic), which he had read at Brandeis, entitled "The Folk Tradition." Hyman's piece, for all its flaws, represented an admirable attempt by a white literary critic to explore the varied traditions of African American writing. Besides discussing the familiar quartet of Wright, Hughes, Ellison, and Baldwin, Hyman's observations took in an impressive array of African American writers and performers—writers whom few other white critics ever mentioned or in some cases even knew: J. Saunders Redding, Rudolph Fisher, Zora Neale Hurston, Melvin Tolson, Jean Toomer, Claude McKay, Margaret Butcher, Billie Holliday, Pine Top Smith, Ma Rainey, the Barbadian writer George Lamming, and the Nigerian writer Amos Tutuola.

But Hyman came under the critical scrutiny of his friend Ralph Ellison in the latter's well-known response, "Change the Joke and Slip the Yoke," the second part of a *Partisan Review* feature on aspects of black writing (rather grandiosely entitled "The Negro Writer in America"). Ellison claimed that Hyman's piece reduced African American writing by overemphasizing certain archetypes within the folk tradition, namely that of the "trickster," which according to Ellison was a construction of white, not black, culture. Ellison noted his differences with Hyman tactfully, save for the equivocal remark near the beginning of his discussion, "if we are to discuss *Negro* American folklore let us not be led astray by interlopers." Presumably Ellison was referring to the "trickster" figure here, but it is difficult to ignore the possibility that Hyman himself is the referent.[27] If my reading of the text's political unconscious is accurate, surely this sort of remark was not likely to encourage white critics to depart from their defined areas of expertise to explore the unfamiliar terrain of African American writing. We can sympathize with the critic who undertakes such a journey with trepidation, not least because a white writer who undertakes to write sympathetically about black views and black writing will likely care about what blacks think of her. But just as surely would it be a mistake to attribute any reluctance to make such a journey to apparently unwelcome remarks like Ellison's, if indeed that is what it was; as uncomfortable as these sorts of remarks may make white critics feel, the most rational among them will understand that decades of systematic neglect of black writing by white critics certainly has more profound causes than white fear of criticism.

In the late 1960s *Partisan Review*, like many other magazines, could hardly avoid responding to the national crisis over the race question. The accumulating number of urban upheavals, the spread of the black power movement, the rise of the Black Panther Party, the white backlash—all combined to create a situation in which white liberals and leftists were forced at least to register the importance and difficulty of the issue of racism in the United States, which they did. They were not so successful, however, at offering long-term solutions, or at providing a beginning basis of interracial cooperation grounded in a serious, rigorous educational effort among sympathetic whites. Writing in *Partisan Review* as a contributor to the 1967 symposium "What's Happening to America," Richard Schlatter was one of the few willing to question the nature of the commitment of sympathetic whites:

> Are those white liberals, students, intellectuals, and religious leaders who have shown some willingness in the last decade to follow the moral leadership of the Supreme Court going to keep on? At the present all these groups seem to be confused by the cry of "Black Power" and frustrated by failure.[28]

Partisan Review did what was expected: throughout the crisis it printed relevant articles on a fairly regular basis. Following the 1967 symposium, which took up the bulk of an entire issue, there was a review of Frantz Fanon's *Towards an African Revolution* (Spring 1967); a review of William Styron's *The Confessions of Nat Turner* (Winter 1967); an essay by Martin Duberman entitled "Black Power in America" (Winter 1968); a symposium on black power (Spring 1968); a debate over the issue of whether Styron, a white novelist, successfully rendered a story narrated by the black insurrectionist Nat Turner; a brief review of James Baldwin's *Tell Me How Long the Train's Been Gone* (Fall 1968); a review of Harold Cruse's

The Crisis of the Negro Intellectual (Fall 1968); "Writing and Teaching," a brief commentary on race and teaching by June Jordan, one of the few African American women to contribute to the magazine (vol. 36, no. 3, 1969); and Peter Brooks's "'Panthers at Yale" (vol. 37, no. 3, 1970). These contributions showed *Partisan Review* to be keeping abreast of current thinking on some aspects of the race question; they did not show the magazine to be part of a cultural avant-garde insistent on reconsidering forms, ideas, and attitudes concerning race.

IV

One publication among the New York intellectuals that began as though it might make a radical departure in its coverage of African Americans and race relations was Dwight Macdonald's iconoclastic *Politics* magazine. Initiated in February 1944 following Macdonald's break from *Partisan Review* over the issue of support for the war effort (Macdonald remained in opposition), the magazine promised to treat race as fundamentally important to any transformation of American society. Introducing "Free & Equal," a regular department "devoted to the most dynamic social issue of today," Macdonald explained that "The Negro in America is the great proletarian. The white worker can dream of rising to middleclass status, but the Negro is a worker in uniform, so to speak, a uniform he cannot take off: his skin." Macdonald went on to underscore the strategic importance of black liberation: "When such a group, deliberately kept for generations at the bottom of the social structure, begins to stir and raise its head, the whole edifice feels the shock."[29]

Immediately Macdonald took up the matter of discrimination against African Americans in the military and in industry. The first issue included a piece on Jim Crow in the military ("Brief in the Lynn Case"), and a review of two articles, one that had appeared in the *Journal of Business* entitled "Negroes in a War Industry: The Case of Shipbuilding," the other entitled "The Negro Automobile Worker." By responding to a trial and to two recent articles, Macdonald's *Politics* immediately took a more political, more topical, and more aggressive stance than *Partisan Review*, which had in Macdonald's view lost a bit of its nerve during the war by cautiously opting to concentrate on literary rather than political matters.[30] Macdonald apparently meant to enter the fray with little of the restraint that characterized the older publication.

The March 1944 issue of *Politics* contained even more material on African Americans. There was a review article by Kenneth Stampp entitled "Our Historians on Slavery," in which Herbert Aptheker's *Negro Slave Revolts* was favorably assessed. In addition there were reviews of Harold Orlansky's pamphlet *The Harlem Riot: A Study in Mass Frustration*, Earl Brown's "The Truth about the Detroit Riot," and J. S. Slotkin's "Jazz and Its Forerunners as an Example of Acculturation." These were the first sustained discussions of African American life to be found in the publications of the New York intellectuals, and together in a single early issue they seemed to promise a serious effort by Macdonald to explore African American experience with some degree of rigor, thereby breaking with the long-standing practice among white leftists of substituting lofty statements of principle opposing discrimination for familiarity with its victims. Yet even *Politics*, with its better coverage of the African American community and its concerns, did not make room in its pages for any significant number of African American writers.

Subsequent issues tended to confirm Macdonald's intention to expose his largely white audience to this new perspective. In the April 1944 number Macdonald himself wrote a piece on the Lynn Case (to date the only court test of Jim Crow in the military), and in the May issue he published a letter from Walter Kerr entitled "Negro Leaders and the Harlem Riot," with a reply by the young Roy Wilkins. The June 1944 issue further pursued the issue of Jim Crow in the military with two pieces written by black servicemen describing the insults and unequal treatment they were forced to endure, entitled "I Was a Seabee" by Isaac McNatt and "The Story of the 477th Bombardment Group" by "Bombardier." In the July 1944 issue the eminent black publisher and writer George S. Schuyler reviewed Gunnar Myrdal's monumental study of race relations in the United States, *An American Dilemma*. In the course of his favorable review, Schuyler addressed himself to the very question under discussion here. "There is something profoundly depressing," he wrote to his largely white audience, "about the universal ignorance of Negroes by white people and, what is worse, the childish zeal to obscure this ignorance and evade any responsibility."[31]

Thus far in the short life of *Politics* Macdonald seemed to be doing everything possible to accept responsibility for educating his white audience. In the same issue he assailed Jim Crow still more vehemently in his editorial "How 'Practical' Is a Racially Segregated Army?" In the August 1944 issue he exposed his audience to differences within the African American community, with which very few whites were likely to be familiar. Although Wilfred Kerr's "Negroism: Strange Fruit of Segregation" was not a particularly insightful discussion of nationalist and separatist trends within the African American community, it did instigate a discussion that Macdonald allowed to continue in letters published in subsequent issues. A common complaint among black leaders, then and now, was that whites rarely looked closely at discussions taking place within the black community, preferring instead of to designate a single spokesman for all African Americans. Walter A. Jackson has recently pointed out that this would soon become a serious problem in the postwar years, when the government repressed black Communists and when black nationalism was on the decline, making it appear to whites that black opinion was more or less uniform. Moreover, Jackson reminds us that the problem of an allegedly homogeneous black community was exacerbated by the fact that many African American leaders cleaved to the traditional strategy of maintaining a strict united front in order to win white liberal support for desegregation, full equality, and equal opportunity.[32] With Kerr's article, despite its bias, Macdonald risked charges of divisiveness and made an important gesture toward a fuller understanding of the actual differences within black political and cultural thought at the time.

Macdonald closed out 1944 with two pieces on African Americans in the labor movement, "The Tucson Strike," part of a pamphlet produced by the Tucson Committee for Inter-Racial Understanding (October 1944), and Nat Glazer and Frederick Hoffman's "Behind the Philadelphia Strike" (November 1944). The latter was a detailed investigation of alleged racism among white workers, which the author claimed was actually the result of union-busting tactics by the company during the Philadelphia Transit Company Strike of that year. These pieces suggested that Macdonald might be taking seriously the need to link considerations of race and class in an effort to provide an analytical basis for biracial cooperation.

But Macdonald's interest in the African American community began to wane in 1945, especially in the latter half. "Free & Equal," a "regular department," became irregular after its first year, and as the year progressed the number of pieces diminished. "Free and Equal: Jim Crow in Uniform" (May 1945) continued Macdonald's campaign against a segregated military; a second piece, "F.D.R.," was authored by George S. Schuyler, and mentioned Roosevelt's failure to ease racial tensions and act on behalf of African Americans (May 1945); another was a republication of Justice John Harlan's eloquent dissent in the 1896 *Plessy* v. *Ferguson* decision which upheld the doctrine of "separate but equal" (June 1945). The July 1945 issue contained three contributions on the matter of race: "My Friends" by William Worthy, Jr., on the military, a lengthy and favorable review of Richard Wright's *Black Boy* by Wilfred H. Kerr, and a review of three books on the Japanese. The August 1945 issue included Harold Orlansky's "A Note on Anti-Semitism among Negroes," and "Jazz, Clock and Song of Our Anxiety" by Arthur Steig, in which African Americans are given credit for inventing the musical form for two essential(ist) reasons: (1) "they are the least bourgeois and hence the least divorced from life and feeling [a]nd (2) they suffer most intensely the spiritual disenfranchisement by our world" (247). Finally, the October 1945 issue included an extended review of Bronislaw Malinowski's *The Dynamics of Culture Change: An Inquiry into Race Relations in Africa* by Helen Constas. Nothing of significance appeared in the final two issues of the year.

During the final three years of *Politics*' short life, the magazine abandoned its maverick identity, at least with regard to race, and reverted to the more characteristic mode of nearly complete neglect. In 1946 only one major entry on race appeared, Macdonald's "Counter-Attack," a passionate condemnation of lynching and other brutality in the South. After this piece not a single article on race appeared in any of the ten remaining issues of the magazine! Certainly Macdonald's drift away from Marxism, and his embrace of anarchism, pacifism, and moralism during these years—combined with the controlled passion and congenial intelligence of his incomparable prose—expanded the possibilities for radicalism in the postwar period. But, as was the case with his fellow New York intellectuals, in the end he broke little new ground when it came to the dilemma of race. This confirms Manning Marable's contention that the cold war, with its disproportionate emphasis on foreign policy and developments in Europe, did significant damage to the struggle for equality at home.[33]

V

Certainly *Commentary* magazine owed a good part of its wide influence during the 1960s to its willingness to address the controversial subject of race relations, especially relations between African Americans and Jews. Quantitatively, there was no comparison between *Commentary* and other New York publications. For that matter few white publications in the country matched *Commentary* in terms of the proportion or amount of coverage devoted to race. Beginning in the mid-1950s and continuing through the early 1970s, *Commentary* ran either an article or a review touching on race in nearly every issue; often a single issue would contain several such pieces. In addition, the range of coverage was unusually broad. The magazine reported upon conditions in the South, conditions in the urban North, developments within the civil rights movement, discrimination in housing and education, the legal battles in the courts, blacks and labor, the activities of the Klan, black-Jewish relations, the psychological and cultural dimensions of race, urban riots and rebellions, and the black power movement. This is an impressive list of subjects; indeed I think it fair to say that the magazine under Podhoretz's editorship (Podhoretz succeeded Elliot Cohen as editor in 1960 and followed his lead) became one of the foremost voices of white left liberalism on the question of race during the 1960s.

Given the magazine's wide influence, there is good reason to inquire further into some of the attitudes and assumptions that shaped its outlook. One of the more consequential aspects of that outlook, only occasionally acknowledged by the magazine's contributors, was how little many liberals actually knew of African American experience, despite their eagerness to espouse detailed strategies and policies meant to solve the race problem. One finds scattered throughout left liberal treatments of race odd moments of self-reflection in which it is acknowledged that for all their sympathy for blacks, leftists and liberals after all had very little contact with them, were not particularly interested in greater contact, and were not well informed about life inside the black community. Characteristically, Norman Podhoretz put the problem more directly than anyone else in "My Negro Problems—and Ours":

> [Thus] everywhere we look today in the North, we find the curious phenomenon of white middle-class liberals with no previous personal experience of Negroes—people to whom Negroes have always been faceless in virtue rather than faceless in vice—discovering that their abstract commitment to the cause of Negro rights will not stand the test of a direct confrontation.

The consequences for race relations and for politics, according to Podhoretz, were disturbing indeed, for they included white flight, an exodus from the public schools, opposition to redistricting, a belief that "Negroes (for their own good, of course) are not perhaps pushing too hard," skepticism about blacks ever attaining equality, and a growing fear of and impatience with inner-city violence and black militancy, especially black nationalism.[34]

We can see that as early as 1963 Podhoretz was suggesting a direct correlation between white liberals' lack of familiarity with African Americans and the increasing reluctance of liberals to support far-reaching reform. Interestingly, the apparent determinism of these sociological changes was mitigated somewhat by Podhoretz's irony, by which he seemed to distinguish himself from the middle-class liberals he described. No doubt what enabled him to make this distinction was a firm commitment to change, but also a background that included numerous encounters with blacks. Nonetheless, as Podhoretz painfully admitted, these encounters were nearly always unpleasant and mutually misunderstood, the results having been that "like all whites," he too was "sick in his feelings" toward blacks. Podhoretz's honesty revealed the depths of the race problem in the United States: either whites were fleeing from direct contact with blacks and becoming cynical, or they were enduring troubled, painful contact. Yet without painful, intensive, and long-term efforts at improving the quality of the national discourse on race, and without far-reaching structural change, simple contact would do little to improve race relations.

Podhoretz's response to this dilemma is well known; by the late 1960s he and his magazine went the way of the white liberals he described, and later became a bulwark of neoconservatism. He too came to believe blacks were making unreasonable demands (such as affirmative action), and that increased militancy in the form of black nationalism or a desire for the equality of results was destroying any hope of equal opportunity for African Americans. This change in perspective, part of the general "white backlash" that began in the late 1960s and is to this day one of the reasons for the nation's tragic failure to address the legacy of slavery, was more than a rational reconsideration of policy in response to threatening developments within the black movement.[35] As Podhoretz himself pointed out, the sources of liberalism's increasingly moderate reforms were part of an ideological shift with distinct social and demographic causes—causes touched on by Podhoretz in the preceding quotation. White flight helped change liberal attitudes by distancing them further from the realities of African American life; consequently it helped to fuel the frustrations of young African Americans, thereby enhancing the appeal of cultural and political forms of nationalism, progressive and reactionary alike.

If Podhoretz astutely attributed changing white attitudes to greater social and ideological distance between blacks and white liberals, the problem he and his fellow white liberals failed to solve was how to develop an interracial and multicultural discourse that would build consensus for the need for a comprehensive program of structural reform. Such a discourse would have to perform several crucial tasks in the struggle for full equality. It would have to lay a foundation for mutual understanding between races by, above all, addressing whites' ignorance of African American experience. It would have to develop an appreciation for the profound historical and psychological intransigence of racism and thereby encourage a more durable commitment to far-reaching change among whites. It would have to go beyond the familiar constituency of educated white progressives to include working-class whites. Finally, it would have to sustain and enlarge itself to the point that white fear and black frustration begin to subside, allowing for further mutual efforts at healing the wounds of racism.

If *Commentary* failed to create a discourse on race equally shaped by whites and African Americans, what sort of discourse *did* it create? It was very much a social scientific, an academic, and a white discourse focused on public policy. Despite a handful of articles by critics exploring culture, literature, and the arts, most contributors were white political scientists, social psychologists, sociologists, or historians who were interested in offering analyses helpful to government policymakers.[36] Naturally these experts often referred to the work of their colleagues as they engaged in the important debates taking place in the academy and among policy makers. In so doing, however, they tended to neglect debates and debaters within the African American community, especially the nonacademic community, and thus their discourse did not adequately register the complexity of attitudes and values found in the black community. Clearly an important way of exploring black subjectivity would have been to give space to African American writers and African American literature. Once again it is Ellison who provided the important insight here. In a remark to Richard Wright regarding the "scientists" of the Communist Party, he once observed that Marx, Engels, Lenin, and Stalin won't help them unless they understand the theoretical world made flesh through literature. The sociology-oriented critics, he noted elsewhere, "rate literature far below politics and ideology" at their peril.[37] Like *Partisan Review* and *Politics* before it, *Commentary* neglected the very writers who might have opened the African American world to its contributors and its audience. Granted, the magazine described itself as a political and not a literary publication; nonetheless, it might have understood that the politics of race in the United States is not normal politics and never has been—it encompasses a whole way of life, or better, two ways of life in fateful embrace. More than any other enterprise, it is literature that possesses the power to illuminate whole ways of life.

Commentary's discourse was also one that stressed the pathologies within the black community. This is not to suggest that writers for *Commentary* necessarily endorsed an etiology based upon nihilism and dysfunction the way later neoconservative "culture of poverty" advocates did; rather, *Commentary* emphasized the degree to which racial prejudice and discrimination had damaged the black community—its family and cultural life, its schools, its self-esteem, the physical and mental health of its members. Articles such as "The Harlem Ghetto: Winter 1948," "The Vicious Circle of Frustration and Prejudice," "Portrait of the Inauthentic Negro: How Prejudice Distorts the Victim's Personality," and "The American Negro in Search of Identity" made explicit what was an underlying assumption of many pieces. A review of *Commentary*'s contributions regarding race confirms Walter Jackson's assessment of liberal approaches to race in the 1940s and 1950s, including Gunnar Myrdal's (although Myrdal was unique insofar as he also looked at the damaging effects of prejudice on *whites*, a subject ignored by most whites, including the New York intellectuals). "When social scientists did look at the effects of prejudice on blacks," observes Jackson,

> they focused on the damage that racial hostility inflicted on Afro-Americans. Myrdal . . . had drawn on a social science literature that charted the deleterious effects of segregation and discrimination on black institutions, family life, personality, and culture. While no serious student of Afro-American life could afford to ignore the manifold injuries of racism, Myrdal . . . had overemphasized social pathologies and had neglected the mechanisms, strategies, beliefs, symbols, and institutions that gave coherence and meaning to Afro-American communities and enabled blacks to resist oppression.[38]

Jackson goes on to suggest that many liberal social scientists, including some African Americans, offered images of a devastated black community in order to jolt whites out of their indifference and compel them to support reform legislation. But, of course, such a rhetorical strategy, though politically expedient, perpetuated a distorted view of the black community, and moreover would not have been necessary had progressive white intellectuals, in cooperation with their African American counterparts, been doing the job of insistently educating themselves and others about the realities of African American life. It is not surprising, therefore, that these images would come under increasing attack by African Americans during the 1960s, precisely because they seemed to provide the ideological basis for liberalism's insufficient program of reforms, as well as its technocratic disdain for movements and leaders arising out of the very slums assumed to lack any coherent, creative, or useful political and cultural impulses.

If we look closely at one of the more important articles published by *Commentary*, "Liberalism and the Negro: A Round-Table Discussion" (March 1964), we find these attitudes and perspectives quite prevalent. The article was an edited version of the important Town Hall discussion by the social scientist Gunnar Myrdal; two prominent New York intellectuals, sociologist Nathan Glazer and philosopher Sidney Hook; and the author James Baldwin, the only African American on the program. Norman Podhoretz served as moderator, and from the audience Charles Silberman, Kenneth Clark, Lionel Abel, Shlomo Katz, and William Phillips each posed a question.

What is striking—and indeed dismaying—about the symposium is just how well it holds up today. This speaks to the acumen of the participants, but more emphatically to the subsequent deterioration in the quality of the discussion of race. Despite having good reason to admire the insights of the participants, the reader who encounters the transcript today will likely find the experience dispiriting, for both the all-too-familiar litany of social problems and the sometimes sharp, tension-filled exchanges remind us that relatively little has changed. Thirty years later, the appalling relevance of nearly every portion of the discussion is chastening indeed. If anything of consequence *has* changed, it is that fewer policy makers and commentators today advocate the kind of far-reaching economic reforms that, for all of their differences, each of the *Commentary* panelists agreed were essential if the legal victories of the civil rights movement were to be expanded. Tragically, not only have comprehensive structural reforms not been implemented, but they are no longer taken seriously in much of the current discussion on race and inequity. One sorely misses in contemporary discussions the earlier assumption that fundamental economic change spearheaded by government intervention was a necessary, though not a sufficient requirement for securing genuine equality for African Americans and making serious inroads into the multifaceted legacy of slavery.

Despite the consensus in 1964 concerning the need for far-reaching economic reform—of the magnitude of the Marshall Plan according to one of the participants—other matters were dealt with less satisfactorily and more irritably. To its credit, the *Commentary* discussion ranged widely, touching on liberalism, quotas, economic assistance, jobs, housing education, ethics, ethnicity and race, pluralism, and racism. But what put the greatest strains on the participants were not policy disagreements as much as the less tangible issues of who speaks for the African American community, what role the African American community should play in the emancipatory process, what values and attitudes within the African American community should shape that process, and what historically determined differences distinguish the African American community from ethnic (and racial) groups also struggling for equality. These matters lent the proceedings a palpable air of tension only partially disguised by the perfunctory politeness displayed by the participants. Indeed the *Commentary* discussion was one of the first prominent public events at which the smoldering differences between African Americans and their white, often Jewish, allies began to surface. Within a few years the alliance between blacks and white liberals, and blacks and Jews, would be severely tested.[39]

It would be but a slight exaggeration to say that the *Commentary* discussion was haunted by the question of representation—that is, by the vexing question of who speaks for the black community, and how much efficacy his words should be granted. I deliberately present the matter this way in order to underscore the power relations that shaped the dialogue, for efficacy was not so much attained from the black community and thereby *possessed* by the chosen representative as it was *granted* by the influential arbiters of the established left-liberal discourse. The decision to convene a panel at which Baldwin was the sole African American was fully consistent with a long-established pattern among the New York intellectuals in which concern for African Americans did not translate into amplifying a representative range of African American voices.

James Baldwin had, of course, fulfilled the role of spokesperson for some time, but the *Commentary* roundtable was the first instance in which Baldwin's status was publicly questioned. Ironically, it was challenged not by Baldwin himself, who had long held private reservations about his role, but by his fellow panelists, who on several occasions questioned whether Baldwin's highly skeptical attitude concerning the values of mainstream America reflected the attitudes of the majority of African Americans. Thus, at an important juncture in the discussion, just after Baldwin had made the controversial statement "From my own point of view, my personal point of view, there is much in th[e] American pie that isn't worth eating," Norman Podhoretz was moved to question the representativeness of Baldwin's claim and his status as spokesman, despite Baldwin's repeated disclaimer:

> O.K., that's you, but what about your fellow Negroes? Wouldn't they be perfectly happy to eat everything in that pie? You said in *The Fire Next Time* that Negroes wonder whether they want to be integrated into a burning house, and at that point—as someone observed—you were speaking as an American social critic rather than as a Negro. Now the question is, is it the Negroes who feel that the house is burning and the pie is rotten, or is it only James Baldwin?[40]

A heated exchange between Baldwin and Glazer followed over whether most blacks distrusted the police; this in turn was followed by Sidney Hook who, after referring to an opinion of Baldwin that distressed him even more than Baldwin's observations about the police, returned to the issue of the novelist's fitness as spokesperson for the black community:

> I'd like to pursue the question that Mr. Podhoretz asked as to whether James Baldwin is speaking for himself or for most Negroes. If I read his most recent book properly, he actually believes that all Negroes hate white people; at least that's how some people have interpreted him. But that's not true for the Negroes I know, and it can't be true for Negroes who have married whites.[41]

How fitting, actually, that the only African American asked to contribute to the New York intellectuals' discussion of race should be attacked for not fulfilling the impossible task assigned. At a time when many frustrated African Americans were reconsidering the beliefs and tactics of the civil rights movement and turning to new leaders such as Malcolm X, white liberals were well aware that approaches they had long favored were losing their foothold. From liberal quarters the pressure on Baldwin somehow to hold the line against this new militancy was mounting; when he seemed to deviate from what was thought by white progressives to be the consensus within the African American community, they openly made their challenge. Perhaps it was only inevitable that they eventually became displeased with their spokesman, who was neither deaf to, nor unchanged by, the multitudinous and increasingly resounding voices of the black community.

Needless to say, many of these voices—provocatively, militantly, and sometimes angrily making claims for black power and revolution—were not voices the New Yorkers were eager or prepared to hear. More than anything else, *Commentary*'s roundtable discussion revealed the white panelists' obvious uncertainty about how the absent black community felt concerning the critical matters on the liberal agenda. The suddenly unreliable Baldwin, and unidentified "friends," were all the white participants seemed to have as resources. When Baldwin was asked to respond, typically it was "as a negro," and the expectation, or at least the hope, had been that there was no significant discrepancy between the writer's individual perspective and that of "his people." Surely Baldwin encouraged this identification by often prefacing his remarks with phrases such as "looking at it as a black American citizen." But Baldwin was less and less constrained by the expectations of his white colleagues. With the appearance of *The Fire Next Time* in 1963, sharp differences in perspective began to emerge. Baldwin's political analysis cum personal memoir included passages that were bluntly critical of white liberals. His famous encounter with Elijah Muhammad, for example, actually took place because of Baldwin's frustration with his white liberal colleagues. "In a way," recalled Baldwin,

> I owe the invitation to the incredible, abysmal, and really cowardly obtuseness of white liberals. Whether in private debate or in public, any attempt I made to explain how the Black Muslim movement came about, and how it has achieved such force, was met with a blankness that revealed the little connection that the liberals' attitudes have with their perceptions or their lives, or even their knowledge—revealed, in fact, that they could deal with the Negro as a symbol or a victim but had no sense of him as a man.[42]

Baldwin immediately went on to defend Malcolm X's controversial endorsement of violence as a means of self-defense by likening it to the Israelis' fight to regain their ancient homeland, to which he appended the provocative remark "I . . . certainly refuse to be put in the position of denying the truth of Malcolm's statements simply because I disagree with his conclusions, or in order to pacify the liberal conscience." As if these comments were not challenging enough to the New Yorkers, the circumstances surrounding their publication added to the tension. For as some readers may recall, it was Podhoretz who proposed the idea for the piece to Baldwin (Baldwin in turn, according to Podhoretz, suggested the idea for "My Negro Problem—and Ours"), for which Baldwin promised the results to *Commentary*, only to renege and publish with the *New Yorker* for more money.

What stands out rather dramatically in the roundtable discussion, then, is a crisis of representation manifested by the absence of a plurality of African American voices. The result was the continuing estrangement of whites from important elements of African American life. The roundtable discussion was the product of over two decades of left-liberal discourse in which precious few African Americans were given the opportunity to supply a sympathetic but largely uninformed group of readers with steady coverage of black life. Such coverage would certainly have been the basis for a more adequate awareness of race in American life, and would have provided the foundation of an education adequate to the tasks facing the nation. Unfortunately, despite their actual uncertainty about African American life, the New York intellectuals insisted on forging ahead with their policy prescriptions in their characteristically Olympian tone, which admitted of no lapse of confidence or authoritativeness. For several decades their style had worked effectively as an audacious assault on safe, parochial writing. But what had once been an expression of chutzpah in response to the dominant culture became an expression of hubris in response to a constituency possessing less power than they. As Jews, perhaps many of them heard African Americans more readily than most Americans, but as whites their capacity for listening remained limited.

What went unheard? In his well-known exchange with Irving Howe on the nature of African American writing, Ralph Ellison answered most eloquently of all. Responding to Howe's sympathetic but constricting understanding of African American writing and experience as necessarily a protest of the harsh historical conditions of oppression, the author of *Invisible Man* responded by pointing out how multifarious and irreducible African American experience actually was. "Being a Negro American," he wrote in "The World and the Jug,"

> has to do with a special perspective on the national ideals and the national conduct, and with a tragicomic attitude toward the universe. It has to do with special emotions evoked by the details of cities and countrysides, with forms of labor and forms of pleasure; with sex and with love, with food and with drink, with machines and with animals; with climates and with dwellings, with places of worship and places of entertainment; with garments and dreams and idioms of speech; with manners and customs, with religion and art, with life styles and hoping, and with that special sense of predicament and fate which gives direction and resonance to the Freedom Movement. It involves a rugged initiation into the mysteries and rites of color which make it possible for Negro Americans to suffer the injustice which race and color are used to excuse without losing sight of either the humanity of those who inflict the injustice or the motives, rational or irrational, out of which they act. It imposes the uneasy burden and occasional joy of a complex double vision, a fluid, ambivalent response to men and events which represents, at its finest, a profoundly civilized adjustment to the cost of being human in this modern world.[43]

These words bear repeating because few white Americans, even today, are familiar with the world Ellison describes. The absence of such familiarity, as I have been implicitly arguing, is due only partly to blatantly racist attitudes on the part of whites. It is also the outcome of several decades of white progressive sympathy from afar, which did not involve sustained contact with a representative range of black experience.[44] Leftists, liberals, and certainly neoconservatives failed to cultivate the ideological grounds for comprehensive socioeconomic transformation, which requires both a knowledge of the full extent of the troubles endured by the black community as well as its capacity for resilient and creative alternatives. And when we consider that with regard to the New York intellectuals the problem of distance from the black community was compounded by distance from the white working class, we can see that they were therefore ill-positioned to undertake a serious and progressive racial project with a national impact. We recall the Faustian bargain the New York intellectuals felt they were forced to make in the early 1940s, by which they sacrificed whatever connections they had to the working class for a degree of intellectual freedom. As I have shown, this paid dividends to be sure, but not necessarily where race was involved. Had they maintained some of their connections to the labor movement, not to speak of to the lives of ordinary white Americans—Richard Nixon's so-called silent majority—they might have contributed to a closer relationship between these two great constituencies, white and black, that constitute the bulk of the American working class. As it turned out, the New York intellectual formation provided the basis for the neoconservative and neoliberal retreat from race and the compromise with the politics of *ressentiment*, which poisoned race relations in the 1980s.[45] It must be added that neither were openly socialist expressions of the New York intellectuals such as Irving Howe's *Dissent* in a position to otherwise affect the white working class by fortifying its members against the ideological seductions of first the far right and the neoconservatives and, more recently, the neoliberals.

David Roediger is right—"the wages of whiteness" have long maintained the rift between the white and black working class. So have the salaries of whiteness. To heal that rift remains the central challenge for those who would take the next essential step toward solving America's enduring dilemma of race. For whites especially who would help build a radical democratic movement with majoritarian hopes, the need to interact with these broader constituencies regarding racial matters is inescapable. There is no more important service to be rendered in the struggle for social justice.

10

"Preserving Living Culture": The 1960s and Beyond

It is very easy to see the flaws in movements if you abolish sentimentality and you question everything that bothers you.

V. S. NAIPAUL

I

Sal Paradise rediscovers a nation in Jack Kerouac's *On the Road*, the signature novel of the Beats and inspiration to a generation that came of age in the 1960s. In all that has been written about the novel, no one has mentioned that Paradise "finds" America as a fugitive from the New York intellectual milieu.

A student and aspiring writer from New York City's western frontier just beyond the Lincoln Tunnel—Paterson to be exact—Paradise leaves home to become a disciple of the manic Dean Moriarty as the two live out the mythos of the Old West beyond the Passaic. Everyone knows what they ostensibly find there: jazz, drugs, alcohol, women—in short, spontaneity, authenticity, America. Together, these elements make up an ineffable transcendence called "IT," roughly the equivalent of YHWH to the God-seeking Dean. What remains much less well known, even today, is what Paradise was escaping. The familiar answer is that he and his friends were fleeing the stupefying banality of the 1950s with its odious political and sexual repressions. True enough, but they were also escaping the dominant modes of dissent from the decade's pervasive conformity as well, and they were doing so when this dissent was seeming less radical and more respectable than ever before. To the Beats, the New York intellectuals' dissidence was compromised by their insufficiently vocal opposition to McCarthyism and by their newfound respectability, which would shortly become manifest as influence in academia and in government policy-making circles as well. To the generation that came of age in the 1950s, the New York intellectuals and their world suddenly seemed conventional and suffocating. This is underscored in the opening pages of *On the Road*, in which Dean Moriarty's "criminality"—described as a "wild yea-saying overburst of American joy . . . Western, the west wind, an ode from the Plains, something new, long prophesied, long a-coming"—is pointedly contrasted with "the tedious intellectualness" of his friends in New York City, all of them still "sniffing at the *New Yorker*" and "in the negative, nightmare position of putting down society and giving their tired bookish or political or psychoanalytical reasons."[1] Paradise sets out not only to find America, but to put the world of the New York intellectuals behind him.

Eventually many of the New York intellectuals became aware that they were being abandoned by a new generation, and they made their displeasure known. Morris Dickstein correctly characterizes the rivalry between the counterculture and the old guard (which included the New York intellectuals) during the 1960s as part of an Oedipal psychodrama.[2] I might add that in some case, especially during the 1950s, the rivalry was not between father and son so much as between sons. It is revealing that the earliest and most censorious response to the Beats was made by a critic who was close to them in age. In 1958, when Kerouac was thirty-six and Ginsberg thirty-two, the twenty-eight-year-old Norman Podhoretz published his scathing review of *On the Road*, "The Know-Nothing Bohemians." And the prelapsarian Leslie Fiedler, the author of the ambivalent "The New Mutants" who would later embrace the Beats and the counterculture, was only five years older than Kerouac. Indeed, the sardonic, sometimes bitter tone of Podhoretz's piece suggests that more was at stake for these young critics than the Beats' alleged unreason, their disdain for authority, their spontaneous style, or their allegedly obscene obsessions with sex, drugs, and popular music. The implacable tone of the piece suggests its author sensed that nothing less than a cultural inheritance was at stake, and he was not about to relinquish his birthright over a cultural and critical heritage that clearly had been bequeathed to him and his young colleagues.

In retrospect we know that Podhoretz had reached one of those much-discussed crossroads where literature and politics meet. It may not have been a dark and bloody one—after all, what flowed in abundance was ink, not blood. Nonetheless, "The Know-Nothing Bohemians" appeared not only at the moment when the old left began to give way to the new, and older sensibilities clashed with newer ones, but it represented one of the first barrages in the battle between modernists and postmodernists. We recall that it was Harry Levin who in 1960 first applied Arnold Toynbee's term the "Post-Modern" to postwar literary and cultural developments, and other New Yorkers were the first to give the term currency. There is something to be learned, I believe, about the ensuing debates, which continue to this day, if we focus our attention on the antinomies that marked their highly polemical origins beginning in the late 1950s.

If "The Know-Nothing Beohemians" didn't set the tone for every New York intellectual responding to the Beats (we have seen that Susan Sontag, for instance, was sympathetic to what she called "the new sensibility"), its author shared a thoroughly contemptuous attitude with others such as Alfred Kazin, Daniel Bell, Lionel Trilling, and Irving Howe, all of whom attacked the counterculture and its Beat precursers during the 1960s. In "The Know-Nothing Bohemians" Podhoretz's witty assaults were unrelenting. There was disdain, and also a hint of resentment as well, in the initial observation that Kerouac's "photogenic countenance" was increasingly visible in "various mass-circulation magazines," he was being interviewed "earnestly" on television, he was part of a Greenwich Village "nightclub act," and his novel had become a best seller.[3] Podhoretz went on to explain the popularity of the "image of Bohemia," which he distinguished from the genuine bohemianism of early decades, as essentially a suburban phenomenon owing to the fascination of conformists with what they imagined as "the heroic road." The new bohemianism "is hostile to civilization," Podhoretz claimed, "it worships primitivism, instinct, energy, 'blood.' To the extent that it has intellectual interests at all, they run to mystical doctrines, irrationalist philosophies, and Left-wing Reichianism." Podhoretz maintained that the only art the Beats were interested in was jazz, whose "primitive vitality" they emulated in their "bop language," meant to undermine "coherent, rational discourse."[4] Podhoretz seemed entirely forgetful of the fact that the genuine bohemianism he valued also worshiped primitivism, and also tried to undermine civilized, rational discourse. To claim as he did that writers such as Hemingway and Pound (not to mention writers he omitted like Lawrence and Yeats) were part of a movement "created in the name of civilization: its ideals were intelligence, cultivation, spiritual refinement," was strangely to remake modernism in the image of Arnold, thereby ignoring the very elements of the movement that would later give him (and Trilling) pause.

Podhoretz continued by rebuking Kerouac for his exclusive concern for forms of spiritual renewal that efficiently relieved suburban angst yet were powerless to challenge the moral and political corruption of middle-class respectability. He derided Kerouac for his sentimental attachment to marginal individuals, whom he (Kerouac) insisted on defining exclusively as representatives of beatified groups such as bums, prostitutes, Mexicans, and African-Americans. Podhoretz in essence accused Kerouac of racism in his depiction of blacks, who were capable of nothing less than graceful, sublime, and tragic artistry in everything they did. In a discerning passage Podhoretz wrote,

> It will be news to the Negroes to learn that they are so happy and ecstatic; I doubt if a more idyllic picture of Negro life has been painted since certain Southern ideologues tried to convince the world that things were just as fine as fine could be for the slaves of the old plantation. Be that as it may, Kerouac's love for Negroes and other dark-skinned groups is tied up with his worship of primitivism, not with any radical social attitudes.[5]

Podhoretz further castigated Kerouac for assigning significant metaphysical meaning to promiscuous sex between men and women, of which, he was careful to point out, there was a surfeit. Not surprisingly, Podhoretz was silent about the ways women were often portrayed as hapless prey in these encounters, instead making the rather incredible claim that women were as sexually predatory as the men. To explain the conventional talk of the men during these encounters—about love and commitment and such—Podhoretz hypothesized that it disguised a deadly fear of sex and of not performing it well. According to the admittedly "square" onlooker Podhoretz, these sexual Olympians (and their rivals?) suffered a "sexual anxiety of enormous proportions."

But Podhoretz's contempt was most palpably felt when he attacked Kerouac for his anti-intellectualism. Here the stakes seemed to be highest and the cuts intended to go deepest. "The plain truth," Podhoretz began one of his most mordant passages, "is that the primitivism of the Beat Generation serves . . . as a cover for an anti-intellectualism so bitter that it makes the ordinary American's hatred of eggheads seem positively benign. Kerouac and his friends like to think of themselves as intellectuals . . . but this is only a form of newspeak." Kerouac's so-called "spontaneous bop prosody," maintained Podhoretz, "has such a limited vocabulary (Basic English is a verbal treasure-house by comparison) that you couldn't write a note to the milkman in it, much less a novel." In sum, concluded Podhoretz,

> Being for or against what the Beat Generation stands for has to do with denying that incoherence is superior to precision, that ignorance is superior to knowledge; that the exercise of mind and discrimination is a form of death. It has to do with fighting the notion that sordid acts of violence are justified so long as they are committed in the name of instinct. It even has to do with fighting the poisonous glorification of the adolescent in American pop culture. It has to do, in other words, with being for or against intelligence itself.[6]

Somewhat less contentious than Podhoretz was Leslie Fiedler in his 1965 essay "The New Mutants." Fiedler openly acknowledged his generation's hostility to the "new mutants," a label meant to communicate both distance from the new sensibility but also irony toward those who would demonize it. To the ambivalent Fiedler, the rising generation was full of "dropouts from history" who have utterly disavowed the past, especially the tradition of Humanism. They embraced the irrational in order to subvert even the most revolutionary apostles of reasons, Marx and Freud. Student protests made it clear to Fiedler that "the new irrationalists" would "prolongue adolescence to the grave," in order to avoid lives of "rationality, work, duty, vocation, maturity, [and] success." To the institutional setting of the university and the relative freedom it afforded, "they prefer the demonstration, the sit-in, the riot: the mindless unity of an impassioned crowd (with guitars beating out the rhythm in the background) whose immediate cause is felt rather than thought out, whose ultimate cause is itself."[7] In turning from the *polis* to *thiasos*, this "Dionysiac Pack" flouted the conventions of gender formation, particularly when it came to the making of masculinity and the conventions of traditional family morality. They did so by adopting "the esthetic of pornography" and producing a new literature whose archetypal text was Michael McClure's "Fuck Ode," and whose chief organ was Ed Sanders's journal *Fuck You*.

There was more—I haven't registered Fiedler's response to the cult of the orgasm or to the drug culture—but the basic tone, I think, is audible. Fiedler was unlike many of his fellow New York intellectuals insofar as he was willing to make a case for the new sensibility and "post-Modernist literature" as a revolt against the past: against reason, masculinity, whiteness—in short, against "humanity" as it had been traditionally defined. But in the end Fiedler's aroused powers of description combined with a peculiar reluctance to offer judgments to produce not acceptance or support but rather a strange sort of abdication before a rising tide he clearly had little in common with.[8] Fiedler's response is certainly not to be compared with that of Norman Mailer and his uncritical enthusiasm for the new sensibility, for Fiedler's explanations dwelled almost entirely on the antinomian and rejectionist aspect of the new sensibility rather than upon its affirmative impulses. But neither was his response comparable with those of his fellow middle-aged critics Philip Rahv and Stanley Edgar Hyman, which in fact he distinguished from his own by calling "irrelevant, incoherent, misleading, and [as] fundamentally scared as the most philistine responses."[9] But in the end he was closer to them than to Mailer, because his gestures toward understanding ended up being just that—gestures.

Both Fiedler and Podhoretz lacked the quality of imaginative identification that must be the staple of liberal criticism—what David Bromwich has called "a habit of sustained attention to things outside the familiar circuit of interests," and in a slightly different context Richard Rorty has described as "a matter of detailed description of what unfamiliar people are like and of redescription of what we ourselves are like."[10] Put another way, Podhoretz's and Fiedler's subjects were described with little of the imaginative range and sympathy one finds, for example, in superior ethnography—an appropriate field by which to judge their work because clearly for both critics the "mutants" constituted another culture entirely.

Another lapse in what Rorty has called "solidarity" can be found in Alfred Kazin's lyrical memoir *New York Jew* (1978), in which the distinguished critic told of confronting the new sensibility as a somewhat bewildered yet acerbic father stung by the condescension of the prodigal son: "The sons were attacking the fathers where we lived. They attacked our attachment to libraries; to books uselessly piled on more books; to our fondest belief that violence had nothing proper to do with sex and sex nothing to do with politics."[11] Kazin's son offered him little comfort or understanding, responding "scornfully" when a beleaguered Kazin visited him at college:

> Twenty years old; army fatigues, of course; the sweeping mustache; an impatient, loving, slightly pitying expression. The young were no longer waiting to be noticed. They were on stage! And as for "Amerika," it is not and never was what you timid old Social Democrats thought it was. [Richard] Hofstadter and I were what Blake called horses of instruction; the sons were tigers of wrath. Blake said that the tigers were wiser in their wrath than the horses of instruction. For we had "made it"; had turned into an Establishment; were compliant intellectuals who lent comfort to the same old system just by our being liberal, moderate—"wise."[12]

The patricidal generation seemed to have accomplished its goal when the estimable Hofstadter was reduced to pleading with his young audience at Columbia's 1968 commencement: "Some of us, even *us*, used to be radicals as well. Had we changed so much? Cannot you recognize that the America we suffered for in the thirties is yours?"[13]

Irving Howe in that same eventful year of 1968 had not pleaded. The latter portion of his well-known essay "The New York Intellectuals" was nothing less than a broadside directed at the counterculture and the new left. A few passages will suffice to evince the tone of Howe's obloquy:

> This generation matters, thus far, not so much for its leading figures and their meager accomplishments, but for the political-cultural style—what I shall call the "new sensibility"—it thrusts into absolute opposition against both the New York writers and other groups. It claims not to seek penetration into, or accommodation with, our cultural and academic institutions; it fancies the prospect of a harsh generational fight; and given the premise with which it begins—that very thing touched by the older writers reeks of betrayal—its claims and fancies have a sort of propriety. It proposes a revolution—I would call it a counterrevolution—in sensibility. Though linked to New Left politics, it goes beyond any politics, making itself felt, like a spreading blot of anti-intellectualism, in every area of intellectual life. (267–68)

> The new intellectual style, insofar as it approximates a politics, mixes sentiments of anarchism with apologies for authoritarianism; bubbling hopes for "participatory democracy" with manipulative elitism; unqualified populist majoritarianism with the reign of the cadres. (268)

> We are confronting, then, a new phase in our culture, which in motive and spring represents a wish to shake off the bleeding heritage of modernism and reinstate one of those periods of the collective naif which seem endemic to American experience. The new sensibility is impatient with ideas. It is impatient with literary structures of complexity and coherence. . . . It wants instead works of literature . . . that will be as absolute as the sun, as unarguable as orgasm, and as delicious as a lollipop. It schemes to throw off the weight of nuance and ambiguity, legacies of high consciousness and tired blood. It is weary of the habit of reflection, the making of distinctions, the squareness of dialectic, the tarnished gold of inherited wisdom. It cares nothing for the haunted memories of old Jews. It has no taste for the ethical nailbiting of those writers on the left who suffered defeat and could never again accept the narcotic of certainty. . . . It breathes contempt for rationality, impatience with mind, and a hostility to the artifices and decorums of high culture. It despises liberal values, liberal cautions, liberal virtues. It is bored with the past: for the past is a fink. (273)[14]

Lest my purpose in quoting these lengthy passages be misunderstood, I want to make it clear that I am not gainsaying the truth of some of Howe's criticisms—indeed, I shall attend to the incisiveness and contemporary relevance of his and other New Yorkers' critiques in due time. Again, it is the vituperative sound of the critique that concerns me, especially insofar as it suggests an illiberal defense of liberalism at a time when whatever embattled liberal tendencies still existed under the sway of the new sensibility would have been strengthened by a more balanced approach from critics. But once again, as they had during the period of the Popular Front, the New Yorkers considered themselves at the proverbial dark and bloody crossroads where literature and politics meet. In other words, they thought themselves in a war, and they opted for carpet bombs rather than delivering smart bombs. Given such a framework, little effort was expended upon differentiating among the so-called enemies—all were characterized by the worst excesses of the worst among them. Little moral imagination was evident here, if by this we mean the kind of solicitous attitude toward one's enemies that Lionel Trilling so admired in John Stuart Mill's response to the conservative Coleridge.[15] One might of course argue that none of the lights of the new sensibility was in any way as illuminating as was Coleridge—not Allen Ginsberg, Herbert Marcuse, C. Wright Mills, or Norman Brown. But we recall it was precisely Trilling's point that during the 1940s there was no formidable conservative critic or tradition outside of liberalism which could put it under pressure; thus liberalism would have to put itself under pressure. With his broad definition of liberalism—it would have included advocates of the new sensibility—Trilling's words should have encouraged a far more substantive and self-critical exchange with opponents during the 1950s and 1960s. Howe's definitive essay shows little of the restraint required of a suitably self-correcting endeavor. In the rare instance where he acknowledged some legitimate appeal or some value among his opponents, he stopped short and returned to the offensive. To quote his influential essay one last time,

> Perhaps because it is new, some of the new style has its charms—mainly along the margins of social life, in dress, music, and slang. In that it captures the yearnings of a younger generation, the new style has more than charm: a vibration of moral desire, a desire for goodness of heart. Still, we had better not deceive ourselves. Some of those shiny-cheeked darlings adorned with flowers and tokens of love can also be campus *enragés* screaming "Up Against the Wall, Motherfuckers, This is a Stickup" (a slogan than does not strike one as a notable improvement over "Workers of the World, Unite").[16]

It was only much later, in his memoir *A Margin of Hope* (1982), that Howe partially reassessed his attitude toward the younger generation during the 1960s by expressing regret at his and others' impatience. Referring to a 1962 visit to the *Dissent* editorial board by several leaders of Students for a Democratic Society (including Paul Potter, Paul Booth, and Tom Hayden), Howe at first rehearsed old complaints regarding the disdain for representative democracy and attraction to Fidel Castro. But then he confessed

> Wise after the event, I now see that we mishandled the meeting badly. Unable to contain an impatience with SDS susceptibility to charismatic dictators like Castro, several *Dissent* people, I among them, went off on long windy speeches. Might it not have been better to be a little more tactful, a little readier to engage in give-and-take rather than just to pronounce opinions?[17]

Yet even with this admission, Howe's next sentence suggested that a gentler response would not have made a difference:

> [A] clash was inevitable. We simply could not remain quiet about our deepest and costliest conviction: that if society still has any meaning, it must be set strictly apart from all dictatorships, whether by frigid Russians or hot Cubans. There is the value of tact, but also the value of candor. (292)

The fact that a decade and a half later a divided Howe still felt compelled to rationalize behavior that some part of him regretted speaks volumes about the profound difficulties of overcoming differences that have become hardened in polemical battle.

Howe's experiences with younger radicals were very similar to those of his political colleague Michael Harrington, a third-generation New York intellectual from the Midwest. Harrington, of course, became well known as the author of the influential *The Other America* (1962); less well known is the fact that, through years of combining intellectual labor with patient organizing work made possible by his evolved skills at reconciling differences, Harrington was in Maxine Phillips's apt words "the only socialist after 1960 whom many Americans could identify and trust."[18] I refer to Harrington's blessed skills as evolved because early in his career, when he was thirty-four to be exact, he tragically missed an opportunity to help bring the old and new lefts togethers. It is in fact quite likely that due to this failure he thereafter dedicated himself to honing his rare and valuable ecumenical talents.

The lost opportunity to which I refer occurred in the same year as the debacle at *Dissent* headquarters. As a leading member of the Student Activities Committee of SDS's parent organization the League for Industrial Democracy (LID), and as "the oldest young socialist alive" as he jokingly called himself, Harrington attended SDS's convention at the United Autoworkers camp in Port Huron, Michigan, where the fledgling organization debated Tom Hayden's draft of a declaration of principles, to be known subsequently as the Port Huron Statement. Harrington played a large role in the early discussions of the draft manifesto, making clear his opposition to what he regarded as a one-sided analysis in which the United States was portrayed as largely responsible for the cold war and the Soviet Union was insufficiently criticized. Harrington was opposed by several people during these early discussions, but subsequently his remarks bore fruit in the revised version of the original as some activists were won over to a position much closer to his. As Todd Gitlin has put it, "Harrington proved more persuasive than he knew."[19] Sadly, however, a frustrated Harrington was not present at these later workshops because he left Port Huron precipitously, according to James Miller because he had a previous speaking engagement.[20] At any rate, Harrington immediately reported the transgressions of his youthful, erstwhile allies to LID leaders.

The response was immediate, and harsh. Without bothering to read the final document produced by SDS, LID concluded that the young radicals had "disagreed with us on basic principles and adopted a popular front position."[21] Whereupon several of SDS's leaders were summoned to New York for a "hearing," presided over by Harrington himself, playing the role, according to SDSer Al Haber, of "Grand Inquisitor."[22] Measures were taken: SDS staffers were taken off salary, LID insisted on the right to approve all future SDS documents, all funds for SDS were cut off, the mailing list was seized, and the lock to the SDS office was changed.[23]

The terribly irony of this episode was that Harrington had recently written an article for *Dissent* on the need for caution and forbearance regarding the new left. According to Gitlin, Harrington warned democratic socialists

> not to come crashing down on the just-radicalized New Left—to remember that their romanticism toward Fidel, however "fuzzy," was not a finished ideology but a "complex feeling" that had to be "faced and changed" but could not be done so "from a lecture platform," through a recital of the old categories or by a magisterial act." "The persuasion must come," he wrote, "from someone who is actually involved in changing the status quo here, and from someone who has a sympathy for the genuine and good emotions which are just behind the bad theories."[24]

Gitlin has suggested that Harrington's unreconstructed anticommunism, forged under the ever vigilant eye of the "brilliant and bitter" Max Shachtman, proved too burdensome an ideological load even given Harrington's appeal for tolerance and flexibility in the early 1960s. Moreover, not unlike Podhoretz and other New York intellectuals, Harrington was perhaps threatened by young mavericks who seemed uninterested in the past, unfazed by the left's long journey through the wilderness of factionalism and sectarianism, and unabashed in their idealistic, youthful vigor. But however determinant these factors were in 1962, we have the example of Harrington's subsequent triumph over them to buttress our hopes today. By incorporating the conciliatory methods and attitudes that Harrington cultivated after this early failure, we may further honor him despite the belatedness of his rehabilitation.

My final example of a failure to imagine fully enough the subtleties of the Beats, the counterculture, and new left—which together I shall designate by the more convenient term postmodernism—was Lionel Trilling's unjustly neglected work of cultural criticism, *Sincerity and Authenticity* (1972). A lightly edited collection of his Charles Eliot Norton lectures delivered at Harvard in 1969–1970, *Sincerity and Authenticity* remains Trilling's most sustained, considered response to the upheavals of the 1960s. We recall, of course, that one of the bitterest of these upheavals occurred at Columbia University, where Trilling labored diligently but with only partial success to mediate between militant students and the administration. That these were extremely trying years for Trilling is apparent in several essays written at the time—"Mind in the Modern World" and "The Uncertain Future of the Humanistic Educational Ideal" immediately come to mind. Recently his travails have been elaborated in Diana Trilling's bracing memoir *The Beginning of the Journey*, where it is revealed, among other things, that at the height of the conflict at Columbia Lionel's photo appeared on posters above the appeal WANTED, DEAD OR ALIVE, FOR CRIMES AGAINST HUMANITY. But despite appeals from old colleagues such as Podhoretz who were already on the road toward neoconservatism, Trilling never repudiated his liberalism, and he refrained from making harsh, apocalyptic denunciations of the new left or the counterculture. When faced with what he frankly considered the cultural emergency of the 1960s, his articulated response was to try to moderate its worst effects by plunging into the past in search of the origins of the crisis. Whereas Podhoretz denounced "the counterrevolution of the 60's" as a "repudiation of the modernist canon and indeed of rationality itself," and called upon intellectuals openly to oppose this "resurgence of philistinism" and "simple cultural barbarism," Trilling's restrained answer to

Podhoretz's entreaties was modestly to suggest that in times of great cultural difficulties "[y]ou become historical-minded."[25]

Sincerity and Authenticity was a historical inquiry into the changing experience of the self as it pertained to the moral life during the past four hundred years in the West. Trilling's narrative began with the ideal of sincerity, which he dated to the early sixteenth century and which was best described by Polonius's advice to Laertes: "to thine own self be true / And it doth follow, as the night the day, / Thou canst not then be false to any man." Sincerity involved "a congruence between avowal and actual feelings" arising from a close relationship between the individual and society, although not so close that the individual was subordinated to external social pressures.[26] During the nineteenth century, however, the ideal of sincerity began to give way to that of authenticity, which severed honest self-expression (the expression of "the honest soul") from any necessary congruence with others in the social world, and instead linked it to self-realization, the goal of the Hegelian "disintegrated consciousness." Bereft of binding social codes and values, and compelled to grapple with unpredictable inward energies both destructive and creative, the authentic self of modernism held sway throughout the first half of the twentieth century.

Trilling credited both the ideals of sincerity and authenticity for having given rise to monumental cultural achievements. With the advent of a third ideal, however, a contemporary and debased form of authenticity that resulted in a condition of "weightlessness," it became increasingly apparent that all along Trilling had been narrating a story of decline. A word used by Nietzsche to describe the vertiginous feeling of loss and freedom in the wake of God's death, weightlessness occurred when the gravity of religious, moral, and social necessity diminished and the individual drifted, free to choose among roles and identities that kept him light and afloat. Whereas the modernist typically regarded this state of being with ambivalence, registering both fear and joy at its sublime potential and often tragic results, the postmodernist seemed altogether untroubled. In the name of play, difference, arbitrariness, and otherness, he offered descriptions of individual existence lacking a sense of wholeness and without a stake in a stable, unavoidably contingent relationship with the world. As Mark Krupnick has observed, in 1970 Trilling was unfamiliar with the key works of Lacan, Foucault, and Derrida, which would soon define the poststructuralist canon, but he did offer a critique of Norman O. Brown, Herbert Marcuse, and R. D. Laing that continues to have relevance for our understanding of poststructuralism. Brown and Laing, for example, were taken to task for the ease with which they dismissed not only narrow but also "democratic" forms of reason available in critical discourse, science, and in the very nature of language itself. This led them, according to Trilling, to valorize forms of madness that presumably negated the prison house that bourgeois society was said to be. Marcuse too, recommended a way to abolish what he took to be the bourgeoisie's repression of our vital instinctual energies, especially when it came to sexuality. While Marcuse did not celebrate the liberating potential of madness, his utopian strategy for establishing a new socialist state founded on the free play and regular gratification of instinctual desire left no room whatever, according to Trilling, for the enabling capacity for "renunciation and sublimation" that are indispensable to any form of political activism.

One needn't gainsay the cogency of much of what Trilling had to say about these Freudian revisionists (and, indirectly, their poststructuralist counterparts) to note as well the limits of both the tone and the scope of his critique. Mark Krupnick has remarked of Trilling's treatment of R. D. Laing that not since his polemics with the critics of the Popular Front had he been so passionate in his denunciations.[27] Perhaps so, but Trilling was no Podhoretz, certainly not in his written work, and even in his most pointed remarks he rarely, if ever, did his adversaries the discourtesy of caricaturing or mocking them. The final sentences of *Sincerity and Authenticity* registered as strong a tone of disapproval as there was to be found in that work, and yet Trilling always respected the constraints that the demands of plausibility placed upon him. "The falsities of an alienated social reality," concluded Trilling with regard to versions of postmodern antinomianism,

> are rejected in favor of an upward psychopathic mobility to the point of divinity, each one of us a Christ—but with none of the inconveniences of undertaking to intercede, of being a sacrifice, of reasoning with rabbis, of making sermons, of having disciples, of going to weddings and to funerals, of beginning something and at a certain point remarking that it is finished.[28]

Trilling is nothing if not concentrated and composed in expressing his adverse judgment here, a judgment that applies equally well to many contemporary examples of critical theory.

Nonetheless, it is difficult to avoid the judgment that, taken together with his critique of Brown, Marcuse, and Laing, Trilling's assessment of postmodernism displayed palpable lapses when it came to the liberal habits of mind he displayed elsewhere. We may contrast, for example, the degree of cognizance of the finest accomplishments of his adversaries shown in *Sincerity and Authenticity* with that shown in one of his most discerning essays, "The Sense of the Past," in which he articulated in brilliant fashion the achievements of the New Critics even as he incisively laid bare their limitations. In Trilling's response to the 1960s and its postmodern legacy, one searches in vain for the dialectical turn of mind that placed one's own position under pressure as well as that of one's adversary. Almost entirely absent from Trilling's critique was an appreciation for the complexity and multiplicity of his postmodern subject. Nowhere, for example, did he carefully distinguish between legitimate and illegitimate skepticism toward government, military, or corporate institutions. For the most part he averted his attention from defining events like the Vietnam War, counterinsurgency at home (COINTELPRO) and abroad (Chile), the civil rights and black power movements, and Watergate. He had little of substance to say concerning democratic, egalitarian, anti-acquisitive, anti-imperialist, or anti-authoritarian impulses within dissenting movements. And save for a lone footnote in *Sincerity and Authenticity* acknowledging the newly emerging black history, Trilling registered scant awareness of other emerging subjects and scholarly fields with roots in the movements for social change of the 1960s: feminist criticism, women's studies, gay and lesbian studies, the new social history, environmental studies, minority and ethnic studies, and cultural studies.

Elsewhere as well there is evidence of an immoderation, or better, a lack of proportion, in Trilling's assessment of postmodernism. In a 1974 symposium on *Sincerity and Authenticity* organized by *Salmagundi*, Trilling responded to a query about "authentic" authenticity, as it were, and its deteriorated form that had allegedly become prevalent only recently. Perhaps because of the format of this particular public setting, which required a less nuanced response than would a formal address or an essay, Trilling's description of how the counterculture had become a respectable part of the postmodern ethos seems reductive, even banal. Contrasting the heroic modernist who had no alternative but "to take things and tear them apart and put them together in a new way" to 1960s-style rebels, Trilling contended that meaningful aggression "against bourgeois respectability and philistinism . . . have become clichés."

> Authenticity and sincerity become confused with a kind of dress, with faded denims, perhaps, with a "look" that's down to earth. As soon as you begin attaching moral meanings to a fashion, to a symbol of that kind and as soon as those meanings become accepted, become the mark of a certain kind of person who asks for a certain kind of respect, then I feel no, this is of no use. . . . [W]hen a revolt becomes an affection, a cliché, I must withdraw from it. I think this is the ground on which my complication, my contradiction, exists.[29]

By any account this must be considered a surprising statement, not because it failed to capture an aspect of the postmodern ethos—after all, as the author of "Sherwood Anderson" well knew, *every* avant-garde or adversary movement in culture had its share of pretenders and fashions. It failed because it did nothing more than capture an aspect of a complex cultural development.

II

An understanding of the limits to the scope of the New York intellectuals, of the bounds of their curiosity and sympathy, is essential today, especially in light of the spreading desire among academics to end their relative isolation and address a wider, public audience. Sometimes this call for the reemergence of the public intellectual has included praise, and a bit of nostalgia, for the New York intellectuals, the last group of critics to reach a general, educated, and middle-class audience. But as I have taken pains to show throughout this book, the New York intellectuals' public was really only a small part of the actual public, important though it was. Their experiences with the broad public demonstrates that the more difficult task is not achieving the crossover from an academic to a nonacademic audience, but rather the securing of a public audience that is drawn from the *whole* public—that is made up, in other words, of readers representing all political and ideological tendencies, all ethnic and racial groups, and all classes. This is the challenge facing public-minded academics today.

My argument thus far has been that many of the New York intellectuals critiqued the counterculture and new left in an illiberal manner. In dwelling upon their animadversions, on examples where patience, comprehensiveness, or generosity were lacking, I have tried to judge them by the exacting standards that they themselves embraced, given their commitment to the liberal cast of mind. I have *not* judged them by the standards now reigning in American politics and in certain academic precincts. These standards are the standards of partisanship and ideological expediency, by which success is overwhelmingly measured by the ability of a politician or critic to establish an impervious line of demarcation between his own political or critical positions and those of his enemies, combined with a willingness to enter into uncritical alliances with one's ideological friends. This has led to no small amount of posturing and Manichaeanism among critics across the spectrum, not to mention a renewed respect for the skills of the caricaturist. Any sentient survey of Dinesh D'Souza's *Illiberal Education* or Allan Bloom's *Closing of the American Mind*, for example, will reveal among other things a striking deficiency of poise, of balance, and of empathy. Neither does one find a repository of liberal values when one conducts a similar survey among poststructuralist critics on the left or among the New Yorkers' detractors on the left.[30] Certainly by these contemporary standards of partisanship, the New York intellectuals hardly rank below the average. In their worst moments they responded in kind to their adversaries, nothing more, nothing less. Naturally our judgment of them should reflect the fact that, as respected members of the nation's cultural elite for three decades, they bear some of the responsibility for what remains factional and superficial in our cultural discourse. But in no way should this conclusion prevent us from judging their less powerful adversaries on the left with comparable rigor. Not surprisingly, this is a task the contemporary left—both academic and nonacademic—has been reluctant to undertake.

Throughout this book I have claimed both tacitly and openly that the New York intellectual critique of progressive thought is directly relevant to literary, cultural, and political discussions today. Thus, for example, I have attempted to identify the fundamental rightness of *Partisan Review*'s break with doctrinaire criticism as early as the mid-1930s. This was a criticism with characteristics still to be found in current academic left criticism: an understanding of culture largely as a manifestation of power and the needs of conflicting groups; a tendency to equate suffering with merit; an undifferentiated, largely rejectionist relation to tradition; an overemphasis of ideology in literature and art; a neglect of human agency in favor of social determinism; a skepticism of aberrant, "unincorporated" thinking; and a disdain for the "frivolities" of craft, aesthetics, and standards of value. I want to suggest that the New York intellectuals, for all their flaws, offer us alternatives to these persistent errors. These alternatives include an understanding of culture's relative autonomy; an unsentimental (albeit often inattentive) view of oppressed groups; a constitutive relation to the past by which tradition can be used against authority; a belief in the distinction between ideological (systematic and coercive) and nonideological thought both inside and outside of literature; an appreciation for acts of will in light of the pressures and limits imposed by social reality; a robust response to craft and the need for making aesthetic judgments; and a respect for unaffiliated intelligence.

In addition, when one considers the New York intellectuals' response to postmodernism, a response whose limitations I have taken pains to identify, one nonetheless finds related elements that are of value. Here I am thinking of the critique of certain anti-intellectual and antiacademic tendencies during the 1960s; of the profound skepticism of rational thought to the point of negation; of the rejection of humanism, liberalism, and consensus building in favor of mechanical materialism and power politics; of a ruthlessness toward the past and tradition; of the complete subsumption of aesthetic value by power and ideology; and of political correctness—what Trilling referred to as "forms of assent which do not involve actual credence."

Recently a number of liberal and leftist scholars, in criticizing conservative and neoconservative views, have had the temerity also to critique key aspects of the left's discourse. In challenging the academic left's unwritten rule of avoiding self-criticism when battling the more powerful forces of the right, critics and historians such as Joyce Appleby, Lynn Hunt, Margaret Jacob, David Bromwich, Frederick Crews, Morris Dickstein, John Diggins, Russell Jacoby, Richard Rorty, and Cornel West have taken on some of the same habits of mind that the New York intellectuals challenged, even to the point of espousing some the same values they espoused.[31] I should like to add to this emerging reassessment by looking closely at several recently published books that seem to me to be typical of the more politicized products of the academic left today. In many ways these efforts are part of the new tide of literary histories that incorporate insights from poststructuralist theory, even as they distance themselves from theoretical modes that have drawn attention away from history and historical scholarship, from traditions of dissent, from considerations of class, and from connections with the workaday world. Yet in some cases these new literary histories perpetuate questionable habits of mind that are long-standing and continue to beset the American left.

III

Cary Nelson's *Repression and Recovery: Modern American Poetry and the Politics of Cultural Memory, 1910–1945* (1989) undertakes to draw attention to several dozen poets and hundreds of poems, most of them exhibiting a radical sensibility if not explicit radical politics (only brief mention is made of the conservative Southern Agrarians and experimentalists like Eugene Jolas who were only peripherally political). Beyond doing justice to neglected individuals, *Repression and Recovery* recommends an entire subculture that flourished in magazines often too little even for Frederick Hoffman's standard account, *The Little Magazine* (1946). This was a vibrant, dynamic group of contending realists, modernists, and realist-modernists whose identity was effaced through the combined pressure of McCarthyism, professionalization, and the New Criticism. (The reader may recall, for example, Cleanth Brooks's use of proletarian literature as his foil in his widely influential *Modern Poetry and the Tradition* [1939]). Nelson describes the resulting critical practices as follows:

> The lives and experiences of disenfranchised populations, insofar as they intersect with poetry, [have been] rejected as inferior or irrelevant to the best of cultural history. English professors should be pressed to explain why, for example, the poetry sung by striking coal miners in the 1920s is so much less important than the appearance of *The Waste Land* in *The Dial* in 1922.[32]

Nelson suggests that extending the canon to include such texts will contribute to a diverse and democratic society in need of an "oppositional language, . . . socially critical perspectives of anger and idealization, . . . [and] subject positions we can take up consciously."[33] Scholars and critics further such goals through acts of countermemory that erode the narrow interest, limited tolerance, and constraining power that the established canon is said to represent.

Nelson's survey takes in a great deal. Its most significant achievement is to have recovered considerable numbers of neglected American poets, songwriters, graphic artists, and magazines through impressive bibliographic and archival work. There are the expected names: Archibald MacLeish, Edna St. Vincent Millay, Langston Hughes, Vachel Lindsay, Mike Gold, Tillie Olsen, Louis Zukofsky; and the expected magazines: *Liberator*, *Masses*, *Partisan Review*, *New Masses*, *Anvil*. But there are many other names that will be wholly unfamiliar to all but a few readers, in some cases unjustifiably: H. H. Lewis, Baroness Else von Freytag-Loringhoven, Marius De Zayas, Sol Funaroff, Angelina Weld Grimke, Abraham Lincoln Gillespie, Covington Hall, Lola Ridge, John Beecher, Joy Davidman, *The Comrade*, *Contact*, *Fire!!*, *Fantasy*, *Dynamo*, *291*, to name a few. There are omissions too: Stanley Burnshaw (whose poems aren't mentioned), Eve Merriam, Josephine Johnson, Aunt Molly Jackson, *Symposium*, *Directions*, *Harlem Quarterly*, *Mainstream*, *Jewish Life*. But no matter—Nelson's study is meant to be provocative and prolegomenous, and its effect has been to put others on their trail. Surely these scholars are appreciative of the seventy-odd pages of notes that contain basic biographical and bibliographical information. The importance of the notes is aptly and subtly underscored by the decision to place the last of the text's many wonderful graphics in their midst.

What is supplied in the notes and text is an alternative history of twentieth-century poetry, in which "a fluid field of both fulfilled and unfulfilled possibilities, a continuing site of unresolved struggle and rich discursive stimulus"[34] replaces the traditional emphasis on the poetic monuments of Frost, Eliot, Stevens, et al. In addition, the antinomies that inform so much thinking about twentieth-century poetry—traditional versus experimental, genteel versus modern, poet versus public—break down once we immerse ourselves in the heterogeneous world of poetic production that was centered around the little magazines and was reflective of larger causes and constituencies. Nelson reconstructs, as well as anyone has, the diverse American cultural environment that produced Zukofsky's oblique verse and Joe Hill's socialist songs, Sterling Brown's marvelous dialect poems and the stunning typography of Marius De Zayas, Anna Louis Strong's free verse narrative and Gladys Hynes's illuminated letters for Pound's *Draft of the Cantos XVII–XXVII.*

More than a survey and revision, *Repression and Recovery* is also a sustained meditation on the problems of history- and canon-making. In a chapterless text, and chronological only in the loosest sense, Nelson patches rather than weaves together history and theory. The form of the narrative thus reinforces the author's fundamental commitment to the close but vexed connections between empirical fact and the historian's shaping, contingent narrative, between literary valuation and the constellation of social relations that produces canons. Nelson's larger argument that our histories and canons are made and not found will by now be familiar—indeed the force and frequency with which he makes the point is a bit surprising given the critical trends of the past decade. The theoretical framework is advanced through a cluster of arguments attesting to the interestedness if not partisanship of all versions of literature and its past: "Nothing we can say or think about a poem is free of social construction. . . . There is no perceptible, unmediated, unconstructed zero degree of literary materiality that serves as a consensual basis for interpretation." "Literary history is never an innocent process of recovery. We recover what we are culturally and psychologically prepared to recover and what we 'recover' we necessarily rewrite." "Even if a text remains in the foreground of the literary history we continue to retell and revise, it will be reinterpreted and embedded in new cultural contexts. . . . [T]ested against any demonstrable notion of identity, it will have become another text." "No purely literary categories exist; there are merely different ways of constructing the textual and social domain of literariness."[35]

In these quotations we detect one persistent problem with Nelson's study: a tendency to debunk manifestly untenable positions. Nelson spends an inordinate amount of time exploding simplistic and unattributed views rather than constructing useful ones by refining compelling but imperfect critical work. It hardly seems necessary to rebut the by-now discredited notions that literature is free of social construction, that literary history is innocent, that the internal attributes of the text are wholly responsible for that text's reception and value, or that a purely literary category exists. By taking his bearings from such views, Nelson's own views merely mirror their distortions in reverse. In the case of the last quotation given, for example, if we omit the word "purely," the statement loses much of its force; for the discovery that literary categories are historically and discursively constructed hardly obliterates them. In fact, a claim that literary categories collapse when confronted with the circumstances of their existence works in favor of the New Critical tenet that literature is best served by being protected from history, which of course the New Critics eventually quarantined. To be fair to Nelson, I should point out that in the discussion from which this quotation is taken he refuses to regard the canon as an extraliterary category, as some do who see it exclusively as a category signifying power. Nonetheless it must be said that he never clearly establishes what is literary about the canon.

Elsewhere Nelson is conspicuously reticent about the related matters of the literary text and literary value. As for the text, Nelson credits it with no determinate existence, owing to the profound changes in interpretation that occur over the course of time. However, since interpretive dissonance occurs synchronically as well as diachronically, Nelson cannot remain consistent unless he is willing to concede that contending interpreters read different texts. But then to speak of different interpretations of different texts is a rather unremarkable corrective to prevailing wisdom. Nelson's unwillingness to regard the single text differentially—or to acknowledge that internal textual constraints, themselves socially constituted, set limits and exert pressures on the responses of readers—plays havoc with the ensemble of internal textual relations. In his model the text is not substantial or autonomous enough to absorb and transform extratextual material. When confronted with external reality, the text succumbs; it cannot participate in shaping so much as the outlines of its own reception and revision.

By contrast, I would contend that, when pushed, a dynamic text pushes back with a force depending at least partly on the destabilizing content of its technical proficiency and formal innovation (that is, its means of production). That some texts are pushier than others in this respect seems to trouble and elude Nelson, as it does many contemporary critics promoting an extension or rejection of the canon on egalitarian grounds. Thus Nelson wishes to make room in an expanded canon for some very resistible poems, one among them being Lucia Trent's satire on academia entitled "Parade the Narrow Turrets": "Thumb over your well-worn classics with clammy and accurate / eyes, / Teach Freshmen to scan Homer and Horace and look wise . . . / And at official dinners kowtow to fat trustees."[36]

Of course, Nelson judges poems, and does so often if not explicitly. The poems presented here are meant to be valued, and Nelson's criteria are those of historical interest and social knowledge as they correspond to contemporary needs for egalitarian values. Moreover, Nelson makes the ability to appreciate these poems depend on the reader's familiarity with the "vital cultural work" they did within the environment in which they were produced. But there are problems with these criteria. Making an appropriately qualified hermeneutical plunge a necessary requirement for responsive reading may indeed encourage a more sympathetic disposition toward a given set of poems, but it will not obviate the need to determine which poems are more and which are less successful. Nelson may lament the fact that, for mainstream critics, the songs of coal miners are "*so much* less important" (emphasis added) than *The Waste Land*, but his very formulation suggests that nonetheless these songs *are* less important. If they are less important, why is this? Unfortunately, no answer is provided. A major difficulty here is Nelson's sporadic attention to matters of poetic composition: although he is concerned to enumerate some of the formal innovations found in these poems, such innovations are dealt with epiphenomenally. On several occasions he condemns poems for their mawkishness, forced rhymes, histrionic poses, and genteel notions, but for the most part he gives only passing notice to the relative success of various incorporations of experimental techniques. Typically a writer is represented by several lines of verse, often not the poet's best work, accompanied by a paragraph of text. In some instances no poetry at all is quoted. Certainly Nelson is constrained by the inclusive aims of his project, but he might have overcome these constraints by including more samples of poetry, and a chapter or two devoted to individual writers.

Nelson's failure to address the complexities of the value question emerges plainly in his discussion of the relation of Pound's fascism, sexism, and anti-Semitism to his poetry. Although he argues persuasively against separating Pound's personality and politics from *The Cantos*, and also against the view that *The Cantos* has a "consistent or finally determinable" politics,[37] Nelson nevertheless breaks off his discussion at precisely the moment when he confronts the central difficulty. As he himself puts it, "Of what value is an elegant literary sensibility that is contaminated by exhortations to genocide and murder? . . . Once we recognize how Pound's economic and social theories and his gestures of historical recovery are tied up in these prejudices, what is left of the 'radiant gist' that one of Pound's major critics finds at the core of his enterprise?"[38] Good questions! But Nelson falters here, for his mode of asking is suspiciously rhetorical, as if self-evident answers eliminate the need to investigate whether objectionable elements of Pound's ideological content are mitigated, neutralized, or even overwhelmed by countercurrents arising out of the text's multiple array of forces and forms. Once we acknowledge that *The Cantos* neither exudes an "elegant literary sensibility" nor constitutes a political document, it remains for us to examine how such a text containing old prejudices and pernicious beliefs can at the same time be a rude assault on moribund forms, a source of innovative and productive modes of representation, and an expression of nuanced thoughts and feelings. I wish Nelson had seriously entertained the possibility that certain textual forces at play in *The Cantos* might be inimical to aspects of the text's manifest content—for example, that the text's subtle self-reflexiveness, its sometimes playful ambience and verbal horseplay, its self-limiting, makeshift quality, and its many examples of shifting, empathetic points of view (to name only a few sources of countervailing exertion) might actually work against Pound's brazen and sometimes obnoxious pronouncements.

I want to test this hypothesis, but first a brief look at how Pound's critics have faced the problem of his politics. Despite Nelson's claim that American literary criticism has systematically avoided history and politics, these matters have in fact occasioned pertinent discussions which he fails to mention, namely, those surrounding the 1949 award of the Bollingen Prize in Poetry to Pound for *The Pisan Cantos*. Obliged to declare publicly whether they supported awarding the prize to Pound or not, critics such as W. H. Auden, Clement Greenberg, Irving Howe, George Orwell, Karl Schapiro, and Allen Tate were not only divided as a group but tended to divide the poems into their aesthetic and their ideological components. On the whole supporters and opponents alike offered dualistic analyses: supporters called attention to the poems' surpassing technique and ignored their propositional content, whereas opponents condemned their opinions and overlooked their formal accomplishments.

Contemporary critics may not find these solutions satisfactory in retrospect, but they provide a better frame of reference for discussing literary politics than Nelson's history of criticism's systematic depoliticization. These debates show that the difficulty is rather more intractable than Nelson suggests because they reveal the weaknesses of politicized as well as depoliticized criticism. Ironically, political criticism in the United States, having been long suppressed by one kind of politics of cultural memory, now seems to elude another. The bane of the dominant modes of postwar American criticism may very well have been their obsessive exclusion of history and politics, but the bane of dominant and subordinate modes alike has been their shared inability to combine political and aesthetic intelligence.

A good deal of the more recent Pound criticism, some of which is cited approvingly by Nelson, has attempted to overcome the form/content dichotomy by insisting on a full examination of Pound's most odious beliefs as constitutive elements of language, form, and technique, rather than as elements superimposed from without. Paul Smith, for example, argues in *Pound Revised* (1983) that beginning with his Imagist phase Pound's language devolves into something "intrinsically totalitarian" due to his subordination of the signifier to the truth-bestowing signified. For Smith, Pound's "revolutionary" poetic line turns out to be oppressive because its small syntactic units and precise images effectively block speculation and force the reader to contemplate contained factual meanings. In "Ezra Pound and the 'Economy' of Anti-Semitism" Andrew Parker has made a related argument, in which he claims that Pound's hatred of Jews is expressed as an animus against whatever is excessive, fugitive, and unassimilable in his own language. Thus Pound's signal resistance to rhetoric and metaphor, themselves analogues of usury and nomadic Jews, represents an economy of language that is hostile both to Jews and to writing itself.[39] Finally, Robert Casillo's *Genealogy of Demons: Anti-Semitism, Fascism, and the Myths of Ezra Pound*, a study Nelson cites as "undertak[ing] a difficult, probably unresolvable, but altogether necessary project," promises "to weigh Pound the poet and Pound the anti-Semite and fascist in a single balance."[40] But although it performs the valuable task of exhaustively identifying every aspect of Pound's reactionary politics, it postpones the task of balancing that it promises. Only in the final paragraphs of his book does Casillo mention the possibility that in spite of Pound's reactionary political aims, *The Cantos* might nonetheless offer, as Fredric Jameson has put it, "renewed access to some essential source of life," perhaps in part through "the fascination of their verbal texture."[41] Coming at the close of his study, however, this view is unfortunately not elaborated, and Casillo's book ends up doing more inveighing than weighing.

By identifying various forms of essentialism in Pound, these approaches manage to avoid the traditional form/content binarism I have mentioned. But they tend to import into their analyses a certain reductiveness that was confined in earlier criticism to the political analysis of literary content. Whereas the traditional weakness of politicized criticism has been its eagerness to distill a coherent political perspective out of a welter of partially discordant beliefs and attitudes, current tendencies attribute a specific political identity to a given formal device or complex of formal strategies. The form/content binarism is thus only apparently overriden;

in actuality forms are wholly overcome and become merely the contaminated carriers of infectious ideologies.

Another approach suggests that, although formal devices are enmeshed with political content, they do not themselves possess (or otherwise derive) determinate political meanings that transcend their specific use within a text. As I have argued, the political valences of literary form depend on the interplay between embedded formal and ideological configurations on the one hand, and groups of readers responding both to these textual pressures and the pressures of an ensemble of semantic and social relations on the other. In other words, no form is doomed to communicate a single political point of view. This is true even within the confines of a single text, which invariably puts form to different uses. Everything depends on how a form or device is deployed, and how the public at a given time receives it. This is especially true for an eight-hundred-page collection like *The Cantos*.

To illustrate, let us compare two passages from Canto 74, the first being a particularly inflammatory passage:

> doubtless conditioned by the spawn of the gt. Meyer Anselm
> That old H. had heard from the ass eared militarist in Byzantium:
> "Why stop?" "To begin again when we are stronger."
> and young H/ the tip from the augean stables in Paris
>
> ....
>
> thus conditioning.
> Meyer Anselm, a rrromance, yes, yes certainly
> But more fool you if you fall for it two centuries later
> . . . [Pound's ellipsis]
> from their seats the blond bastards, and cast 'em.
> the yidd is a stimulant, and the goyim are cattle
> in gt/proportion and go to saleable slaughter
> with the maximum of docility.[42]

Here we are told that Mayer Amschel Rothschild (1743–1812), the founder of the house of Rothschild and progenitor of modern Jewish finance, is preposterously romanticized by the very Christians who continue to be exploited by his successors. The passage contains key elements of Pound's virulent anti-Semitism as sketched out by Casillo: contempt for Jews as a group as opposed to contempt for some Jews, the belief that Jews as a group have systematically oppressed non-Jews through usury and their control of financial networks and a political agenda, only hinted at here, in which non-Jews end their passivity by attacking at least some Jews and effacing the Jewish influence.

If it is fair to say that little value resides in such beliefs, is there perhaps value to be found in certain formal elements that work against these beliefs? To frame the question this way already eliminates the aestheticist response, for which there are only poetic meanings and thus no real political views to be worked against. One politically informed response to this question may be to weigh the effect of certain formal devices against the more obvious political effects. Thus one might point to

Pound's heavy-stressed, pentameter-killing lines ("but more fool you if you fall for it two centuries later"); his abundant, shifting historical allusions (Rothschild, the two Henry Morgenthaus, father and son); his concrete, economical language ("the yidd is a stimulant and the goyim are cattle"); his unique combination of high and low diction ("the tip from the augean stables in Paris . . . blond bastards . . . saleable slaughter"); or his self-mocking rhetoric ("with the maximum of docility") as examples of revolutionary formal elements that subvert the rigid, reactionary sentiments they express. Such elements would be said to represent a new and diverse range of expressive modes that are incommensurate with a politics bent on destroying novelty and diversity.

Another kind of politically informed response, quite common today, asks us to shift our attention from the realm of conscious poetic strategy to that of subtext and linguistic play. Here we find that the text yields unintended examples of countervailing influence. Thus, briefly, inasmuch as Pound's poem is meant precisely to arouse and "stimulate" Christians to defend themselves against Jews, the figure of Jew as "stimulant" (instead of the more logical "tranquilizer") and Christians as "cattle" suggest Pound's repressed admiration for and even identification with the worldly Jews. The text's political unconscious works to unravel the truth of Pound's manifest pronouncements.

Still another kind of politically informed response would acknowledge that formal strategies and/or linguistic play complicate the politics of the passage and may even provide valuable insight into its subtle inconsistencies. But rather than assign value to the passage as a whole, this response emphasizes a manifest political content so overwhelmingly repugnant that poetic elements which normally have an illuminating effect function only to enhance the subtlety, sophistication, prestige, and ultimately the appeal of anti-Semitism. In this view, only when anti-Semitism becomes a historical curiosity with no power to cause harm will intelligent readers who are also informed citizens be willing to look beyond by-then-moribund beliefs and give priority to subtler textual dynamics.

I believe this last response is ultimately the most responsive to aesthetic and social contingencies, so long as we apply it only to Pound's most egregious passages. Once we apply it to *The Cantos* in general, however, the argument shades into reductiveness and determinism, echoing earlier doctrinaire leftist treatments of Pound (and modernism as a whole), which Nelson unfortunately ignores. Thus, even when we take Robert Casillo's findings into account, we must be prepared to grant that many passages in *The Cantos* remain connected to Pound's most pernicious beliefs in a manner slender and remote enough to warrant considered appreciation.

Which brings me to my second illustration from Canto 74, a passage whose highly personal subject matter and supreme artistry not only eclipse Pound's politics (which they, of course, continue to interact with), but make the view that the passage is dangerous by virtue of its excellence seem thuggish. Pound the political prisoner writes from his outdoor cage at the American Disciplinary Training Center north of Pisa:

Butterflies, mint and Lesbia's sparrows,
the voiceless with bumm drum and banners,
and the ideogram of the guard roosts
el triste pensier si volge
ad Ussel. . . .
I have forgotten which city
But the caverns are less enchanting to the unskilled explorer
than the Urochs as shown on the postals,
we will see those old roads again, question
possibly
but nothing appear much less likely,
Mme Pujol,
and there was a smell of mint under the tent flaps
especially after the rain
and a white ox on the road toward Pisa
as if facing the tower,
dark sheep in the drill field and on wet days were clouds
in the mountain as if under the guard roosts.
A lizard upheld me
the wild birds wd not eat the white bread.[43]

If it is worthwhile for politicized critics to explain what they mean when they say a text like *The Cantos* offers "renewed access to some essential source of life"—and given the needs of their purported constituencies for pleasure and for new modes of cultural expression, it would seem to be very worthwhile—they would find in this passage a particularly abundant source of both. Among the manifold sources of pleasure, one might mention the rhythmic and alliterative precision of the first two juxtaposed lines; the magnificently terse "Mme Pujol," coming as it does at the end of a silent line; the simple, poignant declarative "A lizard upheld me"; the extraordinary shift from "wild birds" to "white bread," made without invidious comparison; and the prevailing elegiac tone, which, lacking self-pity or bitterness, preserves a moment of intense self-reflection in the midst of ideological battle. This latter accomplishment is admittedly rare in Pound, but insofar as it is rare in many politicized writers it ought to be valued highly. As to the uses of Pound's techniques, one need not speculate about their value—Pound's toolbox has been ransacked by a host of dissident poets, and it is no small irony that indebted writers as diverse as Louis Zukofsky, George Oppen, William Carlos Williams, Robert Lowell, and Allan Ginsberg share political agendas with contemporary critics who believe the techniques these poets have put to such good use are reactionary.

To return to Nelson's book, many of his aesthetic judgments, judicious though they are, remain unexamined and untheorized. I have already remarked on a principal source of difficulty: the question as to whether innovative formal strategies comprise a thin veneer or are incorporated into a text with strong effect is not addressed. Nelson concentrates on the social force field to the neglect of the textual force field—the interplay of dynamic linguistic elements that in successful texts are every bit as resonant and consequential as the complex social forces shaping them. Such imbalance does not affect all or even most of the examples given by Nelson, but his unwillingness to deal adequately with the value question does an injustice to such poets as Muriel Rukeyser, Kenneth Fearing, and Sterling Brown, whose relative excellence continues to go unrecognized. The imbalance between social and textual force fields raises the further issue of whether Nelson has taken adequate account of the needs of his audience who, regardless of gender, class, or race, are not well served by the political oversimplifications, the narrow notions of praxis, the ubiquitous noble gestures, the insistently powerful physiques, and especially the formal obtuseness that more than occasionally make their way into the images and texts proffered here.

IV

Repression and Recovery is part of a growing endeavor to rediscover, reinterpret, and revitalize twentieth-century literary leftism in the United States. To some, of course, literary leftism remains an oxymoron. Traditionalists claim the left has produced little literature of note and has given scant support to good literature produced by others. Similarly, leftist organizers (as distinguished from academic leftists) have not shown great interest in literature, authors, or critics. Thus, for example, despite the unprecedented example of Edward Bellamy's enormously influential novel *Looking Backward* (1888), the Socialist Party's *Appeal to Reason* at the turn of the century suspended serial publication of that movement's only enduring classic, Upton Sinclair's *The Jungle* (1906), in order to make room for more pressing political matters. And thus, during the 1950s Max Shachtman rebuffed Irving Howe and Stanley Plastrik's request to have the Independent Socialist League sponsor a little magazine devoted to nonsectarian political and cultural discussion. This forced them to found *Dissent* on a wholly independent basis. Other examples abound. Contrary to the common claim, it was not only the cold war, the New Criticism, or the anti-Communism of the New York intellectuals that was responsible for depoliticizing literature and suppressing the literary left. Given the beating the left has been taking of late, it is understandable why left academics have explained the absence of a compelling and popular radical literary tradition by citing causes outside the left as responsible. On the other hand, now that leftists have been unburdened of the heavy Soviet load that the media and public have forced them to carry, they have the chance to remake themselves anew by reconsidering with ruthless honesty the depredations of their past. Thus far the ruthlessness has been reserved for traditional enemies. As for evaluations of the left itself, a comforting solicitude prevails.

Two examples shall suffice. The first is Walter Kalaidjian's *American Culture between the Wars*. Many will find this a useful book, and well they should because it is a readable survey of key literary and artistic projects on the left from the 1910s to the present. A short list includes the *Masses* and Greenwich village radicalism, the Russian and Soviet avant-garde, the Harlem Renaissance (including little magazines like *Fire!!*, and such neglected women poets as alice Dunbar-Nelson, Angelina Weld Grimke, and Gladys May Casely Hayford), homoeroticism and androcentrism on the left, the proletarian literary movement, Scottsboro, Diego Rivera, Muriel Rukeyeser (the most detailed and best part of the book), the LANGUAGE poets (who are compared to Kenneth Fearing!), Barbara Kruger, Jenny Holzer, Nancy Spero, Hans Haacke, and the NAMES Project. What makes the book especially appealing is that, like *Repression and Recovery*, it is full of interesting reproductions of rare posters, little magazine covers, paintings, murals, quilts, museum installations, and the like.

Kalaidjian deploys this medley of projects against what he regards as the dominant discourse and canon of high modernism. He condemns traditional scholarship's "incredibly narrow focus on a select group of seminal careers," which he claims has erased the memory of "modernism's unsettled social text, crosscut as it is by a plurality of transnational, racial, sexual, and class representations."[44] Kalaidjian would reverse the set of antitheses that have operated within "canonical institutions to privilege the literary over the social, aesthetics over politics, formalism over popular culture, and domestic nationalism over global multiculturalism."[45]

Kalaidjian's study succeeds much better on the empirical than on the theoretical level, for several reasons. Given his chosen subject matter, his interests are wide-ranging: even Nelson's *Repression and Recovery* is not as broad in scope, being less international in its purview and stopping at the end of World War II. Also Kalaidjian engages his reader by comparing "modernist" projects of the past with contemporary "postmodern" projects. Generally he offers accurate, informative accounts that situate the projects he describes both historically and ideologically. He manages at times to temper his enthusiasm by pointing out the limitations of the projects he describes. Last but certainly not least, he writes lucidly.

The book's weaknesses are evident in its larger, theoretical claims. A fundamental problem is that the very concept of modernism is muddled. Although at times Kalaidjian would extend the term to encompass the experimental and politicized projects of the left, at other times he ignores or denounces canonical modernism and suggests that the modernist canon must be replaced. Thus he makes no effort to distinguish between the original meaning of canonical modernists, on the one hand, and their appropriation, either by the academy or by advertising, on the other. In a book about the perils of forgetting, he forgets that the canonical modernists were too controversial to be published anywhere but in the alternative press, that they were sued for obscenity and had their books banned, that authoritarian critics denounced them right and left as decadents, aesthetes, antisocial types, psychotics, revolutionaries, and reactionaries, that they were implacable foes of the spiritual detritus of capitalism, and that they utterly transformed the cultural landscape in the West. Moreover, certain shibboleths are repeated: high modernism is limited by its allegedly "impersonal poetics," postwar scholarship has repressed modernism's social context, and T. S. Eliot's value is reduced because of his elitist Eurocentrism. What is repressed by Kalaidjian is the lyrical and personal side of modernism, which continues to represent an antidote to the impersonality of bad leftist politics. Also repressed is the fact that the New Critics and New York intellectuals overcompensated for the repression of modernism by such progressive or leftist critics as V. L. Parrington, Granville Hicks, and Bernard Smith. With regard to Eliot's Eurocentrism, it should be pointed out that Kalaidjian is only slightly more global in scope than Eliot despite his internationalist sentiments. When all is said and done, Kalaidjian has replaced Eliot's idea of the "mind of Europe" with a few individual minds of Europe. Diego Rivera is arguably the key exception, but even here Kalaidjian stresses the muralist's links to European artistic movements.

A few other historical items are forgotten as well. The 2nd International has been blamed for many things, but to blame it for leading leftist writers and artists into the muscular arms of Proletcult is not tenable.[46] On the other hand, to speak of international socialism's antiwar stance in 1917 is rather misleading since most of the parties of the 2nd International were supporting their respective nations during World War I.[47] Regarding the extraordinary Randolph Bourne, it is all well and good to point out his desire to socialize the self, but it is equally true that Bourne and others writing cultural criticism for *Seven Arts* also endeavored to individualize socialism by calling on radicals to build movements capable of satisfying the subtler needs of private life. In a nation where the "pursuit of happiness" is preeminently a private affair, it will not do simply to condemn ad infinitum the bourgeois ideology of the imperious, narcissistic self. Finally, I have heard heroic communism and socialist realism called many things, some positive, but until now I have never heard them called elegant and witty![48] In fact a number of the poetic passages Kalaidjian regards as "powerful" are not. Jack Haynes's "Scottsboro Boys Chant" is one of several examples of inflated doggerel. The chant begins "Innocent men, innocent men / They shall not burn them down there in the pen / White men, red men, yellow men, black / Workers sweat and die, bosses get the jack." Such a chant may or may not serve specific agitational purposes: in any event on these grounds it should be compared with other chants and songs, some of which have done the job of inspiring and cohering more effectively (Shelley's "A Song: 'Men of England,'" and the civil rights hymn "We Shall Oversome" readily come to mind). But to subordinate one's powers of discrimination to such a narrow utilitarian rationale as Kalaidjian often does amounts to a double betrayal—of his own skills as a reader honed over the course of many years, and of the masses he is interested in liberating. Ordinary people, after all, make daily judgments of craft and quality, whether at the supermarket, the record store, the movie theater, when embroidering or doing carpentry. These judgments, both the good and the bad, are often complexly made, and they often include an aesthetic dimension. The irony of so many left academics eschewing aesthetic judgment because it is allegedly elitist and ultimately ideological must not continue to go unregistered. Any sustained encounter with ordinary Americans will reveal how indispensable aesthetic judgments are for dignifying otherwise trying lives. A writer who understood this well was James Agee, who wrote *Let Us Now Praise Famous Men* to dispense with the pieties of mechanical materialists in his own day. Although at times he slipped into sentimentalizing his sharecropper subjects, his response to the very modest decorations above the Gudger's mantel, for instance, remains an alert, respectful, and unreserved meditation on the saving graces of the quest for beauty. It is a recognition altogether lacking in the discourse of the current academic left, and the costs are great.

Another recent study of the literary left, Barbara Foley's *Radical Representations*, does little to remedy the situation. She also avoids the issue of quality in her formidable study of proletarian fiction. Fortunately, this has not prevented her from producing what is by far the most comprehensive and serious treatment of the subject to date, supplanting Walter Rideout's estimable (and recently republished) *The Radical Novel in the U.S.* (1956). Foley's book now becomes the one to which others will need to respond, and it will likely remain so for some time.

It is both an ambitious and provocative book. Part 1 gives a historical account of the reception of proletarian literature from the 1930s to the present, it surveys nearly all of the recent scholarship on the subject, it reviews the major statements made by the critics of the movement, it describes developments within the Soviet Union and assesses their influence in the United States, it details the role of the American Communist Party (CPUSA), and it explores in depth the interaction of class, gender, and race within the proletarian movement. In part 2, Foley employs an eclectic blend of ideology critique, narratology, and genre criticism in offering readings of over fifty proletarian novels. She includes chapters on fictional autobiography, Bildungsroman, the social novel, and the collective novel.

To understand Foley's overall perspective, it is well to recall that American historians, including literary historians, have been involved in a contentious debate about the nature of Communist-led movements. On the one side are the anti-Communists who emphasize the Communist Party's control and manipulation of the American scene as the agent of the Soviet Union. A more recent school of revisionist historians at times acknowledges this control from above and abroad, but places more emphasis on the creative independence of rank-and-file activists, writers, and artists, and the ability of the Party to respond effectively to American political and cultural realities by building influential mass movements. Foley is clearly in the latter camp. She argues that the New Critics and especially the New York intellectuals (in the person of Philip Rahv) denigrated, defamed, and banished proletarian literature and the entire tradition of democratic, social protest literature in the United States. She argues that, by contrast, in its day proletarian fiction won the praise of a great many critics outside the movement, and that, contrary to the accepted view, it was an extremely diverse literature largely unaffected by the few examples of critical or ideological dogmatism that cropped up from time to time. She maintains that, in fact, many proletarian critics were overly influenced by "bourgeois" empiricism and pragmatism, and this caused them to eschew the very didacticism that she claims often made for powerful examples of propagandistic, revolutionary literature. She admits the Soviet influence was significant, but devotes most of her attention to discrepancies between examples of willful control and actual results. Finally, her narratological, generic readings, while duly noting failings, generally stress the accomplishments of this body of work. Although quite critical of the handling of race and gender, Foley emphasizes the social insights rendered by way of interesting, sometimes innovative narrative strategies, whose similarity to those of the canonical modernists belie the strictly enforced literary categories established by the hegemonic discourse.

In reading Foley I am reminded of Engels' well-known remark about Balzac's inability not to see the progressive march of history despite his Legitimist politics. With Foley, of course, the politics are reversed, but there is always an element of drama in the partially unreconstructed Marxist gamely pressing on past old attachments in order, as Engels put it, to see. It is mostly an admirable and sometimes a fascinating process, and it yields important insights. Two chapters in particular, "Race, Class, and the 'Negro Question'" and "Women and the Left in the 1930s," are major contributions to our understanding of the politics of race and gender in Communist Party politics and in proletarian fiction. Here Foley's tone seems perfectly modulated, and the orientation toward the Party's and the writers' handling of these issues impeccably balanced. These are, after all, delicate matters. That there was racism and sexism within this leftist milieu there can be no doubt. On the other hand, compared with traditional milieus, it afforded unprecedented opportunities for women and minorities. It also allowed for a far more serious battle against white chauvinism. Foley nicely registers this ambivalence and shows how it affected many of the writers of proletarian fiction.

But stubborn allegiances do sometimes mitigate Foley's pursuit of the truth. I found alarming, for instance, her disingenuous description of Andrei Zhdanov, Stalin's notorious literary hatchet man, who as a Politburo member denounced Anna Akhmatova as "a nun and a whore" and one of various "literary hangers-on and swindlers" for whom "there is no room in Leningrad."[49] Foley coolly describes Zhdanov as "an active participant in debates about politics and culture during the 1930s and the USSR's cultural commissar in the postwar period."[50] Imagine referring to Senator Joseph McCarthy, who also did enormous damage to the causes of democracy and socialism, as "an active participant in the debates about politics and culture during the 1950s and a U.S. senator." Foley's less than rigorous pursuit of the truth is also evident with regard to the question of the Communist Party's control of the American scene as an agent of the Soviet Union. On this key matter she equivocates:

> Soviet cultural developments exerted significant influences upon Depression-era literary radicalism. Indeed, 1930s literary radicalism would not have existed at all—or would have existed in dramatically different form—if the Bolshevik Revolution had not set in motion the vast changes, including cultural changes, that altered permanently the face of twentieth-century life. At the same time, a close examination of the precise nature of the Soviet influence upon American literary radicalism calls into question the thesis that the American Marxists simply attempted to clone themselves off the Soviet example.[51]

Foley makes two other dubious distinctions: between a nonbinding and presumably noncoercive Party "criticism" and a binding, potentially coercive Party "directive"; and between the Party's "definitive sponsorship and guidance" (loose control) and "control [of] the specific contents or outcomes of debates over aesthetics" (strict control). As to the "criticism/directive" distinction, Foley must know that any criticism proffered by an authority inevitably represented a consensus and was backed by the Party's prestige and apparatus. Its ideological message was clear and its psychological effect was profound; thus it had the same conforming effect as a directive. As to Foley's other distinctions, no anti-Communist historian has made the claim that Soviet hegemony amounted to absolute control over all matters large and small—not Theodore Draper, Harvey Klehr, or the socialist Irving Howe. The question is whether there was room for disagreement with the Party line as determined by the Comintern, or whether the CPUSA ever dissented from policies fashioned in Moscow. The answer to both questions is no. Nothing important ever happened unless initiated and approved of by the Party. This was the basis of its hegemony, and Foley herself supplies ample evidence proving its existence. Of course, there were those who managed a degree of autonomy, and others who allowed it. And certainly the Party itself made contributions to American culture and politics. These are worth our full understanding and appreciation, but they did not break the stranglehold that the Party as Moscow's proxy had on the head and heart of the American left.

Foley's treatment of proletarian fiction returns us to the value question. There is no question but that her detailed analysis of the political effects of narrative strategies adds a good deal to our understanding of this neglected body of work. How proletarian novels begin and end, how time is compressed or extended, how focalization (akin to point of view) is handled, what the relationship is between the narrator and protagonist and how it affects the rendering of speech (the uses of free indirect discourse, for example), whether and how authoritative statements are made and what their content is—Foley explores the politics of each one of these questions and more. Her method is an intriguing blend of formalist, Marxist, feminist, and reader-response criticism, which allows her to avoid the common pitfalls of formal determinism on the one hand and ideological determinism on the other. Foley is aware that identical narrative forms and strategies can have different effects given different circumstances; concomitantly, she shows how the same ideological perspective can be generated by different forms. What makes her analyses persuasive is her alertness to the complex and ever changing contexts that inform literary transactions.

The word "transaction" is appropriate here, for Foley, like Nelson and Kalaidjian, approaches literature in a largely instrumental way. But is anything lost when political and ideological function become the overriding criteria for discussing literature? Partly it depends on how broadly political and ideological function is considered. More often than not, contemporary critics stress projected effects in the realm of belief and social action. In their zeal to overcome the idealist notions of self that inform bourgeois individualism, they often neglect the affective and the personal side of subjectivity or at best subsume it under more public, determinate ideological forces. There are very few instances in any of these books when the precise quality of writing and the quality of the experience it evokes are described. The curious eye, the power of description, the fitness of a chosen word, the adequacy of tone—in short, the elements of craft that contribute as much or more to the effects of reading as do structure, strategy, device, and influences—are sadly unattended. Someday the story will be told how a generation of academics, convinced that the current social arrangement precludes quality lives for the majority, could have become so blind to the rare examples of quality that survive so many morbid symptoms today.

I strongly agree with these critics that the disenfranchised and their supporters, intent on gaining cultural and social emancipation, need to be familiar with the models of resistance and languages of dissent that are supplied by suppressed oppositional literatures from the past. In this connection these books and others like them will continue to prove valuable. But I do not think critics have sufficiently emphasized the degree to which the interests of the oppressed, or ordinary people, are served by an ability to extend critical consciousness to their own traditions—to discriminate among partisan poems between the spurious and conscientious, the clumsy and dexterous, and the hypnotic and animating. A well-made poem holds out the promise that close scrutiny of its construction will yield examples of idealization—but this is a peculiar sort of idealization that calls attention to the vicissitudes, imperfections, and bitter struggles that have gone into its own making. Reading such a poem is politically useful for two reasons. It imparts the kind of affirmative but self-reflexive and self-limiting knowledge that any emancipatory political movement must possess. And it discourages readers from measuring worth solely on the basis of political conviction by inducing a wide enough range of pleasurable sensations and emotions to make readers want to maximize the extent to which politics can include these. In this respect crafted poems empower readers to invent with care new forms of personal and political life, and thus they can give us something of what is needed to build democratic and vital movements for social change. It is this understanding, above all, which the New York intellectuals have bequeathed us.

11

What's Left of Lionel Trilling?

In a nation of critics of literature and critics of culture, no secure place has yet been found for Lionel Trilling, a consummate critic of literature *and* culture. This is not to gainsay Trilling's reputation as one of the country's foremost "men of letters" during the twentieth century—indeed, along with Edmund Wilson and T. S. Eliot, Trilling was among a small number of powerful and public cultural arbiters who fought successfully to canonize certain romantic and modernist works and thus profoundly shape literary taste in the postwar period. But Trilling was not a literary critic in the usual sense of the word: he was not primarily interested in exegesis and the formal complexities of the literary work as were Eliot, Blackmur, or Ransom; nor was he, like Wilson and Leavis, mainly interested in generating brilliant but miscellaneous moral and social insights whose apparent sources were lively encounters with individual works. Trilling was unique because his chief concern was always to illuminate a contemporary cultural, moral, and ultimately political problem. His relationship to individual works was consciously generated by his sense of his historical surroundings and his role as citizen within a flawed industrial democracy with cultural and spiritual difficulties. For Trilling the critical enterprise could be explained only partially as an engagement between a reader and a text; in actuality this was merely one pairing, albeit an essential one, among many in the ongoing processes of acculturation and change involving critic, text, and an array of social forces. Characterizing his highly contextual approach to literature, he wrote "[M]y own interests lead me to see literary situations as cultural situations, and cultural situations as great elaborate fights about moral issues, and moral issues as having something to do with gratuitously chosen images of personal being, and images of personal being as having something to do with literary style."[1] His meticulously crafted essays move easily from one level of analysis to another, and nearly always compensate for the itineracy of interdisciplinary work with arresting juxtapositions and syntheses.

Trilling was very much a pragmatic, situational critic, which makes it difficult to generalize about the conclusions he reached. Cultural needs, as Trilling assessed them, shifted over time, and he adjusted his critical emphases accordingly. Many of his apparently inconsistent judgments can very often be counted as altered responses to changed circumstances. History may not have vindicated Trilling in all of these judgments, and several of the poorer ones will be identified here, but what is remarkable in retrospect is how impressively flexible his responses remain when understood in context. If we allow ourselves to see how thoroughly history impinged on his criticism, we are bound to take note of the care expressed through nuance, subtle qualification, and dialectical precision, which made him the foremost contextual critic of his time. His life's work nicely exemplifies Brecht's remark that the survival of cultural artifacts depends on the degree to which they are immersed in their time, not the degree to which they transcend it. He belongs in the line of the worldly citizen-critics, which extends from Plato and Aristotle to Hazlitt and Arnold, critics who gave pronounced attention to literature's connection to the well-being of the polity, and who portrayed that connection, as well as the nature and interests of the polity, with unusual specificity and insight.

Trilling considered himself part of the liberal culture that he spent a lifetime trying to strengthen. The point bears emphasis because among fellow inhabitants of this culture Trilling's position as ally came under increasing suspicion during the 1960s and 1970s as his critique of the counterculture intensified. Since the 1970s, in fact, his legacy has been abandoned by many progressives and claimed by neoconservatives such as Hilton Kramer and Norman Podhoretz. But his work will not remain compelling if, ironically, it becomes associated with traditional values—not because traditional values cannot be compelling, but because such an association diminishes Trilling's role as social critic, as well as the actual complexity and challenge of renewing the liberal-radical culture, which was at the heart of his engaged criticism. Broadly speaking, Trilling's lifelong critical work can best be understood as a critical humanist challenge to the culture and politics of liberalism, or, as he occasionally and more accurately referred to it, the "liberal-radical culture" of the 1930s and the postwar period. He defined liberalism as "a large tendency rather than a concise body of doctrine,"[2] held by an "educated class" whose Enlightenment legacy of overly rational, prosaic habits of mind and utilitarian attitudes was manifested in "a ready if mild suspiciousness of the profit motive, a belief in progress, science, social legislation, planning, and international cooperation, perhaps especially where Russia is in question."[3] Trilling did not mean by such a broad definition the political viewpoint of a section of the Democratic Party; rather, he meant the whole spectrum of viewpoints which then went under the category of "progressive" and more or less still does today, and included liberals, social-democrats, socialists, and Communists. Although it is certainly true that for much of his career his immediate target was Stalinism, his insights had, and continue to have, direct relevance to wide areas of social and ideological life in any capitalist democracy.

He described this larger social realm, and literature's relation to it, in the preface to *The Liberal Imagination*:

> It is one of the tendencies of liberalism to simplify, and this tendency is natural in view of the effort which liberalism makes to organize the elements of life in a rational way. And when we approach liberalism in a critical spirit, we shall fail in critical completeness if we do not take into account the value and necessity of its organizational impulse. But at the same time we must understand that organization means delegation, and agencies, and bureaus, and technicians, and that the ideas that can survive delegation, that can be passed onto agencies and bureaus and technicians, incline to be ideas of a certain kind and of a certain simplicity: they give up something of their largeness and modulation and complexity in order to survive. The lively sense of contingency and possibility, and of those exceptions to the rule which may be the beginning of the end of the rule—this sense does not suit well with the impulse of organization.[4]

The function of criticism in such a society must be "to recall liberalism to its first essential imagination of variousness and possibility, which implies the awareness of complexity and difficulty."[5] It does so through its reliance on literature, which is uniquely relevant "not merely because so much of modern literature has explicitly directed itself upon politics, but more importantly because literature is the human activity that takes the fullest and most precise account of variousness, possibility, complexity, and difficulty."[6]

Trilling's own political experiences were decisive in establishing the framework and direction of his criticism. Soon after receiving his B.A. from Columbia, he began contributing to the *Menorah Journal*, a secular and humanist journal dedicated to the advancement of Jewish culture. It had been founded in 1915 by Henry Hurwitz of the Menorah Society, and placed under the managing editorship of Elliot Cohen in 1926, at which time it began moving steadily in the direction of cosmopolitanism and later, in the 1930s, of internationalism. From Cohen, a loquacious, witty, and charismatic figure, Trilling and others within the Menorah group learned to "live [their] intellectual lives under the aspects of a complex and vivid idea of culture and society. . . . I have long thought of him as the greatest teacher I have ever known."[7] Trilling published his first piece, the short story "Impediments," in the *Menorah Journal* in 1925, and subsequently contributed book reviews, essays, and stories. None of the pieces reveal evidence of direct influence by traditional Jewish religious and intellectual thought, a point Trilling later emphasized when discussing his career:

> I cannot discover anything in my professional intellectual life which I can specifically trace back to my Jewish birth and rearing. I do not think of myself as a "Jewish writer." I do not have in mind to serve by my writing any Jewish purpose. I should resent it if a critic of my work were to discover it in either faults or virtues which he called Jewish.[8]

Nevertheless Trilling's refined Anglophilia, which he derived from his London-born mother, was supplemented in some significant measure by the moral seriousness and respect for the word that emanated outward into the Jewish community from the talmudic tradition. This background equipped Trilling and a legion of young Jewish writers for the strenuous, unprecedented journey toward respectability and eventually authority within the dominant American culture, notwithstanding the degree to which they may have suppressed aspects of their ethnic heritage as part of the established processes of assimilation.

Like others within the Menorah circle, Trilling was drawn to the political radicalism of the Communist Party in the early 1930s. Although he never joined the Party, for a relatively brief period he was involved with the National Committee for the Defense of Political Prisoners, an affiliate of the Party's International Labor Defense League, and during the 1932 presidential campaign he signed a statement, along with fifty-two other intellectuals, supporting the Party's ticket. In May 1933, Trilling and other Menorah intellectuals resigned from the NCDPP, and less than a year later he signed an open letter opposing the Communist Party's violent disruption of a Socialist rally at Madison Square Garden (the letter also rejected reformism, capitalism, and fascism, and affirmed the signers' support for the working-class movement). Although Trilling's involvement with the Party and its auxiliary organizations was relatively short-lived, he continued to write for the liberal and left-wing press, including the *Nation*, *New Republic*, *Partisan Review*, and V. F. Calverton's *Modern Monthly*. From the mid-1930s to the early 1940s, his anti-Stalinist leftism focused on the intellectual bankruptcy of the Party, and the lively and often overlooked polemical writing of the period represents some of his best work. Trilling's polemical work has been overlooked partly through his own devices. More than once he excluded incendiary passages from later versions of his essays. One may compare, for example, "Mr. Parrington, Mr. Smith and Reality"[9] with its later, tamer incarnation as "Reality in America" in *The Liberal Imagination*. Trilling's polemical skills were at their delightfully destructive peak in his 1937 review of Robert Briffault's Marxist novel, *Europa in Limbo*. Having disposed of the author's historical account of the bourgeoisie's oppression of the masses through sexual depravity, which Trilling labeled "the license and rape theory of social upheaval," he turned to Briffault in the realm of thought, where he was said to be "quite as seminal." Here, observed Trilling, "his great enemies are defunct ideas. Under his lash every extinct notion of the nineteenth century lies perfectly still. Ruthlessly he banishes our last stubbornly-held illusions about the survival of the fittest, the immutability of human nature, liberal democracy, the idealistic philosophies of reaction, Fabian socialism, the aesthetics of Ruskin. Supererogation, though dull, is not dangerous."[10]

Trilling opposed political litmus tests in "Hemingway and His Critics" (1939), attributing the inferiority of *To Have and Have Not* and *The Fifth Column* to a critical community that had encouraged Hemingway "the man" to usurp Hemingway "the artist":

> Upon Hemingway were turned all the fine social feelings of the now passing decade, all the noble sentiments, all the desperate optimism, all the extreme rationalism, all

> the contempt of irony and indirection—all the attitudes which, in the full tide of the liberal-radical movement, became dominant in our thought about literature. There was demanded of him earnestness and pity, social consciousness, as it was called, something "positive" and "constructive" and literal. . . . One almost wishes to say to an author like Hemingway, "You have no duty, no responsibility. Literature in a political sense, is not in the least important."[11]

Trilling went on to discuss Hemingway's virtues as a writer, offering perhaps the best account of his style we have, and putting to rest the charge of mindlessness by drawing the important distinction between resisting reason and resisting rationalization: "[I]n the long romantic tradition it never really *is* mind that is in question but rather a dull overlay of mechanical negative proper feeling, of a falseness of feeling which people believe to be reasonableness and reasonable virtue."[12] In a particularly audacious essay, an appreciation of Eliot entitled "Elements That Are Wanted," Trilling scored the elitism of *The Idea of a Christian Society*, yet maintained that Eliot was asking pertinent questions ignored by the left, such as what the good life might consist in, what the morality of politics should be, what "the spiritual and complex elements in life" might yield to politics. What Eliot and other modernist writers seemed to be recommending to the liberal culture of both bourgeois capitalism and its left opposition, was "the sense of complication and possibility, of surprise, intensification, variety, unfoldment, worth. These are the things whose more or less abstract expressions we recognize in the arts; in our inability to admit them in social matters lies a great significance."[13]

Trilling's willingness to enter the opposition's camp and admire some of what he found there was certainly an act of dissent among the dissenters, but he felt that liberalism would be doomed if it gave credence only to its own advocates. He was fond of quoting John Stuart Mill's famous essay on Coleridge in which he urged his fellow liberals to welcome the critical insights of this "powerful conservative mind." What made such solicitude absolutely essential for contemporary liberals was the fact that arrayed against them was every major writer of the modern period. "Our liberal ideology has produced a large literature of social and political protest," Trilling noted, "but not, for several decades, a single writer who commands our real literary imagination." To "the monumental figures of our time"—Trilling listed Proust, Joyce, Lawrence, Eliot, Yeats, Mann, Kafka, Rilke, and Gide—"our liberal ideology has been at best a matter of indifference."[14] The lack of a connection between liberalism and "the best literary minds of our time" meant that "there is no connection between the political ideas of our educated class and the deep places of the imagination."[15] As did the entire circle of New York intellectuals who contributed to *Partisan Review*, Trilling advocated an uneasy alliance with modern literature because liberalism seemed incapable of sustaining a culture autonomous and imaginative enough to produce incisive self-critique. This cause Trilling and others to diminish the importance of reactionary ideology in modernist works—roughly in proportion to the degree that Popular Front critics exaggerated its importance; instead they emphasized the modernists' adversarial stance. "Any historian of the literature of the modern age," Trilling claimed, "will take virtually for granted the adversarial intention, the actually subversive intention, that characterizes modern writing. . . . [A] primary function of art and thought is to liberate the individual from the tyranny of his culture in the environmental sense and to permit him to stand beyond it in an autonomy of perception and judgment."[16]

This passage is among several from the preface to *Beyond Culture* sometimes cited by Trilling's critics as proof that, like Arnold, he opted ultimately for, a transcendent notion of culture. It is charged that Trilling's culture is a high culture whose achievements are ultimately spiritual, whose perfection is realized to the degree that it escapes society, and whose beneficiaries are a select elite capable of purely aesthetic contemplation. It is, of course, perfectly true that Trilling had little respect for popular culture (at least in print–Diana Trilling has testified that he was an avid moviegoer), and that, like so many of his generation, he did entertain uninformed, undifferentiated, and frankly elitist opinions alleging its relative worthlessness. Thus he had nothing to say about achievements in film, television, theater, or popular forms of literature: he was simply unwilling to challenge the Popular Front critics and their successors by staking out his own claims in these areas. Figures such as Hitchcock, Kovacs, or Blitzstein elicited no response whatsoever. The furthest Trilling went in this direction was no negligible distance, but it was, finally, inadequate: he did chastize Arnold for his exclusion of Chaucer from the first rank, observing that "if Chaucer is not serious, then Mozart is not serious and Molière is not serious and seriousness becomes a matter of pince-nez glasses and a sepia print of the Parthenon."[17] He also defended Howells and Orwell precisely for their responsiveness to the details of actuality and the attributes of those who must get along in it–in other words, for their *lack* of solemn seriousness, greatness, and genius.

Nevertheless, the charge that Trilling held to a transcendent notion of culture is simply false, as a careful reading of the preface will reveal. Moving "beyond culture" for Trilling could never mean transcending "a people's technology, its manners and customs, its religious beliefs and organization, its systems of valuation, whether expressed or implicit."[18] Culture taken in this sense, and this is the sense in which we take it today, can never be left behind. No person may escape his or her culture–on this matter Trilling is unequivocal:

> [I]t is not possible to conceive of a person standing beyond his culture. His culture has brought him into being in every respect except the physical, has given him his categories and habits of thought, his range of feeling, his idiom and tones of speech. No aberration can effect a real separation: even the forms that madness takes . . . are controlled by the culture in which it occurs. No personal superiority can place one beyond these influences. . . . Even when a person rejects his culture (as the phrase goes) and rebels against it, he does so in a culturally determined way.[19]

It is only when we think of culture as exclusive rather than inclusive, as "that complex of activities which includes the practice of the arts and of certain intellectual disciplines,"[20] do we see what Trilling wished to acknowledge by speaking of modern literature's adversarial stance. The ability to go beyond culture was nothing other than the continuing possibility, against great odds, of human agency manifesting itself in intelligent revolt against existing artistic and intellectual practices. This view was consistent with Trilling's cultural materialist perspective: critics may demur when it comes to endorsing the particular forms of revolt that appealed to Trilling, but it seems illogical for advocates of social change to condemn him as an idealist because he thought it possible for literature to be a locus of rejection and innovation. Trilling did privilege literature, of this there can be little doubt, but if one wants to avoid such privileging, it would mean identifying adversarial impulses in nonliterary realms of experience rather than denying their existence in literature by regarding literature or subjective experience in general as wholly imprisoned by established conventions, power, and authority.

Trilling's first book, one of only two full-length studies he produced, was the intellectual biography *Matthew Arnold* (1939), still the definitive work on Arnold. Here Trilling's direct engagements with sectors of the left were significantly amplified, a fact that belies the common assumption that Trilling took from Arnold's "gospel of culture" an unsavory mandarinism. It is much closer to the truth that Trilling valued Arnold for his refusal to remain aloof, for his aggressive political interventions, for his wish to spread superiority, as it were, and thereby ennoble the masses. If in theory Arnold's definition of literature as a "criticism of life" meant that it "illuminates and refines" Reason, for Trilling its function in practice was ultimately political because it prepared Reason by bringing to mind "some notion of what is the right condition of the self" so that "man might shape the conditions of his own existence."[21] Following Arnold, Trilling considered literature capable of enabling the fullest imagination of that which the political will then strives to realize. Where Arnold's "touchstones" failed to offer such a vision, or where they seemed too abstracted to help erode social prejudice, Trilling was quick to point it out. He upbraided Arnold, for example, for his intemperate response to mass agitation for the vote, and it is testimony to his own political engagement that the biography was laced with remarks connecting his subject matter with the current struggles against anti-Semitism and fascism.

Although Trilling's admiration for E. M. Forster waned somewhat in later decades, during the 1940s he argued strenuously for the author's major status, again within a very definite political and ideological context. In *E. M. Forster* (1943) he based his claim on the novelist's rare combination of social understanding, particularly where money and class were concerned, and his intimate knowledge of how these conditioned the closely rendered moral lives of his characters. According to Trilling, Forster's intense attachment to tradition allowed him to shun the eschatological belief in the future in favor of "belief" in the present. He was worldly because he accepted "man in the world without the sentimentality of cynicism."[22] He could deal with the momentous changes hastened by industrial capitalism—urbanization, economic and cultural imperialism, skepticism with regard to history and tradition—without the lugubrious piety, which, for Trilling, detracted from so many American efforts at social criticism. Indeed, contrary to what might be expected from a critic noted for his unrelenting seriousness, Trilling praised Forster precisely for his lack of seriousness—for his comic manner, playfulness, and "relaxed will." Forster's ease permitted him to remain content with both human possibility and limitation, and to confront the stark conflicts of his often melodramatic plots with a saving ambivalence. Trilling referred to this ambivalence as "moral realism": "All novelists deal with morality," he explained, "but not all novelists, or even all good novelists, are concerned with moral realism, which is not the awareness of morality itself but of the contradictions, paradoxes and dangers of living the moral life."[23] Forster's ease penetrated all absolutes, casting doubt upon both sides of an issue, and upon that side of character most constrained by social expectation. Montaigne-like, he caught by surprise the subtlest of private hesitations. Such a disposition was of particular use to middle-class liberalism to which Forster appealed "from the left,"[24] because it inured liberalism to the surprise which, for lack of imagination, it habitually faced before giving way to "disillusionment and fatigue."[25] In place of the simple logic of good versus evil, Forster presented a third possibility, an understanding of "good-and-evil."[26] This acceptance of contingency and interconnection might have gone under the name of dialectical understanding had not the term been wrongly appropriated by doctrinaire Marxism, the "intellectual game of antagonistic principles,"[27] which Forster so effectively repudiated.

The sixteen essays composing *The Liberal Imagination* were devoted to the project that Trilling, in "The Function of the Little Magazine," attributed to the journal with which he was most closely associated, *Partisan Review*: its goal was "organiz[ing] a new union between our political ideas and our imagination – in all our cultural purview there is no work more necessary."[28] Trilling opened the volume with his first widely influential essay, "Reality in America." Along with two of Philip Rahv's essays ("Paleface and Redskin" [1939] and "The Cult of Experience in American Writing" [1940]), the essay's withering and at times injudicious attack on progressive critical standards provided the death knell for the liberal-radical version of a usable American past. "Parrington," claimed Trilling, "expressed the chronic American belief that there exists an opposition between reality and mind and that one must enlist oneself in the party of reality."[29] In Parrington writers who refused to join up were denigrated for their elitist concern with private experience and literary form – they included Hawthorne, Poe, Melville, and James; writers who dealt directly with economic, political, and social matters, and who did so, moreover, in a manner compatible with certain instrumental notions of democracy, were profusely praised. Applying a version of Eliot's dissociated sensibility to the American scene, Trilling (and Rahv) argued that the progressive critics were merely sustaining the American bifurcation of experience by insisting on a reality defined as fixed, given, and material. They were meeting "mind" – contemplation, imagination, creativeness, *active* thought – with a settled hostility, as though these experiences were somehow antidemocratic. One of the writers the progressives favored was Dreiser, whom Trilling disparaged in this essay and again in "Manners, Morals, and the Novel," for his inability to portray a single interesting subjective life whose "electric qualities of mind" involve the whole personality in ideas, and for, ironically, his pedantic ignorance of colloquial speech. In Dreiser and writers like him, Trilling maintained, ideas were merely "pellets of intellection," whereas in actuality ideas arose out of emotional responses to social situations and were therefore "living things, inescapably connected with our wills and desires," as he argued in "The Meaning of a Literary Idea."[30]

Not only did Trilling seek a way to link ideas with emotion and action—he also sought mediation on the fundamental question of the relation between subjective experience and the objective world, between being and consciousness as Marx had put it. Trilling claimed that the Stalinists and progressives were wrong in giving absolute priority to material factors and in refusing to grant the reciprocal power of human agency in making history. The progressives' failure to provide a dialectical understanding of a question that went to the heart of a proper understanding of the novel caused Trilling to provide an intermediary category, which he called manners. "What I understand by manners," he wrote in a famous definition,

> is a culture's hum and buzz of implication. I mean the whole evanescent context in which its explicit statements are made. It is that part of a culture which is made up of half-uttered or unuttered or unutterable expressions of value. . . . In this part of culture assumption rules, which is often much stronger than reason.[31]

Manners were the constantly changing, conflicting systems of styles and actions that a society made available to its members and, in turn, its members evolve through conscious and unconscious reproduction and modification. Trilling contended that liberalism's objectivist understanding of reality caused it to respond condescendingly and often with hostility to manners. He challenged the view that the complex of experience "below all the explicit statements that a people makes" could be described by the sociologically top-heavy and overly cognitive term "ideology," by which Trilling meant "the habit or ritual of showing respect to certain formulas . . . [to which] we have very strong ties of whose meaning and consequences in actuality we have no clear understanding."[32] For Trilling it was quintessentially the novel that corrected such a view by demonstrating that the world of explicit statement and "ordinary practicality" was not "reality in its fullness." It did so—and here he applied a notion akin to Marx's commodity fetishism—by exposing the obfuscatory effects of money and class relations on perception and behavior. Trilling acknowledged that the novel's common coin was, precisely, ideology. But he claimed the novel was not *confined* to the analysis of ideology: by virtue of its focus on "quality of character" as expressed by the ideas, attitudes, and styles that accompany action, it also traded in manners. Trilling's analysis of Hyacinth Robinson, the would-be activist and assassin in *The Princess Casamassima*, is instructive here. Robinson's "superbness and arbitrariness," his intense response to art and his heightened sensitivity to the suffering masses that followed, and above all his final, heroic acceptance of responsibility for both the ideal of revolution and the ideal of civilized life—the quality of character that these attributes exemplified may have negated immediate practical necessity (Robinson chose suicide over assassination and, it is less often observed, over life after the refusal to assassinate), but they did reveal the unresolved moral dilemmas, which most adherents of positive social action militantly avoided.

In Trilling's view the American novel had been deficient in attaining such a dialectic between social and individual knowledge. "American writers of genius have not turned their minds to society," he wrote in a famous passage. "Poe and Melville were quite apart from it; the reality they sought was only tangential to society. Hawthorne was acute when he insisted that he did not write novels but romances."[33] If Henry James was the only nineteenth-century American writer capable of complicating the subjective lives of his characters by immersing them in a dense, highly textured class society, nonetheless a writer of the romance such as Hawthorne was to be preferred to the modern American social novelists—to Dreiser, Anderson, Lewis, Steinbeck, Dos Passos, or Wolfe—whose passivity before aspects of material reality attenuated the subjective lives they depicted. Hawthorne, "forever dealing with shadows," was nonetheless dealing with "substantial things" *by virtue* of his aloofness from "'Yankee reality.'" It was his very distance from society that allowed him to raise "those brilliant and serious doubts about the nature and possibility of moral perfection."[34]

These judgments, of course, have been extremely influential: they have provided part of the conceptual framework for the prevailing literary histories from the 1950s to the 1970s, among them Richard Chase's *The American Novel and Its Tradition* (1957), Leslie Fiedler's *Love and Death in the American Novel* (1960), and Leo Marx's *The Machine in the Garden* (1964). For his part, Trilling modified these views in the later essay "Hawthorne in Our Time" (1964), arguing with James and himself that Hawthorne's "hidden, dark, and dangerous" world of internal moral conflict "interpenetrates the world of material circumstance."[35] If "romance" implied a material world that was but "thinly composed," it was nonetheless a world "of iron hardness," which proved entirely "intractable" to those wishing for personal or social transfiguration. Hawthorne was therefore not a writer to be embraced by the 1960s, Trilling claimed, a period in which he believed the inner life had been commercially and publicly appropriated, in the name, paradoxically, of noncontingency and spontaneity.

For some progressive critics, the essays of *The Opposing Self* (1955) and *Beyond Culture* (1965) have represented something of a retreat for Trilling. It has appeared to them as though his interest in broad cultural and political tendencies was replaced by a new concern for individual, relatively private experience, and this has been thought to reflect, and contribute toward, the cold war consensus. There is some basis for this view, surely, when we consider how much more ideologically efficacious Trilling's anti-Communism became in this period, and how relatively narrow, though certainly not shallow, was the "critical non-conformism" that he advocated in the famous *Partisan Review* symposium of 1952, "Our Country and Our Culture."[36] Indeed few of the abuses of American global power, the threats to democracy posed by McCarthyism or examples of the continuing effects of racial oppression, diverted Trilling from his increasingly prominent role as critic of the liberal-radical culture. Throughout his career Trilling paid little heed to conservative opinion, believing that because it had won virtually no assent among intellectuals it was not dangerous or worth refuting. Instead, during the 1950s and 1960s it was becoming increasingly clear to Trilling that the great danger for American culture arose from the left, and although he himself never renounced liberalism, by the 1970s it was nearly impossible in a polarized society for many progressives to consider him an ally.

But Trilling's failure to criticize the abuses of postwar American capitalism does not in itself invalidate his insights into certain of liberalism's deficiencies. As we have seen, Trilling identified as chief among these a profound and historic uncertainty with regard to irrational, intuitive, or otherwise unsociable experience that did not lend itself to organized forms of social cooperation. In the 1940s and early 1950s Trilling sought to correct this problem through complicating liberalism's notion of experience by assimilating to it the insights of romanticism and modernism. The key figure here was Freud, who, despite the inadequacies of his own theory of artistic production and the function of art, nonetheless rightly conceived of mind as something to which poetry was indigenous—in Trilling's words the mind was to Freud "a poetry-making faculty."[37] The mind was, moreover, a thoroughfare of biology, culture, and creativity. But rarely did the traffic flow smoothly. In "Freud: Within and Beyond Culture," Trilling argued for the necessary, and tragic, conjunction of all three, claiming at once that Freud "made it apparent how entirely implicated in culture we all are"; that this principle of culture is "terrible" for the self set against it; and that biology, for all its asociality, may after all provide "a residue of human quality beyond the reach of cultural control."[38] Then, identifying a social phenomenon central to Antonio Gramsci's idea of hegemony, he observed, "In a society like ours, which, despite some appearances to the contrary, tends to be seductive rather than coercive, the individual's old defenses against the domination of the culture become weaker."[39] In order to bolster these defenses, Trilling spoke on behalf of difficult, even inadmissible experience. In what many consider to be his finest essay, "Keats: The Poet As Hero," he credited the poet for delighting in "the infantile wish" despite his culture's fear and repression of "the passive self-reference of infancy."[40] Keats's "geniality toward himself, his bold acceptance of his primitive appetite," extended to his abiding capacity for indolence, which Trilling, following Keats himself, distinguished from what we would call laziness, apathy, or self-indulgence. Keats referred to "diligent indolence," marking the power of passivity as a source for what Trilling called "conception, incubation, gestation,"[41] all of which were constitutive of the active life. Similarly, in "Wordsworth and the Rabbis" Trilling counterposed Wordsworth's quietism, "which is not in the least a negation of life but, on the contrary an affirmation of life so complete that it needed no saying," to "the predilection for the powerful, the fierce, the assertive, the personally militant, [which] is very strong in our culture."[42]

In *Sincerity and Authenticity* (1971) and the major essays of the 1960s and 1970s—"On the Teaching of Modern Literature" (1961), "The Fate of Pleasure" (1963), "Mind in the Modern World" (1972), "Art, Will, and Necessity" (1973)—Trilling's position hardened against what he considered to be ubiquitous systems of mass and middlebrow culture, systems that in his view had devoured the adversary culture of modernism and discharged the ersatz avant-garde of the Beats and then the counterculture. Confronted by the counterculture and the student movement at Columbia, where he taught for some thirty-five years, Trilling sounded notes of increasing despair, even questioning whether the "adversary culture" was not in part responsible for the uses to which it was being put. Only time will tell the extent to which his enemies and their successors, now extensively reevaluating the period, will find it prudent to acknowledge the wisdom of portions of Trilling's attack, especially where he assailed the movement for its anti-intellectualism, its sectarianism, its hedonism, and its irrationalism. As things now stand, reasons other than simple ideological differences conspire to make Trilling's criticism underappreciated today. His work presents a direct challenge to the long-standing separation in the United States between politics and intelligence, to the academicization of postwar American criticism and the ensuing pressure to distance criticism from immediate cultural and social problems, and to the resulting tendency for even historical, ideological, and politicized critics to remain aloof from nonacademic cultural and political movements. Trilling's criticism serves to remind us that this distance has helped cause these critics at times to express their militancy by exaggerating or simplifying the efficacy of material factors on authors, texts, and readers, instead of finding ways to encourage willful acts of "critical non-conformism" on the part of a wider public audience.

To these reasons we must add the particular limitations of Trilling's criticism. These center on his relative neglect of power. Not the power of culture and the word—of these he was an exemplary critic—but power that descends upon middle-class intellectuals and the university from corporate, state, and military sources, and the power of the innumerable small acts of circumvention and dissent in the daily routines of those who do not live in certain neighborhoods of Manhattan. Trilling, for all his social perspective and for all his understanding of the conditioned nature of life, was perhaps a bit possessive of the intellectual life and the authority it generously bestowed on him. One senses that he lacked, as did most intellectuals "withdraw[n] from the ordinary life of the tribe," as Trilling himself put it, the worldliness for which he praised Orwell so profusely, a worldliness rooted in the "passion for the literal actuality of life."[43] Had Trilling been such a man "whose hands and eyes and whole body were part of his thinking apparatus,"[44] he might have understood just how fine is the line between the "populist sentimentality" he deplored and the perspicacious generosity Orwell exhibited toward the culture of ordinary people in an essay such as "The Art of Donald McGill." He might also have acknowledged that in certain important respects the progressive politics and postmodern culture of the 1950s and 1960s challenged dangerous values coupled with great power, and were thus worthy of the name "critical non-conformism."

But for all the claims that have been made for the conservatism of Trilling's cold war liberalism and his adamant anti-utopianism, on the question of the need for and the possibility of radically transforming American liberal-radical culture, he was a visionary. He wished to make thoroughgoing self-criticism the prevailing mode within the liberal-radical culture, and he implored those sharing this culture to consider private, politically passive acts as not only sustaining, but, as Whitman believed, the very criterion of democracy. It is, finally, as close to religion as we might expect a secular cultural critic to come. Trilling placed a very heavy burden on liberalism indeed, a burden that as yet it has refused to bear.

Notes

Introduction

1. Richard Flacks, *Making History: The American Left and the American Mind* (New York: Columbia University Press, 1988), p. 277.

2. Rudolf Bahro, *The Alternative in Eastern Europe* (London: Verso, 1978), pp. 361ff., as quoted in Flacks, *Making History*, pp. 277–78.

3. Flacks, *Making History*, pp. 203–5.

4. Fredric Jameson, *Marxism and Form* (Princeton: Princeton University Press, 1971), p. ix.

5. Ibid., p. 416.

6. Certainly an abundant amount has been written on the New York intellectuals during the past two decades, most of it by the New Yorkers themselves. These include Lionel Abel, *The Intellectual Follies: A Memoir of the Literary Venture in New York and Paris* (New York: Norton, 1984); William Barrett, *The Truants* (New York: Doubleday, 1982); Sidney Hook, *Out of Step: An Unquiet Life in the 20th Century* (New York: Harper and Row, 1987); Irving Howe, *A Margin of Hope* (New York: Harcourt Brace Jovanovich, 1982); Alfred Kazin, *New York Jew* (New York: Random House, 1978); William Phillips, *A Partisan View: Five Decades of the Literary Life* (New York: Stein and Day, 1983); Norman Podhoretz, *Breaking Ranks: A Political Memoir* (New York: Harper and Row, 1979); and Diana Trilling, *The Beginning of the Journey: The Marriage of Diana and Lionel Trilling* (New York: Harcourt Brace Jovanovich, 1993). In addition, there are several general historical accounts, including Alexander Bloom, *Prodigal Sons: The New York Intellectuals and Their World* (New York: Oxford University Press, 1986); Terry Cooney, *The Rise of the New York Intellectuals: Partisan Review and Its Circle, 1934–1945* (Madison: University of Wisconsin Press, 1986); James Gilbert, *Writers and Partisans: A History of Literary Radicalism in America* (New York: Wiley and Sons, 1968); Neil Jumonville, *Critical Crossings* (Berkeley: University of California Press, 1991); and Alan Wald, *The New York Intellectuals: The Rise and Decline of the Anti-Stalinist Left from the 1930s to the 1980s* (Chapel Hill: University of North Carolina Press, 1987). The effort to apply the criticism of the New Yorkers to current critical problems has been a relatively recent development. The best of these attempts has been, in the visual arts: Serge Guilbaut, *How New York Stole the Idea of Modern Art* (Chicago: University of Chicago Press, 1983); and in literature and cultural criticism: Mark Krupnick, *Lionel Trilling and the Fate of Cultural Criticism* (Evanston, Ill: Northwestern University Press, 1986); Russell Reising, *The Unusable Past* (New York: Methuen, 1986); Giles Gunn, *The Culture of Criticism and the Criticism of Culture* (New York: Oxford University Press, 1987); Cornel West, *The American Evasion of*

Philosophy (Madison: University of Wisconsin Press, 1989); and Morris Dickstein, *Double Agent: The Critic and Society* (New York: Oxford University Press, 1992). An essay that has noted the relevance of the New York intellectuals to contemporary theoretical and critical issues is Catherine Gallagher's "Marxism and the New Historicism," in *The New Historicism*, ed. H. Aram Veeser (New York: Routledge, 1989), pp. 37–48, in which the author argues that the new historicism is best viewed as a productive elaboration of certain new left preoccupations formed during the 1960s. Much to her credit, in discussing some of the sources of the new historicism Gallagher rejects the common antithesis between the old and the new left, suggesting that aspects of contemporary "left formalism" can be traced not only to continental Marxism but also to "American left-wing literary critics [who had] developed a strong, optimistic, and politically problematic brand of formalism a generation before the New Left." Gallagher continues: "In its most independent and intelligent sectors, the American left offered its children in the post-war years a politics that had already begun to transfer its hopes from the traditional agent of revolutionary change, the proletariat, to a variety of 'subversive' cultural practices, the most prominent of which was aesthetic modernism" (p. 38). Gallagher doesn't elaborate on either the nature of this critical project or on what our relationship to it might be—in fact, she goes on to suggest that both native and continental left-wing criticism were less important factors in the creation of the new historicism than were specific forms of political practice and knowledge accumulated during the 1960s.

7. Sidney Hook, "The Radical Comedians: Inside *Partisan Review*," *American Scholar* (Winter 1984–1985), pp. 45–61.

8. See Gilbert, *Writers and Partisans*, pp. 191, 212.

Chapter 1

1. *The Liberal Imagination* (1950; repr. New York: Harcourt Brace Jovanovich, 1979), p. 95.

2. Throughout this book I refer to the official ideology of the Comintern and the Communist Party of the United States as "official" or "doctrinaire" Marxism. "Orthodox" Marxism refers to this and other currents of Leninism and Trotskyism adopted by cadre organizations in the wake of the Bolshevik Revolution. These are to be distinguished from "classical" Marxism, the body of work produced by Marx and Engels themselves.

3. Philip Rahv, response to "For Whom Do You Write," *New Quarterly* 1 (Summer 1934), p. 12.

4. For a perspective on the early *Partisan Review* that emphasizes ethnicity, see Terry A. Cooney, *The Rise of the New York Intellectuals: Partisan Review and Its Circle* (Madison: University of Wisconsin Press, 1986). Cooney draws on David Hollinger's notion of cosmopolitanism to explain the Jewish New Yorkers' movement outward from the ghetto to "respectability."

5. Philip Rahv, "The Literary Class War," *New Masses* 8,2 (August 1932), p. 7.

6. Ibid.

7. Ibid. In later years Rahv would retain this essentially tragic view of art.

8. Ibid.

9. Ibid.

10. See, for example, Karl Marx, *Capital* (New York: International Publishers, 1967), pp. 59–60, 408; and Friedrich Engels, *Anti-Duhring* (New York: International Publishers, 1939), p. 26.

11. For Burke's early view of catharsis, see *The Philosophy of Literary Form* (Berkeley:

University of California Press, 1973), p. 320. Brecht's views can be found in *Brecht on Theater* (New York: Hill and Wang, 1957), esp. pp. 78–79, 181.

12. Rahv, "The Literary Class War," p. 8. Rahv's disgust at what he regarded as decadence barely concealed a strain of puritanism common to the Communist left then and since. Although officially committed to the idea of equality for women, the Communist movement generally maintained traditional views concerning femininity, masculinity, and sexual practice. In many proletarian novels, for example, this meant that the proletarian heroine remained pure by resisting the inevitable advances of the boss, and the proletarian hero, in respect for his female comrades, managed to demonstrate his manhood elsewhere with unnamed women in unspecified places. Throughout the proletarian and leftist male corpus, sexuality remained taboo as a legitimate, serious subject, and there was little consciousness of a double standard. Politicized women writers, on the other hand—Josephine Herbst, Meridel LeSueur, Mary McCarthy, Tess Slesinger—wrote perceptively on the matter of sexual politics. Much, of course, remains to be written on the intriguing subject of communism and sexuality. The reader would do well to begin with Mary McCarthy's penetrating rendering of sexual politics on the left in "Portrait of the Intellectual as a Yale Man," in *The Company She Keeps* (1942; repr. New York: Harcourt Brace Jovanovich, 1970), pp. 165–246; Tess Slesinger's *The Unpossessed* (1934; repr. Old Westbury, N.Y.: Feminist Press, 1984); Lionel Trilling's "The Fate of Pleasure," in *Beyond Culture* (1965; repr. New York: Harcourt Brace Jovanovich, 1979); pp. 50–76; Charlotte Nekola and Paula Rabinowitz, eds., *Writing Red: An Anthology of American Women Writers, 1930–1940* (Old Westbury, N.Y.: Feminist Press, 1987); and Barbara Foley, *Radical Representations: Politics and Form in U.S. Proletarian Fiction, 1929–1941* (Durham: Duke University Press, 1993). For a more complete discussion of these matters, see chapter 8.

13. Rahv, "The Literary Class War," p. 8.

14. Ibid., pp. 8–9. In attributing a debilitating asociality to Eliot, Rahv had seriously misread the meaning of Eliot's words. If we attend to Eliot's text, "Imperfect Critics," in *The Sacred Wood*, we note that Rahv substituted an ellipsis for the following: "a man like Wyndham brings several virtues into literature. But there is only one man better and more uncommon than the patrician, and that is the Individual" ([1920; repr. New York: Methuen, 1980], p. 32). Eliot's neglected statement was thus quite pointedly "social"—one might venture to say even "progressive" insofar as it stood against snobbery and undeserved privilege. Rahv's was one of many examples where a critic of the left confused modernism's impressive disdain for philistine arrogance and its egalitarian defense of refined sensibility, if I may put it that way, for elitism. Rahv's opinion eventually changed, and Eliot's influence came to be decisive as Rahv repeatedly returned to his writing as a touchstone of modernism.

15. Rahv, "The Literary Class War," p. 8.

16. It should be noted that the juxtaposition of avant-garde and vanguard was meant to contrast two competing notions of internationalism—it was not an example of literary nationalism feeding off of Europhobia or Francophobia. For proletarian critics such as Rahv, the problems of American life reflected the contradictions of a distinctly global crisis, not a national crisis. Thus a worldwide revolutionary movement would ultimately defeat capitalism—the Communist Party was only a part of this overall offensive. Not until the Popular Front strategy of 1935 would some Party critics join with the advocates of a native tradition to oppose the increasingly reactionary politics of so-called international modernists like Eliot, Lewis, and Pound (see chapter 7). *Partisan Review* maintained its own position in later years, standing against both the reactionary aspects of Eliot's and

Pound's modernism, and the search for a usable American past. Throughout the late 1930s and early 1940s the magazine and its circle defended modernism, but as I will demonstrate, it never abandoned its profoundly ambivalent feelings toward its more radical experimentalists, including the late Joyce, Pound, Zukofsky, Barnes, and West, to name a few.

17. Rahv, "The Literary Class War," p. 10.

18. A. B. Magil, "Pity and Terror," *New Masses* 8, 5 (December 1932), p. 16.

19. Ibid., p. 18. In the strange world of Stalinist politics, Magil's opinions had to conform to the overall strategic line of the Party and the Comintern, which in this period of ultra-leftism meant the condemnation of all social-democrats and liberals as "social fascists." To be consistent, therefore, Magil reinterpreted Rahv's leftism as rightism, the main political danger. The arcane logic here was that by not enlisting all possible allies, Rahv was encouraging capitalism and liberalism to prevail, indicating a right deviation. It is significant that Magil was willing to overlook politics in the case of writers whose work was bitterly critical of capitalism.

20. Philip Rahv, "An Open Letter to Young Writers," *Rebel Poet* 16 (September 1932), p. 3.

21. Ibid.

22. Ibid.

23. Philip Rahv, "T. S. Eliot," *Fantasy* 2, 3 (Winter 1932), p. 17.

24. Ibid.

25. Ibid., pp. 17–18.

26. Ibid., p. 18.

27. Ibid., p. 19.

28. Ibid.

29. Both appeared in January 1933. A third piece, which I will not look at here, was a doctrinaire review of Max Planck's *Where Is Science Going* for the February 1933 issue of the *Communist*.

30. Wallace Phelps (William Phillips), "Class-ical Culture," *Communist* 12, 1 (January 1933), p. 93. Phillips used the pseudonym Wallace Phelps when publishing within the Communist movement until the July–August 1935 installment of *Partisan Review*. He retained the name William Phillips when publishing outside the Party discourse.

31. Ibid., pp. 93–94. Phillips did not identify the school.

32. Ibid., p. 96.

33. Ibid., p. 94. Ironically, when it came time for the Party to denounce Phillips and Rahv four years later, it used similar tactics. V. J. Jerome at that time wrote "Men who were but yesterday unknown to or ignored by the class enemy become luminaries overnight. . . . *Presto*, when they emit their first open attack upon the Soviet Union, the twin tyros are suddenly emblazoned in 'liberal' quarters as 'Marxist' critics, and given financial backing by bourgeois 'angels.'" "No Quarter for Trotzkyists—Literary or Otherwise," *Daily Worker* (October 20, 1937), p. 3.

34. Phelps (Phillips), "Class-ical Culture," p. 95.

35. Ibid., p. 96.

36. This first published work of Phillips has by now been long forgotten, a victim of the subsequent ordering of critical discourse into distinct schools, a process that has regrettably neglected those approaches which belong to the interstices, in this case between the New Criticism and doctrinaire Marxism. Even Phillips's own recollection of his suggestive essay seems to have suffered certain omissions. Writing about the essay in his memoirs, Phillips remarked rather modestly: "Everything I had learned and thought about [at NYU] was poured later into an essay that was printed in *The Symposium* (my first published piece)

under the weighty intellectual title 'Categories for Criticism.' . . . Though the essay contained some original speculations, I must say it was full of heavy thinking and a heavy prose. . . . But it impressed some people. Years later, I came upon a review of American periodicals in T. S. Eliot's magazine *The Criterion* by H. S. D. . . . which singled out my piece as representing a breakthrough that was forging a new critical language. Forging or not, the prose reflected the youth of the writer" (*A Partisan View: Five Decades of the Literary Life* [New York: Stein and Day, 1983], p. 29). On inspection, the reviewer, identified by G. A. M. Janssens as Hugh Sykes Davies, was not quite as complimentary as Phillips remembered, although perhaps he should have been. It turns out that Davies was mainly interested in comparing Phillips's diction to that of an inferior writer. Superior though Phillips's prose was in this regard, "it would be a pity," Davies remarked, "if critical prose should become generally like that of Mr. Phillips. It may be because I am not American, but I find it terribly hard to follow." Elsewhere, Phillips was praised because he had "something new to say" and he was "reaching after some elusive . . . notion" (*Criterion* 12, 48 [April 1933], pp. 542–43). There was no mention of a "breakthrough."

37. William Phillips, "Categories for Criticism," *Symposium* 4, 1 (January 1933), p. 32.

38. Ibid.

39. Ibid., p. 33.

40. Ibid.

41. Ibid.

42. Ibid., p. 37.

43. For more on the official theory of historical development, see Maurice Cornforth, *Materialism and Dialectical Method* (New York: International Publishers, 1953); and Joseph Stalin, "The Foundations of Leninism," in *The Essential Stalin*, ed. Bruce Franklin (New York: Doubleday, 1972), pp. 89–186. The latter was a popular textbook of official Marxism during the 1930s.

44. Phillips's union of judgment, need, and action, not to put too fine a point on it, resembles similar notions of "praxis" and "human interest" advocated by Horkheimer and Habermas. See, for example Max Horkheimer, "Materialismus und Metaphysik," *Zeitschrift für Sozialforschung* 2, 1 (1933), as quoted in Martin Jay, *The Dialectical Imagination* (Boston: Little, Brown, 1973), esp. pp. 53–54; and Jürgen Habermas, "Modernity—An Incomplete Project," in *The Anti-Aesthetic: Essays on Postmodern Culture*, ed. Hal Foster (Port Townsend, Wash.: Bay Press, 1983), pp. 3–15; "An Alternative Way Out of the Philosophy of the Subject: Communication versus Subject-Centered Reason," in *The Philosophical Discourse of Modernity* (Cambridge, Mass.: MIT Press, 1987), pp. 294–326.

45. Phillips, "Categories For Criticism," p. 47.

46. Daniel Aaron has argued persuasively that in spite of a very definite political disposition that writers and critics of the movement shared with the Party, a relative, pale sort of autonomy, as it were, existed for them. Aaron writes: "That *The New Masses*, as well as many other proletarian magazines, puffed up ineptly written revolutionary books is . . . true. Yet it is misleading to say that in 1934 and 1935 the party 'dominated' leftist critics and required them to make literary evaluations repellent to *Partisan Review*. Such a conclusion rests upon two faulty assumptions: that the party, then and after, deeply concerned itself with writers and writing, and that radical writers formed a cohesive and malleable group. Naturally, party functionaries hoped to organize writers and to induce them to support policies in line with the official party position, but party leaders like Browder valued writers for their prestige and popularity rather than for the purity of their Marxism or intrinsic literary merit. Mass movements, labor, unemployment—these were the throbbing issues" (Daniel Aaron, *Writers on the Left* [New York: Oxford University

Press, 1961], p. 299). More recently Cary Nelson in *Repression and Recovery: Modern American Poetry and the Politics of Cultural Memory, 1910–1945* (Madison: University of Wisconsin Press, 1989) and Barbara Foley in *Radical Representations: Politics and Form in U.S. Proletarian Fiction, 1929–1941* (Durham, N.C.: Duke University Press, 1993) have supplied an important corrective to the notion that the Party and its culture were monolithic, but in making their case they exaggerate the actual diversity of content, form, and quality found in the poetry produced within this milieu. For a fuller discussion, see chapter 10.

Chapter 2

1. Schapiro's discussion of the relevance of the city stroller to modern art predated Benjamin's discussion of the flaneur by two years. See Meyer Schapiro, "The Nature of Abstract Art," *Marxist Quarterly* 1, 1 (January–March 1937), pp. 77–98 (reprinted in Meyer Schapiro, *Modern Art, 19th and 20th Centuries: Selected Papers* [New York: George Braziller, 1979], pp. 185–212).

2. Wallace Phelps (William Phillips) and Philip Rahv, Editorial Statement, *Partisan Review* 1, 1 (February–March 1934), p. 2.

3. Ibid.

4. More specifically the period from 1927 to 1934 when the main international danger to the working class was said to be not fascism but liberalism, social democracy, and democratic socialism, all of which were alleged to represent petty bourgeois forces of capitulation within the working-class movement. The Comintern contemptuously labeled this trend "social fascism."

5. See pp. 29–30.

6. Specifically, the number of book reviews would reach an average of about six in each of the four 1935 issues. Starting with the third issue (June–July 1934), the average number of critical essays rose to three. At the same time the number of reportage pieces, poems, and short stories declined slightly.

7. Phillips and Rahv at one point publicly apologized for the inferior quality of the creative work offered by the magazine, admitting that "much of the published material has not measured up . . . at least seventy-five percent of the stories and poems submitted . . . are [ultra] 'leftist' in conception and in execution" ("Problems and Perspectives in Revolutionary Literature," *Partisan Review* 1, 3 [June–July 1934], p. 9). The wide divergence between the high expectations of the criticism and the poor quality of the creative work contributed to Phillips's and Rahv's ultimate disenchantment with the idea of a proletarian literary renaissance.

8. William Phillips and Philip Rahv, "In Retrospect: Ten Years of *Partisan Review*," in *The Partisan Reader* (New York: Dial Press, 1946), p. 680. See also Joseph Freeman to Floyd Dell, April 13, 1952, New York, Dell Papers, Newberry Library, Chicago, p. 8. Freeman also claimed to have written the opening editorial statement.

9. Irving Howe consistently defended democratic versions of socialism in *Dissent*. Dwight Macdonald embraced various forms of oppositional politics, most notably pacifism. Philip Rahv during the 1960s returned to the Marxism of his earlier years—a "born-again Leninist" as one commentator has put it, and Mary McCarthy, among others, was an active opponent of the Vietnam War.

10. William Phillips, *A Partisan View: Five Decades of the Literary Life* (New York: Stein and Day, 1983), p. 29.

11. Terry Cooney argues in *The Rise of the New York Intellectuals: Partisan Review and Its Circle, 1934–1945* (Madison: University of Wisconsin Press, 1986) that Edmund Wilson exerted an influence greater than that of Eliot, especially with *Axel's Castle* (pp. 28–29). Brooks in *America's Coming of Age* (1915) and James T. Farrell in *A Note on Literary Criticism* (1936) both provided the two critics with compelling models of socially conscious criticism.

12. As I will demonstrate, their appropriation of Eliot also had implications for their subsequent drift away from Marxist criticism in the early 1940s.

13. The range of more or less officially sanctioned views on the correct definition of proletarian literature can be found in Henry Hart, ed., *American Writers' Congress* (New York: International Publishers, 1935). Some critics insisted on more than one of these prerequisites if literature was to be considered authentically proletarian. This said, it must be kept in mind that the proletarian literary movement was far from monolithic. Phillips and Rahv represented one example among many of critics and especially writers sympathetic to the idea of proletarian literature who yet dissented from the semiofficial pronouncements and produced committed work of genuine originality. Significantly, a large proportion of writers willing to dissent from the orthodoxies of dissent were women and people of color, whose experiences differed from those of "typical" workers. These included Tillie Olsen, Meridel LeSueur, Muriel Rukeyser, Richard Wright, and Langston Hughes, to name a few. For other exemplary cases, see Constance Coiner, *Better Red: The Writing and Resistance of Tillie Olsen and Meridel LeSueur* (New York: Oxford University Press, 1995); Barbara Foley, *Radical Representations: Politics and Form in U.S. Proletarian Fiction, 1929–1941* (Durham: Duke University Press, 1993); Charlotte Nekola and Paula Rabinowitz, eds., *Writing Red: An Anthology of American Women Writers, 1930–1940* (Old Westbury, N.Y.: Feminist Press, 1987); Cary Nelson, *Repression and Recovery: Modern American Poetry and the Politics of Cultural Memory, 1910–1945* (Madison: University of Wisconsin Press, 1989); and Paula Rabinowitz, *Labor and Desire: Women's Revolutionary Fiction in Depression America* (Chapel Hill: University of North Carolina Press, 1991). Still there were others who attempted to dissent from within the movement and found themselves on the wrong side of the well-patrolled border of ideological acceptability. The example of Kenneth Burke is well known in this regard (see his "Revolutionary Symbolism in America" in Hart, *American Writers' Congress*, pp. 87–94). For suggesting to the Congress that the symbol of "the people" was superior to that of "the worker," Burke was attacked and then ostracized. Ironically, within months the line of the Comintern and the Party shifted to the right, whereupon Burke's perspective was enthusiastically adopted.

14. Janet Todd, *Sensibility: An Introduction* (New York: Methuen, 1987).

15. Wallace Phelps (William Phillips), "Sensibility and Modern Poetry," *Dynamo* 1, 3 (Summer 1934), p. 25.

16. Ibid., p. 20.

17. Ibid., p. 22.

18. Ibid.

19. Ibid.

20. Wallace Phelps (William Phillips), "The Anatomy of Liberalism," review of *Anatomy of Criticism* by Henry Hazlitt, *Partisan Review* 1, 1 (February–March 1934), p. 49. It is worth noting that the art critic John Berger has profitably employed the notion of "ways of seeing" in similar fashion.

21. Wallace Phelps (William Phillips) and Philip Rahv, "Criticism," *Partisan Review* 2, 7 (April–May 1935), p. 20.

22. See Theodor Adorno, *Aesthetic Theory*, trans. C. Lenhardt (1970; New York: Routledge, 1984), pp. 320–69; Louis Althusser, *Lenin and Philosophy*, trans. B. Brewster (New York: Monthly Review Press, 1971), pp. 127–86, 221–27; and Raymond Geuss, *The Idea of a Critical Theory: Habermas and the Frankfurt School* (Cambridge: Cambridge University Press, 1981), pp. 26–44. To suggest further comparisons, like Gramsci, Phillips and Rahv refused to equate consciousness with a single system of signification. They shared his premise that bourgeois culture was heterogeneous and contradictory, and that the orthodox model of a monolithic, economically determined culture failed to explain the complex character and function of "hegemonic" culture in the industrial democracies. Like Raymond Williams, in their comments on Faulkner's *Sanctuary* Phillips and Rahv devalued the importance of ideology. They certainly considered formally held systematic beliefs to be relevant, but like Williams they believed that neither bourgeois culture nor bourgeois literature were reducible to them (see Raymond Williams, *Marxism and Literature* [Oxford: Oxford University Press, 1977], pp. 108–14).

23. Wallace Phelps (William Phillips), "Form and Content," *Partisan Review* 2, 6 (January–February 1935), p. 34.

24. Ibid., pp. 33–34.

25. Ibid., p. 34.

26. I have employed Fredric Jameson's terms here. See *Marxism and Form* (Princeton: Princeton University Press, 1971), pp. 306–416, for his highly influential and still useful discussion of the form/content problem.

27. For Phillips and Rahv, as well as for the rest of the proletarian critics, the "cultural heritage" was acknowledged to include modernism as a major component. The writers and critics of the 1930s were consumed by the 1920s as they struggled to develop workable responses to that productive and innovative decade. The passion with which the proletarian movement, for example, set out both to reject and renew the literature of the 1920s has led some critics to view the movement itself as an epiphenomenon of modernism—as a "rebellion within a revolution," as Marcus Klein has put it. See his "The Roots of Radicals: Experience in the Thirties," in *Proletarian Writings of the Thirties*, ed. David Madden (Carbondale: Southern Illinois University Press, 1968), pp. 134–57. This, I think, is a somewhat exaggerated view: the proletarian writers, in spite of incorporating a surprisingly complete array of modernism's formal innovations (though not often successfully), never shared the modernists' belief in the transformative power of re-forming literature in and of itself.

28. Philip Rahv, review of *Winner Take Nothing* by Ernest Hemingway, *Partisan Review* 1, 1 (February–March 1934), p. 58.

29. Philip Rahv, "How the Waste Land Became a Flower Garden," *Partisan Review* 1, 4 (September–October 1934), p. 41. This was the first of several antinomial distinctions he was to make over the years, the best known of which was the Paleface/Redskin antithesis, about which I will have more to say later.

30. Ibid.

31. Wallace Phelps (William Phillips), "Three Generations," *Partisan Review* 1, 4 (September–October 1934), pp. 52–53.

32. Eliot's dialectical philosophical inquiries in both his dissertation and his early criticism have been explored most recently in Michael Levenson, *A Genealogy of Modernism* (Cambridge: Cambridge University Press, 1984); Walter Benn Michaels, "Philosophy in Kinkanja: Eliot's Pragmatism," *Glyph* 8 (1981), pp. 170–202; Sanford Schwartz, *The Matrix of Modernism* (Princeton: Princeton University Press, 1985); and Richard Shusterman, *T. S. Eliot and the Philosophy of Criticism* (London: Gerald Duckworth, 1988).

Schwartz makes some very interesting comparisons between Eliot's subject/object dialectic and Lukács' analysis of reified subjectivity and objectivity.

33. Rahv, review of *Winner Take Nothing*, p. 60.

34. Philip Rahv, "The Novelist as a Partisan," *Partisan Review* 1, 2 (April–May 1934), p. 50.

35. Philip Rahv, "A Season in Heaven," *Partisan Review* 3, 5 (June 1936), p. 10. Phillips had reviewed the notorious volume when it first appeared in 1934. Although he joined the leftist chorus of denunciation, claiming that "only the blind would hesitate to call Eliot a fascist," he sounded an unusually elegiac note as he mourned the loss of the critic whose early work had "won him a large following among young American and English writers." "Eliot Takes His Stand," *Partisan Review* 1, 2 (April–May 1934), p. 52.

36. T. S. Eliot, *The Sacred Wood* (1920; repr. London: Methuen, 1980), pp. 63–64.

37. Ibid., pp. 170–71.

38. Ibid., p. 165.

39. Specifically, they had stated that "the critic is the ideologist of the literary movement," and followed this with a long passage from Lenin that began thus: any ideologist "is worthy of that name only when he marches ahead of the spontaneous movement, points out the real road [and solves] all the theoretical, political, and tactical questions." "Problems and Perspectives in Revolutionary Literature," *Partisan Review* 1, 3 (June–July 1934), p. 5.

40. Phelps (Phillips) and Rahv, "Criticism," pp. 16–17.

41. Ibid., p. 16.

42. Ibid., p. 17.

43. Ibid.

44. Publicly announced in Rahv's insightful but ultimately reductive attack "Proletarian Literature: A Political Autopsy," in which he claimed that "proletarian literature is a literature of a party disguised as the literature of a class" (*Southern Review* 4, 3 [Winter 1939], p. 623). Rahv only acknowledged in passing that many others, like himself and Phillips, had struggled to reshape the movement from within. See pp. 80–81 and 146–148.

45. The most notable examples are Rahv's "Paleface and Redskin" *Kenyon Review* (Summer 1939; reprinted in Essays in Literature and Politics, 1932–1972 [Boston: Houghton Miflin, 1978]), pp. 3–7, and "The Cult of Experience in American Writing," *Partisan Review* 7,6 (November–December 1940), pp. 412–24, Trilling's "Reality in America," in *The Liberal Imagination* (1950; repr. New York: Harcourt Brace Jovanovich, 1979), pp. 3–20, and Chase's *The American Novel and Its Tradition* (Garden City: Doubleday, 1957). Along with Eliot's notion of the dissociated sensibility, Van Wyck Brooks's highbrow/lowbrow dualism was also influential.

46. Phelps (Phillips) and Rahv, "Criticism," p. 23.

47. See Raymond Williams, *Marxism and Literature* (Oxford: Oxford University Press, 1977), pp. 206–12.

48. One might contrast this prevailing attitude toward modernism with that of the British left, for which modernism generally did not present nearly the same kind of threat. W. H. Auden, Louis MacNeice, Cecil Day-Lewis, and Stephen Spender, among others, were all popular poets of the left whose debt to modernism was profound and openly acknowledged. There were many reasons for the difference, two of the most important having been the relative strength of British cultural and intellectual traditions—hence the greater confidence and autonomy of its writers—and the relative weakness of the British Communist Party compared with the American Communist Party during the 1930s. See Valentine Cunningham, *British Writers of the Thirties* (New York: Oxford University Press, 1988), and Samuel Hynes, *The Auden Generation* (Princeton: Princeton University Press, 1971).

Chapter 3

1. Wallace Phelps (William Phillips) and Philip Rahv, "Problems and Perspectives in Revolutionary Literature," *Partisan Review* 1, 3 (June–July 1934), p. 3.

2. Wallace Phelps (William Phillips) and Philip Rahv, "Criticism," *Partisan Review* 2, 7 (April–May 1935), p. 23.

3. Ibid.

4. Phelps (Phillips) and Rahv, "Problems and Perspectives," p. 8. Those familiar with the New Criticism may recall the name of Don West: he and his poem "Southern Lullaby" provided Cleanth Brooks his foil for the important chapter "Metaphysical Poetry and Propaganda Art" in the widely influential *Modern Poetry and the Tradition* (1939; repr. Chapel Hill: University of North Carolina Press, 1967); see p. 51.

5. Phelps (Phillips) and Rahv, "Problems and Perspectives," p. 8.

6. Ibid.

7. In Europe the Frankfurt Institute was the first serious departure from this institutional arrangement, representing, as Perry Anderson has indicated, a transition from the party to the university in the affiliations of Marxist intellectuals. See Perry Anderson, *Considerations on Western Marxism* (London: Verso, 1976), pp. 49–50.

8. It should be mentioned that the same was not as true for intellectuals and organizations within the Trotskyist movement, which made space, albeit a somewhat constricted space, for such intellectuals as Dwight Macdonald and James Burnham. For more on intellectuals in the Trotskyist movement, see Dwight Macdonald, *Politics Past* (New York: Viking Press, 1957), pp. 272–83; and Alan Wald, *The New York Intellectuals: The Rise and Decline of the Anti-Stalinist Left from the 1930s to the 1980s* (Chapel Hill: University of North Carolina Press, 1987).

9. For more on the Proletarian Artists and Writers League, which was modeled on the better known Clarté group, see Daniel Aaron, *Writers on the Left* (New York: Oxford University Press, 1961), p. 410; and Joseph Freeman, *An American Testament: A Narrative of Rebels and Romantics* (New York: Farrar and Rinehart, 1936), p. 284.

10. *New Masses* 5 (January 1930), as quoted in Aaron, *Writers on the Left*, p. 213. Aaron notes that Gold was opposed by Joseph Kalar, who argued for the more romantic idea of roving reporters in the tradition of John Reed himself.

11. For an account of one ex–John Reed Club member's disillusionment in the face of these ideological vicissitudes and institutional changes, see Richard Wright's contribution to *The God That Failed*, ed. Richard Crossman (New York: Harper, 1949), pp. 115–62.

12. Entitled "The Fascist Offensive and the Tasks of the Communist International in the Struggle of the Working Class against Fascism." An abridged version can be found in Georgi Dimitrov, *Selected Works*, vol. 2 (Sofia: Sofia Press, 1972), pp. 7–85.

13. The Party's reformism deepened in the years that followed. Just as it had abandoned the proletarian point of view in literature, so too in national politics and the labor movement it jettisoned demands on behalf of the working class. All was sacrificed to the Democratic Front, which, once the Democratic Party was included, reflected Roosevelt's policies more than the Communists'. The culmination of this policy was the Party's self-immolation at war's end when it turned itself into a "political association" and relinquished many of its hard-won rank-and-file organizations and leadership positions. For an account of the Party's steady rightward shift, see Harvey Klehr, *The Heyday of American Communism* (New York: Basic Books, 1984), pp. 184–222.

14. James T. Farrell Diary, August 14, 1936, James T. Farrell Collection, Van Pelt

Library, University of Pennsylvania, Philadelphia. As quoted in Terry A. Cooney, *The Rise of the New York Intellectuals: Partisan Review and Its Circle, 1934–1945* (Madison: University of Wisconsin Press, 1986), p. 97.

15. The Trials continued into 1938 when Bukharin, Rykov, and Kretinsky were charged with crimes ranging from treason and sabotage to plotting to reestablish capitalism. By this time Phillips and Rahv had already broken with the Party.

16. Klehr, *The Heyday of American Communism*, p. 358.

17. Farrell Diary, August 20, 1936, and August 21, 1936, as cited by Cooney, *The Rise of the New York Intellectuals*, p. 97.

18. "Falsely Labeled Goods," Editorial, *New Masses* 24, 12 (September 14, 1937), pp. 9–10.

19. William Phillips and Philip Rahv, Letter, *New Masses* 25, 4 (October 19, 1937), p. 21.

20. See V. J. Jerome, "No Quarter for Trotzkyists–Literary or Otherwise," *Daily Worker* (October 20, 1937), p. 3.

21. "Trotzkyist Schemers Exposed," *Daily Worker* (October 19, 1937), p. 3.

22. Appropriately the fiftieth anniversary issue of *Partisan Review* (a combined issue of vol. 51, no. 4, and vol. 52, no. 1, 1984) reprinted the full text of the editorial statement, affording its readers the chance to familiarize themselves with one of the more important documents of twentieth-century American literary history.

23. Editorial Statement, *Partisan Review* 4, 1 (December 1937), p. 4.

24. Ibid., p. 3.

25. Ibid.

26. Ibid.

27. Ibid.

28. Ibid., p. 4.

29. Dwight Macdonald, Letter to Leon Trotsky, July 7, 1937 (bms Russian 13.1 [2836]), Trotsky Archive, Houghton Library, Harvard University, Cambridge, Mass. Macdonald suggested that Trotsky write on Dostoevsky, Freud, or Silone's latest novel *Bread and Wine*, or that he update his classic *Literature and Revolution*.

30. Leon Trotsky, Letter to Dwight Macdonald, July 15, 1937 (bms Russian 13.1 [8951]), Trotsky Archive, Harvard.

31. Dwight Macdonald, Letter to Leon Trotsky, August 23, 1937 (bms Russian 13.1 [2838]), Trotsky Archive, Harvard.

32. Leon Trotsky, Letter to Dwight Macdonald, January 20, 1938 (bms Russian 13.1 [8953]), Trotsky Archive, Harvard.

33. This statement and the following three quotations are from Philip Rahv, Letter to Leon Trotsky, March 1, 1938 (bms Russsian 13.1 [4211]), Trotsky Archive, Harvard.

34. By "literary forces" Rahv most likely meant influential publications and individuals who were promoting the kind of politically and socially relevant works *Partisan Review* was most interested in. By "purely formalist work" he may have been referring to such poems as Wallace Stevens's "The Dwarf," and "Anything Is Beautiful If You Say It Is"; e.e. cummings' "Speech from a Forthcoming Play"; James Agee's "Lyrics"; and Elizabeth Bishop's "Love Lies Sleeping"–all of which were published in one or another of *Partisan Review*'s first three issues.

35. To be sure, this attitude was indicative of another side to *Partisan Review*'s appropriation of modernism. Although the magazine's championing of modernism in a desperate age was enormously liberating, its judgment remained parochial when it came to the more

experimental modernists. Thus Pound, Louis Zukofsky, Djuna Barnes, and Nathanael West, to name only four, were utterly ignored by the magazine, the latter three writers laboring in New York virtually under the noses of its editors. Indeed, the literary manifestations of Breton's surrealism, either in West, the late Joyce, or elsewhere, were never held in very high esteem by *Partisan Review*, testimony to the magazine's limited tolerance of formal innovation. This is of course a vital element in judging *Partisan Review*'s ultimate effect on American literature and culture.

36. Trotsky, Letter to *Partisan Review*, March 21, 1938, Trotsky Archive, Harvard.

37. Clement Greenberg, "Avant-Garde and Kitsch," *Partisan Review* 6, 5 (Fall 1939), pp. 34–49. Reprinted in Clement Greenberg, *Art and Culture* (Boston: Beacon Press, 1961), pp. 3–21.

38. "This Quarter," *Partisan Review* 6, 1 (Fall 1938), p. 7.

39. "No progress" was reported in a letter from Dwight Macdonald to Trotsky dated December 19, 1938 (bms Russian 13.1 [2843], Trotsky Archive, Harvard). Later thirty people turned out for a preliminary meeting of IFIRA, but afterward hopes for an American affiliate expired (Dwight Macdonald, Letter to Trotsky, March 3, 1939 (bms Russian 13.1 [2845]), Trotsky Archive, Harvard).

40. *Socialist Appeal*, December 4, 1937.

41. John Wheelwright, Letter, "To the Editors of the *Socialist Appeal*," *Partisan Review* 4, 3 (February 1938), p. 63.

42. Ibid.

43. Ibid.

44. Ibid. The Trotskyist George Novack has subsequently admitted to possessing these much-maligned body parts.

45. *Poetry* 51, 3 (December 1937), p. 171.

46. "Ripostes," *Partisan Review* 4, 3 (February 1938), p. 62.

47. Ibid.

48. *The Marxist Quarterly* was apparently too "little" to be noticed by Fredrick Hoffman in his study, *The Little Magazine* (Princeton: Princeton University Press, 1947), which devoted an entire chapter to the political magazines of the 1930s. Michael Harrington has written a brief but appreciative account of the magazine in Joseph R. Conlin, ed., *The American Radical Press*, vol. 2 (Westport, Conn.: Greenwood Press, 1974), pp. 514–17.

49. Among the more influential of Wilson's article were "The Economic Interpretation of Wilder" (1930), "The Literary Class War" (1932), "American Critics, left and Right" (1937), all of which are collected in *The Shores of Light* (New York: Farrar, Straus, and Giroux, 1952), pp. 500–503, 534–39, 640–61; "Marxism and Literature," in *The Triple Thinkers* (New York: Harcourt Brace Jovanovich, 1938), pp. 266–89, and "The Myth of the Marxist Dialectic," *Partisan Review* 6, 1 (Fall 1938), pp. 66–81.

50. *Modern Monthly* 8, 3 (April 1934), pp. 143–65.

51. See his introduction to Regis Debray's *Teachers, Writers, Celebrities: The Intellectuals of Modern France* (London: New Left Books, 1981).

52. "Results of the P. R. Questionnaire," *Partisan Review* 8, 4 (July–August 1941), pp. 344–48.

53. From Barbara Ehrenreich and John Ehrenreich, "The Professional-Managerial Class," *Radical America* 11, 2 (March–April 1977), p. 19.

54. William Cain, *F. O. Matthiessen and the Politics of Criticism* (Madison: University of Wisconsin Press, 1988), pp. 44–45. Cain points out the bulk of this extraordinary list comes from a *Boston Herald* article, published just after Matthiessen's suicide, that was meant to smear his reputation.

Chapter 4

1. See pp. 75–76.

2. With one possible exception: according to James Gilbert there was a brief attempt in 1940 to broaden the magazine's readership by distributing free copies to inmates of Sing Sing Prison. See James Gilbert, *Writers and Partisans: A History of Literacy Radicalism in America* (New York: Wiley and Sons, 1968), p. 197.

3. Malcolm Cowley, "Red Ivory Tower," *New Republic* 97 (November 9, 1938), p. 22.

4. Milton Howard, "Esthetics of the Cage," *Mainstream* 1, 1 (Winter 1947), p. 56.

5. William Phillips and Philip Rahv, "Literature in a Political Decade," in *New Letters in America*, ed. Eleanor Clark and Horace Gregory (New York: Norton, 1937), pp. 174–75.

6. Ibid., p. 175.

7. Philip Rahv, "Proletarian Literature: A Political Autopsy," *Southern Review* 4, 3 (Winter 1939), p. 623.

8. Ibid., p. 618.

9. Ibid.

10. Ibid., p. 624.

11. William Phillips, *A Partisan View: Five Decades of the Literary Life* (New York: Stein and Day, 1983), pp. 43–44. See also William Phillips, Letter to Dwight Macdonald, July 27, 1937, Dwight Macdonald Papers, Yale University Library, New Haven, Conn., in which he told Macdonald he had been reading Trotsky.

12. Leon Trotsky, *Literature and Revolution* (Ann Arbor: University of Michigan Press, 1968), p. 193.

13. Ibid., p. 202.

14. Leon Trotsky, "Art and Politics," *Partisan Review* 5, 3 (August–September 1938), p. 9.

15. Ibid., p. 10.

16. Ibid., p. 3.

17. André Breton and Diego Rivera, "Manifesto: Towards a Free Revolutionary Art," *Partisan Review* 6, 1 (Fall 1938), pp. 50–51. According to Breton, Trotsky wrote the manifesto even though he and Rivera were the signatories. See André Breton, *Entretiens* (Paris: Gallimard, 1969), p. 182.

18. Breton and Rivera, "Manifesto," p. 50.

19. Ibid., p. 51.

20. Trotsky, *Literature and Revolution*, p. 221.

21. Ibid.

22. Breton and Rivera, "Manifesto," p. 49.

23. See pp. 69–70.

24. Philip Rahv, "Trials of the Mind," *Partisan Review* 4, 5 (April 1938), p. 3.

25. Ibid., p. 4.

26. Ibid., p. 8.

27. Julien Benda, *The Treason of the Intellectuals* (New York: Norton, 1969), p. 57.

28. Rahv, "Trials of the Mind," p. 9.

29. Ibid.

30. Ibid.

31. See p. 68.

32. In his letter dated April 10, 1938, Rahv informed Trotsky that the symposium had been dropped because Trotsky's objections had proved to be sound. Obviously none of these objections stopped Rahv from using the title anyway. See Philip Rahv, Letter to Trotsky, April 10, 1938, Trotsky Archive, Houghton Library, Harvard University, Cambridge, Mass.

33. Philip Rahv, "What Is Living and What Is Dead," *Partisan Review* 7, 3 (May–June 1940), p. 176.

34. Ibid., p. 177.

35. Ibid.

36. Ibid.

37. Philip Rahv, "Twilight of the Thirties," *Partisan Review* 6, 4 (Summer 1939), p. 5.

38. Ibid.

39. Ibid., p. 12.

40. Ibid.

41. Ibid., pp. 11–12.

42. Trotsky, *Literature and Revolution*, p. 193.

43. Antonio Gramsci, *Prison Notebooks* (New York: International Publishers, 1971), p. 20.

44. Rahv, "Twilight of the Thirties," p. 5.

45. Ibid., p. 15.

46. William Phillips, "The Intellectuals' Tradition," *Partisan Review* 8, 6 (November–December 1941), p. 482.

47. Ibid., p. 484.

48. Ibid., p. 485.

49. Ibid., p. 489.

Chapter 5

1. Dwight Macdonald has given a convenient roll call of groups that were to the left of the liberals on the political spectrum. "Reading from right to left" they were Communists, Social Democrats, Socialists, Lovestoneites, Socialist Workers' Party, Socialist Labor Party, and "a whole brigade of 'ites.'" *Politics Past: Essays in Political Criticism* (New York: Viking, 1957), p. 15.

2. Lionel Trilling, "The America of John Dos Passos," *Partisan Review* 4, 5 (April 1938), p. 27. The essay has been reprinted in Lionel Trilling, *Speaking of Literature and Society* (New York: Harcourt Brace Jovanovich, 1980), pp. 104–12.

3. Lionel Trilling, introduction to *The Partisan Reader: Ten Years of Partisan Review*, ed. William Phillips and Philip Rahv (New York: Dial Press, 1946), p. xii. Trilling's introduction was later reprinted as "The Function of the Little Magazine," in *The Liberal Imagination* (1950; repr. New York: Harcourt Brace Jovanovich, 1979), pp. 89–99.

4. Lionel Trilling, introduction to *The Partisan Reader*, p. xiv.

5. Lionel Trilling, preface to *The Liberal Imagination*, p. iii.

6. Fredric Jameson, *Marxism and Form* (Princeton: Princeton University Press, 1971), p. 306.

7. Trilling, preface to *The Liberal Imagination*, pp. vi–vii.

8. Though Rahv would soon change his mind about James's limitations, as we shall see, and in fact was instrumental in initiating the James revival of the 1940s, at this earlier juncture Rahv believed the novelist was doomed to spend his life wrestling with the prohibitive American legacy of what Rahv termed "playing hide-and-seek with experience."

Though admitting that James's idea that one should "live" one's life was revelatory within the American cultural context, Rahv claimed that the European writers were superior, for whom James's idea "has long been a thoroughly assimilated and natural assumption." "The Cult of Experience in American Writing," *Partisan Review* 7, 6 (November–December 1940), p. 412.

9. Ibid., p. 413.

10. Rahv added this passage in revising his essay. For the revised version, see his *Essays on Literature and Politics, 1932–1972*, ed. Arabel Porter and Andrew Dvosin (Boston: Houghton, Mifflin, 1978), p. 11.

11. Trilling, introduction to *The Partisan Reader*, p. xiii.

12. Trilling, "The America of John Dos Passos," pp. 27–28.

13. Ibid., p. 29.

14. Ibid. p. 30.

15. Ibid.

16. Ibid.

17. Ibid., pp. 30–31.

18. Ibid., p. 32.

19. Lionel Abel, "Ignazio Silone," *Partisan Review* 4, 1 (December 1937), p. 37. This was the first of many pieces Abel would contribute to *Partisan Review*. During the 1960s he moved toward neoconservatism. See Lionel Abel, *The Intellectual Follies: A Memoir of the Literary Venture in New York and Paris* (New York: Norton, 1984).

20. Lionel Trilling, "'Elements That Are Wanted,'" *Partisan Review* 7, 5 (September–October 1940), p. 367. The title of the essay was later changed to "T. S. Eliot's Politics," as if its major purpose was something other than to call attention to the left's lack of concern for Eliot's subject—the quality of one's life. "T. S. Eliot's Politics" is reprinted in *Speaking of Literature and Society*, pp. 156–69. Trilling also quoted Matthew Arnold's title-inspiring dictum that criticism "must be apt to study and praise elements that for the fullness of spiritual perfection are wanted, even though they belong to a power which in the practical sphere may be maleficent."

21. Trilling, "Elements That Are Wanted," p. 367.

22. Ibid., p. 376.

23. Ibid., p. 375.

24. Kenneth Burke, *Counter-Statement* (Berkeley: University of California Press, 1968), pp. 104–5, 106.

25. See Frank Lentricchia's incisive study of Burke, *Criticism and Social Change* (Chicago: University of Chicago Press, 1983).

26. Lionel Trilling, *E. M. Forster* (New York: New Directions, 1964), p. 22.

27. Lionel Trilling, "F. Scott Fitzgerald," in *The Liberal Imagination*, p. 231.

28. Ibid.

29. Trilling dropped the passages devoted to a critique of Bernard Smith, a progressive, fellow-traveling critic. Other changes served to blunt the politicized and highly polemical character of the initial version.

30. See pp. 159–60.

31. Lionel Trilling, "Reality in America," in *The Liberal Imagination*, p. 10.

32. Ibid., p. 5.

33. Ibid., p. 8.

34. Ibid., p. 12.

35. Ibid.

36. Ibid., p. 11.

37. Lionel Trilling, "Hemingway and His Critics," *Partisan Review* 6, 2 (Winter 1939), p. 53.

38. Ibid., p. 54.

39. Ibid., p. 60.

40. F. W. Dupee, "André Malraux," *Partisan Review* 4, 4 (March 1938), p. 24.

41. Ibid., pp. 25–26.

42. Ibid., p. 31.

43. Ibid.

44. By the mid-1940s *Partisan Review* was enamored enough of James that it was running full-page ads featuring the "special combination offer" of a one-year subscription and a free copy of *The Great Short Novels of Henry James*, edited by Rahv.

45. Philip Rahv, "The Heiress of All the Ages," *Partisan Review* 10, 3 (May–June 1943), p. 227.

46. Ibid., p. 231.

47. See p. 145, where I compare *Partisan Review*'s cosmopolitanism and its battle against American "provincialism" with the Bolsheviks' efforts to "Westernize" Russia in opposition to the Slavophiles.

48. Rahv, "The Heiress of All the Ages," p. 232.

49. F. O. Matthiessen, *Henry James: The Major Phase* (London: Oxford University Press, 1944), p. xiv.

50. A revised version of the essay appears under the title "Dostoevsky in *The Possessed*," in *Essays on Literature and Politics: 1932–1972*, pp. 107–28.

51. Hannah Arendt, "Franz Kafka: A Revaluation," *Partisan Review* 11, 4 (Fall 1944), p. 419.

52. Ibid., p. 421.

53. William Troy, "The D. H. Lawrence Myth," *Partisan Review* 4, 2 (January 1938), p. 11.

54. Ibid., p. 12. Troy here echoed Marx on "the present splendid brotherhood of fiction writers in England [Dickens, Thackeray, Gaskell, and "Miss Brontë"] whose graphic and eloquent pages have issued to the world more political and social truths than have been uttered by all the professional politicians, publicists, and moralists put together" ("The English Middle Class," [1854]) and Engels on Balzac, whose "complete history of French Society" taught him "more than . . . all the professed historians, economists, and statisticians of the period together" (Letter to Margaret Harkness [1888]). See Marx and Engels in *Literature & Art*, ed. Lee Baxandall and Stefan Morawski (St. Louis: Telos Press, 1973), pp. 105, 115.

55. Irving Howe, "The Culture of Modernism," in *The Decline of the New* (New York: Harcourt Brace Jovanovich, 1970), p. 16.

56. Fredric Jameson, "'Ulysses' in History," in *James Joyce and Modern Literature*, ed. W. J. McCormack and Alistair Stead (London: Routledge and Kegan Paul, 1982), pp. 126–41.

57. Lionel Trilling, "James Joyce in His Letters," in *The Last Decade: Essays and Reviews, 1965–75* (New York: Harcourt Brace Jovanovich, 1979), p. 33.

58. Ibid., p. 36.

59. Ibid. p. 30.

Chapter 6

1. The poems were "The Dwarf," (1937), "Loneliness in Jersey City" (1938), "Anything Is Beautiful If You Say It Is" (1938), "The Woman That Had More Babies Than

That" (1939), "Life on a Battleship" (1939), "Montrachet-Le-Jardin" (1942), "Someone Puts a Pineapple Together" (1947), and "Of Ideal Time and Choice" (1947). The prose pieces published by *Partisan Review* included "The Situation in American Writing" (1939), "The Realm of Resemblance" (1947), and "The State of American Writing" (1939). The books that were reviewed were *The Man with the Blue Guitar, and Other Poems*, reviewed by Delmore Schwartz (1938); *Notes toward a Supreme Fiction* and *Parts of a World*, reviewed by R. P. Blackmur (1943); *Transport to Summer*, reviewed by Delmore Schwartz (1947); and *Three Academic Pieces*, reviewed by Leslie Fiedler (1948).

2. See his *Ariel and the Police* (Madison: University of Wisconsin Press, 1988), esp. 217. Lentricchia very suggestively elaborates upon the politics of Stevens's post–"Owl's Clover" poetry in "Penelope's Poetry—The Later Wallace Stevens."

3. In Addition to Lentricchia, see Milton Bates, *Wallace Stevens: A Mythology of Self* (Berkeley: University of California Press, 1982), chap. 5; Alan Filreis, *Wallace Stevens and the Actual World* (Princeton: Princeton University Press, 1991), and *Modernism Right and Left: Wallace Stevens, the Thirties, and Literary Radicalism* (New York: Cambridge University Press, 1994); and James Longenbach, *Wallace Stevens: The Plain Sense of Things* (New York: Oxford University Press, 1991).

4. Needless to say I have paraphrased Daniel Bell's highly influential definition from *The End of Ideology* (Glencoe, Ill.: Free Press, 1960) because it accurately reflects the assumptions of Stevens's critics from the 1940s through the 1970s. There are of course broader and in my view better definitions of ideology that effectively link unsystematic and unconscious thought to behavior and to a range of social determinants.

5. Some of the characteristic tropes and polarities of cold war discourse are evident in this passage from Joseph Riddel's *The Clairvoyant Eye* (Baton Rouge: Louisiana State University Press, 1965): "Stevens' *distraction* by Marxist criticism and radical *ideology* . . . is as much poetic strategy as individual pique. The Marxist program becomes for him only the most recent of a long line of conceptual failures to provide a rationally ordered society. Though Stevens understood the dangers of putting poetry in the service of politics, he was not so alert to the consequences of trying to subsume the practical in the ideal. It was he, not his critics, who *made the error of deserting a defensible position to contend with the enemy on its own terms*. Cummings had more wisely ignored the argument; Frost had consistently avoided 'ideas.' Neither did Stevens have the political acumen of an Auden, who knew full well how far poetry could go in the service of a cause, and to what degree it had to be impersonal, aloof from causes and action" (pp. 121–22; emphasis added). I do not mean to single out Mr. Riddel—on the contrary, my point is that he was sharing with many other critics, too numerous to mention, certain widespread beliefs about leftism and an ideological language that reinforced them. Of course, doctrinaire Marxist critics on the other side of the barricades possessed their tropes and polarities, which they too deployed in order to diminish Stevens's poetry. In the last analysis the only question that divided Marxist and mainstream critics had to do with Stevens's nonpolitical poems: Could a poet so aloof be interesting?

6. Stanley Aronowitz, *The Crisis in Historical Materialism: Class, Politics, and Culture in Marxist Theory* (New York: Praeger, 1981), p. 3.

7. Lack of space prevents me from making detailed comparisons; I trust those familiar with the work of revising Marxism will recognize the similarities with Stevens's project as I describe it here. For surveys of these traditions, see Perry Anderson, *Considerations on Western Marxism* (London: Verso, 1979), and *In the Tracks of Historical Materialism* (London: Verso, 1983); Stanley Aronowitz, *The Crisis in Historical Materialism* (New York: Praeger, 1981); Fredric Jameson, *Marxism and Form* (Princeton: Princeton Univer-

sity Press, 1972); Martin Jay, *Marxism and Totality* (Berkeley: University of California Press, 1984); and *Marxism and the Interpretation of Culture*, ed. Cary Nelson and Lawrence Grossberg (Urbana: University of Illinois Press, 1988).

8. Wallace Stevens, *The Letters of Wallace Stevens*, ed. Holly Stevens (New York: Knopf, 1966), pp. 286, 287, 289, 295, 296.

9. See *Wallace Stevens: The Later Years, 1923–1955* (New York: William Morrow, 1988), 110. Richardson argues correctly that in his poems Stevens was interested in exploding polarities within politics, as well as between political and other kinds of experience. Such a project, however, need not have precluded the poet's own Manichaeanism, especially where issues of acute controversy and historic insensitivity were concerned, such as race, class, and gender.

10. Wallace Stevens, "The Noble Rider and the Sound of Words," in *The Necessary Angel* (New York: Knopf, 1951), p. 19.

11. Ibid., p. 17.

12. Ibid., pp. 18, 19.

13. Ibid., p. 20.

14. Ibid., p. 22.

15. "Among the handful of clichés which have crept into left-wing criticism," began Burnshaw, "is the notion that contemporary poets . . . have all tramped off to some escapist limbo where they are joyously gathering moonshine" ("Turmoil in the Middle Ground," *New Masses* 17, 1 [October 1, 1935], pp. 41–42; reprinted in *Wallace Stevens: The Critical Heritage*, ed. Charles Doyle [London: Routledge, 1985], pp. 137–40). As many readers are doubtless aware, not only baleful social developments but also the left's cool response to Stevens's poetry provoked him into reassessing and reshaping his work during the 1930s. I refer not only to Burnshaw's review of *Ideas of Order*, but also to the leftist reviews of Ruth Lechlitner ("Imagination as Reality," *New York Herald Tribune Books* [December 6, 1936], p. 40; repr. Doyle, *Wallace Stevens*, 156–60) and Geoffrey Grigson ("The Stuffed Goldfinch," *New Verse* 19 [February–March 1936], pp. 18–19), both of which Stevens read and remarked upon (see *Letters of Wallace Stevens*, pp. 309, 313, 332). Burnshaw, whose influence was greatest, made the claim that the heightened class struggles of the Depression period had disoriented Stevens, and that his latest poetry had expressed confusion about the poet's proper relationship to economic deprivation and social conflict. Because of the swiftness and vehemence of Stevens's response, it is still generally assumed that the review must have expressed a doctrinaire Marxist position by attacking Stevens for his alleged aestheticism, decadence, or elitism. But Burnshaw's review, although it shared important assumptions with doctrinaire Marxism, was less dogmatic than the usual leftist fare. His judgment of Stevens was based partially on the then familiar model of class struggle, in which the momentous battle between labor and capital at first disorients middle-class artists, and eventually forces them to choose sides as their class—the petty bourgeoisie—gradually erodes and finally disappears. But three factors distinguished Burnshaw's review from doctrinaire Marxism. First, his specific remarks concerning Stevens's poetry, although attenuated, were usually sensitive; they demonstrated reading habits easily distinguishable from those of the often benumbed critics of the left during the 1930s (this was especially telling with regard to their reading of modernist texts). Second, as I have already shown, Burnshaw began his article by assailing precisely this kind of critic. Indeed, Burnshaw's effort compares favorably to that of Grigson, the usually reliable British poet, critic, and editor of *New Verse*, a magazine Stevens otherwise admired. In his review Grigson displayed all of the clichés Burnshaw was alluding to: "In *Harmonium* we had a delicate man, an ironist, an imagist, a modern, a thin-fingered undemocratic American.

Here we have fewer melons and peacocks but still the finicking privateer, prosy Herrick, Klee without rhythm, observing nothing, single artificer of his own world of mannerism, mixer-up of chinoiserie. . . . " In sum, wrote Grigson, "Too much Wallace Stevens, too little everything else." The most important reason Burnshaw avoided dogmatism was that he posed the very interesting possibility that Stevens might become an ally of the left, depending upon how this poet of the middle ground turned out. We must remember that only a minority of leftist critics in the 1930s held out this possibility for such difficult and apparently self-absorbed writers as Stevens. Nonetheless, it must be remarked that the operating assumption was that any such alliance could be made possible only by the *poet* changing. The need or the possibility that the *left* change was not entertained, except only remotely in Burnshaw's disapproving statement about sectarian critics. Self-criticism and change, especially when it came to examining theoretical roots, operating assumptions, habits of mind, attitudes, and the like, were never carried out by the Communist party—and throughout the century have been extremely rare commodities among major socialist parties East and West (see Aronowitz, *The Crisis of Historical Materialism*, p. 133).

16. Wallace Stevens, *Collected Poems* (New York: Knopf, 1955), p. 65 (henceforth designated in the text as *CP*).

17. I am not committed to the belief that Stevens necessarily intended "Owl's Clover" to be a systematic critique of doctrinaire Marxism. I have strong doubts that Stevens would ever have conceived of doing such a thing in a poem, believing Elder Olson right when he observed that Stevens did not argue, he meditated, and that in his poetry there is no such thing as a connected argument. I am quite comfortable with the idea that in the course of imaginatively exploring broad philosophical and aesthetic issues, Stevens invariably dealt with doctrinaire Marxism explicitly and implicitly. I also think that it is worthwhile to draw out the consequences of his exploration of these issues for orthodoxy even where no link, explicit or implicit, exists in the poem.

18. Aronowitz, *The Crisis of Historical Materialism*, p. xvi. Within the Marxist tradition Eduard Bernstein, the author of the much maligned *Evolutionary Socialism* (1909) has come to be seen as the chief spokesperson for the reformist perspective.

19. Wallace Stevens, "Mr. Burnshaw and the Statue," in *Opus Posthumous* ed. Milton Bates (New York: Knopf, 1989), p. 50 (henceforth designated in the text as *OP*).

20. Ibid., p. 49. Elsewhere, in his remarks to Simons concerning "The Old Woman and the Statue," Stevens distinguished his own view of change and difference from that of orthodox dialectics: "When I was a boy I used to think that things progressed by contrasts, that there was a law of contrasts. But this was building the world out of blocks. Afterwards I came to think more of the energizing that comes from mere interplay, interaction. . . . Cross-reflections, modifications, counter-balances, complements, giving and taking are illimitable. They make things inter-dependent, and their inter-dependence sustains them and gives them pleasure" (*Letters of Wallace Stevens*, p. 368).

21. Stevens, Letter to Latimer, in *Letters of Wallace Stevens*, p. 289.

22. Ibid., pp. 291–92.

23. It may be worth pointing out that though the relationship between being and consciousness is characterized by reciprocity in these formulations, elsewhere Marx gave priority to the material factors associated with being, even if only "in the last instance," as Engels put it. No doubt this is ultimately the nub of the difference between Stevens and most varieties of Marxism and materialism.

24. See Aronowitz, *The Crisis of Historical Materialism*, p. 225.

25. Ralph Waldo Emerson, *Nature* (New York: Library of America, 1983), p. 16.

26. *Mike Gold: A Literary Anthology*, ed. Michael Folsom (New York: International

Publishers, 1972), p. 139. Gold was at least modest enough not to have added "I *am* that man."

27. See Adrienne Rich, *Poetry*, ed. Barbara Charlesworth Gelpi and Albert Gelpi (New York: Norton, 1975), esp. p. 116.

28. Stevens, "The Irrational Element in Poetry," in *Opus Posthumous*, p. 225.

29. See Wilhelm Reich, *The Mass Psychology of Fascism* (1933; repr. New York: Farrar, Straus, and Giroux, 1970). It goes without saying that Adorno, Horkheimer, Fromm, Marcuse, and others associated with the Frankfurt Institute also sought to develop a materialist understanding of fascism that incorporated an understanding of the unconscious and subjectivity. For a summary of these efforts, see Martin Jay, *The Dialectical Imagination* (Boston: Little, Brown, 1973), chap. 3.

30. Stevens, Letter to Hi Simons, in *Letters of Wallace Stevens*, p. 373. If this subjective experience is collective and even somehow racial as is indicated, in the earlier poem "The Greenest Continent" Stevens makes it clear that the transhistorical character of "what we knew" is not at all immune from culture and history. In fact, Stevens is interested in gauging the extent to which certain more or less constant features of human experience can be modified. However, he does not minimize nature's intractability, and thus he differs from Marxist orthodoxy, which subordinates nature to the social process, to conscious, revolutionary society, and to the material needs of society. Part of the reason orthodoxy avoided coming to terms with the unconscious and "the feminine principle" as defined by its own masculinist culture is simply because these named realms of human experience were thought to be relatively constant and not subject to immediate change. Thus "nature" and "human nature" became reified as absolute and unchanging categories completely outside the parameters of Marxist thought.

31. G. Dimitrov, *The United Front: The Struggle against Fascism and War* (London: L. Lawrence and Wishart, 1938), p. 9.

32. "Another Way of Looking at A Blackbird," *New Republic* 137, 4 (November 4, 1957), pp. 16–19. Reprinted in Doyle, *Wallace Stevens*, pp. 443–44, 444, 439.

33. A. Walton Litz, *Introspective Voyager: The Poetic Development of Wallace Stevens* (New York: Oxford University Press, 1972), p. 196.

Chapter 7

1. Philip Rahv, "Where the News Ends," *New Leader* 21 (December 10, 1938), p. 8. The G.P.U. was the Soviet secret police.

2. Philip Rahv, Letter to George Orwell, January 11, 1946, *Partisan Review* Archives. As quoted in James Gilbert, *Writers and Partisans* (New York: Wiley and Sons, 1968), p. 257.

3. "War Is the Issue!" *Partisan Review* 6, 5 (Fall 1939), p. 125.

4. See pp. 148–58.

5. Irving Howe and Lewis Coser, *The American Communist Party* (New York: Praeger, 1962), pp. 333–34.

6. Ibid., pp. 385–86.

7. James Dugan, "Changing the Reel," *New Masses* 34, 3 (January 9, 1940), p. 28.

8. "The Great American Film," *New Masses* 34, 7 (February 6, 1940), p. 28.

9. Book Review, *New Masses* 34, 8 (February 13, 1940), p. 25.

10. James Dugan, "Two Brilliant Medical Pictures," *New Masses* 34, 13 (March 19, 1940), p. 28.

11. Emil Pritt, "Citizen Kane," *New Masses* 38, 7 (February 4, 1941), p. 26.

12. Dwight Macdonald, "Soviet Society and Its Cinema," *Partisan Review* 6, 2 (Winter 1939), p. 80.

13. Dwight Macdonald, "A Theory of Popular Culture," *Politics* 1, 1 (February 1944), p. 20.

14. Ibid., p. 21.

15. Ibid.

16. Ibid.

17. Ibid.

18. The irony of the first two examples of Nazi-like "coordination" will not be lost on anyone familiar with Macdonald's own career. He himself had labored on the staff of Luce's *Fortune* magazine before quitting in early 1937, and later in the 1950s he began writing a regular column of "socio-cultural reportage" for the *New Yorker*.

19. Dwight Macdonald, "The Russian Cultural Purge," *Politics* 3, 9 (October 1946), pp. 300, 302.

20. Dwight Macdonald, "Masscult and Midcult," *Partisan Review* 27, 2 (Spring 1960), p. 208. This essay was an expanded and updated version of "A Theory of Popular Culture."

21. Dwight Macdonald, "A Theory of Popular Culture," p. 23.

22. Dwight Macdonald, "Masscult and Midcult," in *Against the American Grain* (New York: Random House, 1962), p. 72.

23. Greenberg's and Macdonald's 1941 statement "10 Propositions on the War" (*Partisan Review* 8, 4 [July–August 1941], p. 277) summed up their views, and provided Philip Rahv, a skilled polemicist, with a relatively easy target. In his "10 Propositions and 8 Errors" (*Partisan Review* 8, 6 [November–December 1941], pp. 499–506), Rahv noted the complete absence of a working-class movement, which might wage such a civil war. He considered their strategy to be dangerously adventurist because, even had it had a historical agent, it would open the doors to an Axis victory. At the time Rahv took the fascist threat to democracy, and thus its differences with capitalist democracy, more seriously than either Greenberg or Macdonald. Nevertheless, like them he responded above all to the alleged totalitarian trends in American society, and identified the Popular Front as their source.

24. Clement Greenberg, "Avant-Garde and Kitsch," *Partisan Review* 6, 5 (Fall 1939), p. 49.

25. Greenberg never specified which writers or artists he was talking about. He did say that the poets of the later avant-garde were Rimbaud, Mallarmé, Valèry, Eluard, Pound, Hart Crane, Stevens, Rilke, and Yeats.

26. Clement Greenberg, "Avant-Garde and Kitsch," p. 36.

27. Ibid., p. 49.

28. Richard D. Altick, *The English Common Reader: A History of the Mass Reading Public* (Chicago: University of Chicago Press, 1957), p. 6.

29. See George Orwell, "The Art of Donald McGill," in *A Collection of Essays by George Orwell* (Garden City, N.Y.: Doubleday, 1954), pp. 111–23.

30. Thus is not to suggest they neglected American literature: the New York intellectuals' focus on postromantic European literature, especially modernist fiction, is well known, but it should not obscure the fact that they had a great deal to say about American literature as well, a point Vincent Leitch has underscored in *American Literary Criticism from the 30s to the 80s* (New York: Columbia University Press, 1988). Aside from the important essays by Phillips, Rahv, and Trilling treated earlier, the New Yorkers made numerous contributions to the study of American literature, including Alfred Kazin's classic *On Native Grounds* (1942), and *Contemporaries* (1962); Chase's *The American Novel and Its Tradition* (1957); Fiedler's *Love and Death and the American Novel* (1960); Howe's *Sherwood*

Anderson (1951) and *William Faulkner* (1952); Elizabeth Hardwick's *Seduction and Betrayal: Women and Literature* (1974) and *Bartleby in Manhattan and Other Essays* (1983); Mary McCarthy's "Theatre Chronicles" and *On the Contrary*; and Rahv's edited versions of *The Great Short Novels of Henry James* (1944), *The Bostonians* (1945), *Discovery of Europe: The Story of American Experience in the Old World* (1947), *Literature in America: An Anthology of Literary Criticism* (1957), and *Eight Great American Short Novels* (1963).

31. William Phillips and Philip Rahv, "Literature in a Political Decade," in *New Letters in America*, ed. Eleanor Clark and Horace Gregory (New York: Norton, 1937), p. 176.

32. Ibid., p. 178.

33. Ibid., p. 179.

34. For assessments of Trilling's influence, see Russell Reising, *The Unusable Past: Theory and the Study of American Literature* (New York: Methuen, 1986), p. 94; David H. Hirsch, "Reality, Manners, and Mr. Trilling," *Sewanee Review* 72, 3 (Summer 1964), p. 420; Nicolaus Mills, *American and English Fiction in the Nineteenth Century* (Bloomington: Indiana University Press, 1973), p. 6; and Leo Marx, *The Machine in the Garden* (New York: Oxford University Press, 1964), pp. 341–42.

35. William Cain, "An Interview with Irving Howe," *American Literary History* 1, 3 (Fall 1989), p. 556.

36. In fact, there was a good deal of overlap. See Marcus Klein, "The Roots of Radicals: Experience in the Thirties," in *Proletarian Writers of the Thirties*, ed. David Madden (Carbondale: Southern Illinois University Press, 1968), pp. 134–57; and Cary Nelson, *Repression and Recovery: Modern American Poetry and the Politics of Cultural Memory, 1910–1945* (Madison: University of Wisconsin Press, 1989).

37. Philip Rahv, "Proletarian Literature: A Political Autopsy," *Southern Review* 4, 3 (Winter 1939), p. 618.

38. Ibid.

39. Ibid., p. 623.

40. See pp. 37, 275.

41. Rahv, "Proletarian Literature: A Political Autopsy," p. 624.

42. Ibid., p. 621.

43. This issue was quite dramatically played out at a panel devoted to the career of Tillie Olsen at an MLA convention several years ago. Following an astute and sympathetic presentation on the difficulties faced by Olsen and other women writers within the patriarchal milieu of the Party, Tillie Olsen herself stood up and proceeded to chastize the panelist for dwelling on the negative and forgetting that, in her words, "the Party taught me everything I knew." A balanced assessment of the proletarian literary movement remains difficult to make precisely because the issue was polarized at the outset and because the ideological stakes remain very high. For an admirable, though finally I think somewhat defensive treatment of the Party's role in these gender matters, see Barbara Foley, "Women and the Left in the 1930s," *American Literary History* 2, 1 (Spring 1990), pp. 150–69.

44. William Phillips and Philip Rahv, Editorial Statement, *Partisan Review* 4, 1 (December 1937), p. 4.

45. Hook went on to compare the Popular Front's "dangerous" and "defensive" strategy with both the social revolutionary regime of Kerensky and the coalition government of the Weimar Republic, asserting that whenever parties representing different classes ally, the far right prevails: "everything else is meaningless. Only when the working class convinces its potential allies—the farmers and the lower middle class (including the intellectuals) that the socialist solution is the only one, will the strategic problems of the revolution be worked

out." Sidney Hook, "Anatomy of the Popular Front," *Partisan Review* 6, 3 (Spring 1939), pp. 29–45. For Lenin's critique of economism, see "Engels on the Importance of the Theoretical Struggle," in *What Is To Be Done* (Beijing: Foreign Language Press, 1973), pp. 26–33.

46. As quoted in James Hoopes, *Van Wyck Brooks: In Search of American Culture* (Amherst: University of Massachusetts Press, 1977), p. 198.

47. F. W. Dupee, "The Americanism of Van Wyck Brooks," *Partisan Review* 6, 4 (Summer 1939), p. 77.

48. Ibid., p. 78.

49. Ibid., p. 85.

50. Archibald MacLeish, "The Irresponsibles," *Nation* 150, 20 (May 18, 1940), p. 619.

51. Ibid., p. 621.

52. Ibid., p. 618.

53. *Nation* 150, 23 (June 8, 1940), p. 718. Waldo Frank, Perry Miller, Max Lerner, and several others also wrote letters expressing strong agreement. See James R. Vitelli, *Van Wyck Brooks* (New York: Twayne, 1969), p. 144.

54. Van Wyck Brooks, "On Literature Today," speech given at Hunter College, collected in *Opinions of Oliver Allston* (New York: E. P. Dutton, 1941), pp. 195–96.

55. Brooks, *The Opinions of Oliver Allston,* pp. 211, 216.

56. Pound had recommended Corbière and Rimbaud in "How To Read." See *The Literary Essays of Ezra Pound* (New York: New Directions, 1968), pp. 33, 38.

57. Brooks, *The Opinions of Oliver Allston*, p. 237.

58. Morton Zabel, "The Poet on Capital Hill," *Partisan Review* 8, 1 (January–February 1941), p. 2.

59. Ibid., pp. 2–3.

60. Ibid., p. 14.

61. Dwight Macdonald, "Kulturbolschewismus Is Here," *Partisan Review* 8, 6 (November–December 1941), p. 449.

62. Ibid., p. 446.

63. Ibid., pp. 448–49.

64. Ibid., p. 451.

65. "On the 'Brooks-MacLeish Thesis,'" *Partisan Review* 9, 1 (January–February 1942), pp. 38–42.

66. Nancy Macdonald for the editors to Glenway Westcott, November 17, 1941, *Partisan Review* Archives. As quoted in James Gilbert, *Writers and Partisans: A History of Literary Radicalism in America* (New York: Wiley and Sons, 1968), p. 229.

67. T. S. Eliot, "Letter to the Editors," *Partisan Review* 9, 2 (March–April 1942), pp. 115–16.

68. Ransom, "On the 'Brooks-MacLeish Thesis,'" p. 40.

69. Clement Greenberg and Dwight Macdonald, "Ten Propositions on the War," *Partisan Review* 8, 4 (July–August 1941), pp. 271–78.

70. Robert Warshow, "The Legacy of the 30's," *Commentary* 4, 4 (December 1947), p. 543.

71. Rahv, "The Cult of Experience in American Writing," *Partisan Review* 7, 6 (November–December 1940), p. 9.

72. For more on this matter, see pp. 266–67.

73. For a full explanation of these matters, see Irving Howe, *Socialism and America* (New York: Harcourt Brace Jovanovich, 1985), esp. pp. 105–75.

74. From 1938 to 1951 Matthiessen contributed one major article to *Partisan Review*, "Henry James's Portrait of the Artist," four book reviews, and had three of his own books reviewed, two of them favorably.

75. See Donald Pease and Walter Benn Michaels, eds., *The American Renaissance Reconsidered* (Baltimore: Johns Hopkins University Press, 1985), esp. the essays by Pease and Jonathan Arac; and Eric Cheyfitz, "Matthiessen's *American Renaissance*: Circumscribing the Revolution," *American Quarterly* 41, 2 (June 1989), pp. 341–81.

76. Myra Jehlen, introduction to *Ideology and Classic American Literature*, ed. Myra Jehlen and Saevan Bercovitch (Cambridge: Cambridge University Press, 1986), p. 2.

77. William Cain, *F. O. Matthiessen and the Politics of Criticism* (Madison: University of Wisconsin Press, 1988), p. 150.

78. Ibid.

79. See Jonathan Arac, "F. O. Matthiessen: Authorizing an American Renaissance," in *The American Renaissance Reconsidered*, ed. Walter Benn Michaels and Donald E. Pease (Baltimore: Johns Hopkins University Press, 1985), pp. 96–97. Arac argues that the neglect of the class struggle and the rhetoric of ameliorism that *The American Renaissance* absorbed from the Popular Front was partly responsible for Matthiessen's conspicuous silence concerning the Civil War, an apocalyptic event whose only counterpart in the 1930s could have been proletarian revolution.

80. William Cain, *F. O. Matthiessen and the Politics of Criticism*, p. 155.

81. W. E. B. DuBois, "Criteria of Negro Art," *Crisis* 33 (October 1926), reprinted in *W. E. B. DuBois: Writings* (New York: LIbrary of America, 1986), p. 1000.

82. F. O. Matthiessen, "The Responsibilities of the Critic," in *The Responsibilities of the Critic* (New York: Oxford University Press, 1952), p. 7.

83. Stanley Edgar Hyman was Matthiessen's immediate target. In a review of a collection of essays gathered in appreciation of the music critic Paul Rosenfeld, Hyman had written that Rosenfeld was being praised for his warm capaciousness, "a thoroughly degraded function." Hyman, who later authored *The Armed Vision* (1955), was a frequent contributor to the *New Republic*, *New Leader*, and the *New Yorker*.

84. Matthiessen anticipated this kind of question with only the slightest hint of defensiveness in his introduction, entitled "Method and Scope." He acknowledged his book was not going to concern itself with "how" the concentration of masterpieces between 1850 and 1855 occurred (that is, he would not offer a "descriptive narrative of literary history"); nor would it concern itself with the economic, social, and ideological causes of the American Renaissance. Rather, it would be a book whose preoccupation would be with the works themselves, with "an author's resources of language and of genres, in a word, . . . with form" (p. xi). But lest his focus be construed as formalist, Matthiessen reminded his readers that his was not the "pathological" obsession of the aesthetes and decadents. Form, according to Matthiessen's axiom, "was nothing else than the entire resolution of the intellectual, sentimental, and emotional material into the concrete reality of the poetic image and word" (p. xi). Thus, his sensitivity to elements of language did not preclude wide-ranging discussions of any number of issues fusing formal and substantive concerns, from Emerson's Oversoul to employer-employee relations.

85. F. O. Matthiessen, *American Renaissance* (New York: Oxford University Press, 1941; repr. 1977), pp. xv–xvi.

86. Ibid., pp. 8–9.

87. Ibid., p. 8.

88. Ibid., p. 4.

89. Ibid., p. 17.

90. Ibid.

91. Ibid., p. 54.

92. Ibid., p. 42.

93. As was Antonio Gramsci in another context, who recommended that socialists wage a political and cultural "war of position" in the capitalist democracies instead of the classic "war of maneuver," the direct assault on state power of the kind the Bolsheviks mounted against the Russian autocracy.

Chapter 8

1. Paula Rabinowitz has recently made this point. "Much of the scholarship of literary radicalism," she observes, "has been institutional; it has focused on what Josephine Herbst characterized as the 'head boys' who edited journals, wrote criticism, and prescribed the ideological and aesthetic content of the movement." *Labor and Desire: Women's Revolutionary Fiction in Depression America* (Chapel Hill: University of North Carolina Press, 1991), p. 4.

2. Raymond Williams, *Marxism and Literature* (Oxford: Oxford University Press, 1977), p. 119. For more on the concept of the "formation," see Raymond Williams, *Culture* (London: Fontana, 1981), pp. 57–86.

3. For a critique of Rabinowitz's views, see Barbara Foley, "Women and the Left in the 1930s," *American Literary History* 2, 1 (Spring 1990), pp. 150–69.

4. Contributors with androgynous names who were unfamiliar to me were not counted. Writers whose last names were given along with initials were counted as male, but given their relatively small number, these figures leave little margin for error. See *Partisan Review: Fifty-Year Cumulative Index* (New York: AMS Press, 1984).

5. See Elaine Showalter, *Sister's Choice: Tradition and Change in American Women's Writing* (Oxford: Oxford University Press, 1991), pp. 22–41.

6. When asked in an interview about the title, Sontag replied with irritation, recalling an encounter she once had with Mary McCarthy:

> She remembers once, some 20 years ago, meeting the novelist Mary McCarthy. . . . Ms. McCarthy told Ms. Sontag, "Oh, you're the imitation me."
>
> "She said it to embarrass me, I suppose," Ms. Sontag said, going on to denounce as misogynous any effort to give her or any another *[sic]* woman a preordained role.

Richard Bernstein, "Susan Sontag, as Image and as Herself," *New York Times* (January 26, 1989), C17 (as quoted in Sohnya Sayres, *Susan Sontag: The Elegiac Modernist* (New York: Routledge, 1990), p. 35.

7. Carol Gelderman, *Mary McCarthy: A Life* (New York: St. Martin's Press, 1988), p. xi.

8. Interview with Elizabeth Hardwick, in *Women Writers at Work: The Paris Review Interviews*, ed. George Plimpton (New York: Viking Penguin, 1989), p. 216. As quoted in Showalter, *Sister's Choice*, pp. 23–24.

9. Interview with Mary McCarthy, in *Women Writers at Work*, p. 189.

10. Gelderman, *Mary McCarthy: A Life*, p. 307.

11. Ibid.

12. Interview with Elizabeth Hardwick, in *Women Writers at Work*, p. 214.

13. Ibid., p. 216.

14. Elizabeth Hardwick, *A View of My Own* (New York: Farrar Straus, 1962), p. 35.

15. In "Are Three Generations of Radicals Enough? Self-Critique in the Novels of Tess

Slesinger, Mary McCarthy, and Marge Piercy," *Review of Politics* 53, 4 (Fall 1991), pp. 602–26, Philip Abbott discusses these novelists' visions of the relationship between politics and personal life, and between the left and the public. Abbott argues that the contradictions they reveal within the organizational life of the left are finally insurmountable. He concludes, "Though unlikely to be followed, the central, though submerged and unintended, political lesson derived from the self-critiques of Slesinger, McCarthy and Piercy would thus involve a moratorium on new magazines, circles, collectives, communes, and cadres during the next wave of 'restless, prying criticism' that flows through American society before these are really the only alternatives left." It remains unclear whether this is a call for an absolutely new beginning for the left, or for its demise.

16. These and other biographical details are derived from Shirley Biagi, "Forgive Me for Dying," *Antioch Review* 35 (1977), pp. 224–36; and Janet Sharistanian, afterword to *The Unpossessed* (Old Westbury, N.Y.: Feminist Press, 1984), pp. 359–86, and "Tess Slesinger's Hollywood Sketches," *Michigan Quarterly Review* 18 (1979), pp. 429–38.

17. Tess Slesinger, "A Hollywood Gallery," ed. Janet Sharistanian, *Michigan Quarterly Review* 18 (1979), pp. 439–54.

18. See Mary McCarthy, *Intellectual Memoirs* (New York: Harcourt Brace Jovanovich, 1992), p. 84.

19. Horace Gregory, review of *The Unpossessed, Books* 10, 36 (May 13, 1934), p. 2.

20. Robert Cantwell, review of *The Unpossessed, New Outlook* 163, 6 (June 1934), p. 53.

21. T. S. Matthews, review of *The Unpossessed, New Republic* 79, 1016 (May 23, 1934), p. 52.

22. J. D. Adams, review of *The Unpossessed, New York Times* (May 20, 1934), p. 6.

23. John Chamberlain, review of *The Unpossessed, New York Times* (May 9, 1934), p. 17.

24. Murray Kempton, *Part of Our Time* (New York: Simon and Schuster, 1955), pp. 121–23.

25. Reprinted in Lionel Trilling, *The Last Decade* (New York: Harcourt Brace Jovanovich, 1981), pp. 3–24.

26. Lionel Trilling, "A Novel of the Thirties," in *The Last Decade*, p. 20.

27. Ibid., p. 23.

28. See Carol Gelderman, *Mary McCarthy: A Life*, p. 73.

29. Mary McCarthy, *Theatre Chronicles, 1937–1962* (New York: Farrar, Straus, 1963), p. ix; interview with Mary McCarthy, in *Women Writers at Work*, p. 183. As quoted in Gelderman, *Mary McCarthy: A Life*, p. 81.

30. There is some irony in this, for as Morris Dickstein has pointed out, during the height of McCarthyite prosecution Mary McCarthy chose to depict in *The Groves of Academe* an unscrupulous faculty member who attempts to hold onto his job by posing as a victim of the Red scare. See Morris Dickstein, *Gates of Eden* (New York: Penguin, 1977), p. 29.

31. As quoted in Gelderman, *Mary McCarthy: A Life*, p. 286.

32. As quoted in ibid., p. 332.

33. See Alan Wald, *The New York Intellectuals: The Rise and Decline of the Anti-Stalinist Left from the 1930s to the 1980s* (Chapel Hill: University of North Carolina Press, 1987), p. 241.

34. See Gelderman, *Mary McCarthy: A Life*, pp. 139–48; and Carol Brightman, *Writing Dangerously: Mary McCarthy and Her World* (New York: Clarkson Potter, 1992), pp. 305–17.

35. Mary McCarthy, *The Oasis* (New York: Avon, 1981), p. 11.

36. Ibid., p. 14.
37. Ibid., p. 46.
38. Ibid., p. 37.
39. Ibid., pp. 38–39.
40. Ibid., p. 44.
41. Interview in Suzanne Horer and Jeanne Socquet, eds., *La Création etouffée* (Paris: P. Horay, 1973). As quoted in Gayle Greene and Coppelia Kahn, *Making a Difference* (New York: Routledge, 1981), p. 100.
42. McCarthy, *The Oasis*, p. 91.
43. Ibid., p. 93.
44. I owe this point to Janet Gray. See her entry, "Elizabeth Hardwick" in the forthcoming *American Writers, Supplement III* (New York: Charles Scribner's). Much of the biographical information on Hardwick that follows is derived from this piece.
45. Interview with Elizabeth Hardwick, in *Women Writers at Work*, p. 212.
46. As quoted in Gray, "Elizabeth Hardwick."
47. The others were the publisher A. Whitney Ellsworth, Robert Silvers, and Barbara and Jason Epstein.
48. As quoted in Gray, "Elizabeth Hardwick."
49. Elizabeth Hardwick, "Domestic Manners," in *Bartleby in Manhattan and Other Essays* (New York: Random House, 1983), pp. 96–97.
50. Elizabeth Hardwick, *Sleepless Nights* (1979; repr. New York: Vintage, 1990), p. 71.
51. Ibid., p. 58.
52. Ibid., p. 59.
53. Ibid., p. 67.
54. Ibid., p. 85.
55. Elizabeth Hardwick, introduction to *A Susan Sontag Reader* (New York: Random House, 1982), p. xi.
56. Susan Sontag, "Against Interpretation," in *Against Interpretation* (New York: Dell, 1979), pp. 14, 7.
57. Susan Sontag, "On Style," in *Against Interpretation*, p. 18.
58. Ibid., pp. 21–22.
59. Ibid., p. 25.
60. Here I must take issue with Wald and Sayres, who suggest that in the 1960s Sontag was going against the grain of the New York intellectuals' postwar emphasis on American culture. As evidence Wald cites the interest in Henry James "as a home-grown modernist," and the championing of the American school of abstract expressionism (see *The New York Intellectuals*, p. 218, as cited by Sayres, *Susan Sontag: The Elegiac Modernist*, p. 4). But the interest in Henry James, as I have shown, preceded the postwar period: Rahv's seminal essay "The Heiress of All the Ages," for example, appeared in 1943. As for the promotion to the fore of the abstract expressionists, Serge Guilbault has made the case for the nationalist impulse behind that effort. But one must keep in mind that these cosmopolitan artists were valued precisely because they transcended American provincialism and could stand up to the highest standards of European modernism. These criteria were fully consistent with the long-standing project of "internationalizing" American culture, which dated from the late 1930s. Thus Rahv, Greenberg, and the rest of the New York intellectuals continued to take their bearings from Europe well into the postwar period, as evidenced by the very substantial number of essays published throughout the period by *Partisian Review* concerning European intellectual and cultural developments. Sontag, unlike her predecessors, was quite willing to explore and appreciate the diverse American cultural

scene. But like them, she too ultimately took her bearings from Europe, albeit from a different set of Europeans.

61. See, for example, Irving Howe's comments on Sontag in "The New York Intellectuals," in *Irving Howe: Selected Writings, 1950–1990* (New York: Harcourt Brace Jovanovich, 1990), pp. 276–77.

62. Susan Sontag, "The *Salmagundi* Interview," in *A Susan Sontag Reader* (New York: Random House, 1982), p. 333.

63. Susan Sontag, "Communism and the Left," *Nation* 234, 8 (February 27, 1982), pp. 230–31.

64. Sontag, "The *Salmagundi* Interview," p. 334.

65. Irving Howe, *Politics and the Novel* (1957; repr. New York: Columbia University Press, 1992), pp. 162–63.

66. Howe, preface to *Politics and the Novel*, pp. 7–8.

Chapter 9

1. Kenneth Clark, "Candor about Negro-Jewish Relations," *Commentary* 1, 4 (February 1946), pp. 8–14.

2. For more on the Party's cultural activities in Harlem, see Mark Naison, *Communists in Harlem during the Depression* (New York: Grove Press, 1983), pp. 193–226.

3. Ralph Ellison, *Shadow and Act* (1964; repr. New York: Vintage, 1972), p. 112.

4. See Woody Guthrie, *American Folksong* (1947; repr. New York: Oak, 1961), pp. 9–14.

5. Amiri Baraka (LeRoi Jones), *Blues People* (New York: Morrow Quill, 1963), pp. 150–51.

6. Leon Dennen, "Negroes and Whites," *Partisan Review* 1, 5 (November–December 1934), pp. 51–52. Angelo Herndon was a Party activist imprisoned for organizing interracial rallies in Birmingham, Alabama. Along with the Scottsboro Boys, Herndon became an international cause célèbre for the Party and its followers.

7. James T. Farrell, "Lynch Patterns," *Partisan Review* 4, 6 (May 1938), pp. 57–58.

8. David Chanler, "Cross Country: Newark, N.J.," *Partisan Review* 7, 2 (May–June 1940), p. 232. In fairness to the author it should be added that the next sentence reads "You can see white people muttering and shaking their heads about what they consider a usurpation, but the negroes are the fastest growing group in the city." I find this to be something of a non sequitur, and the tone here somewhat ambiguous. Chanler seems to want to distinguish himself from these whites even as he uses terms indicating a shared perspective.

9. David Daiches, "The American Scene," *Partisan Review* 7, 2 (May–June, 1940), pp. 244–47. The other novels under review were Meyer Levin's *Citizens*, Erskine Caldwell's *Trouble in July*, James Still's *River of Earth*, Caroline Slade's *The Triumph of Willie Pond*, and Kay Boyle's *The Hunter*.

10. James Agee, "Variety: Pseudo-Folk," *Partisan Review* 11, 2 (Spring 1944), pp. 219–23.

11. "Statement of the L.C.F.S. [League for Cultural Freedom and Socialism]," *Partisan Review* 6, 4 (Summer 1939), p. 126. The statement was signed by nearly all members of *Partisan Review*'s staff. As quoted in S. A. Longstaff, "*Partisan Review* and the Second World War," *Salmagundi* 43 (Winter 1979), p. 112.

12. See Clement Greenberg and Dwight Macdonald, "10 Propositions on the War,"

Partisan Review 8, 4 (July–August 1941), p. 277. In his otherwise effective response, "10 Propositions and 8 Errors" (*Partisan Review* 8, 6 [November–December], pp. 499–506), Philip Rahv did not mention the struggle for equal rights.

13. Elizabeth Hardwick, "Writer and Spokesman," *Partisan Review* 12, 3 (Summer 1945), p. 407.

14. Delmore Schwartz, "Fiction Chronicle: The Wrongs of Innocence and Experience," *Partisan Review* 19, 3 (May–June 1952), p. 358.

15. We shall shortly encounter another example of idealization, a not infrequent response among whites for whom it offers, like indifference, as excuse for withholding one's serious attention.

16. None of the mass circulation magazines such as *Life*, *Look*, the *Saturday Evening Post*, or *Reader's Digest* published work by African American writers on a regular basis. Black writers did not fare much better in middlebrow magazines such as *Harper's*, *Esquire*, *Saturday Review*, and *Vanity Fair*. I have not examined all of the many special interest and avant-garde little magazines of the period, but among those I have encountered I have not seen any serious commitment to publishing black writers. As for the left press, the *Monthly Review* devoted many articles to vital developments within the international anti-imperialist movement, but it had much less to say about the internal struggle for equality. The Party's *Science & Society* also presented a strongly internationalist perspective, but its philosophical and theoretical approach left little room for coverage of the idioms of African American life or writing. Neither of these magazines featured an unusual number of black writers. I have not examined the contents of other little magazines on the left which may have had a better record in dealing with these issues: *Jewish Life*, *Masses and Mainstream*, *New Foundations*, *Contemporary Writer*, *Venture*, *Promethean Review*, or *California Quarterly*.

17. James Campbell, *Talking at the Gates* (New York: Penguin, 1991), p. 40.

18. James Baldwin, "Smaller Than Life," *Nation* 165, 3 (July 19, 1947) pp. 78–79.

19. As quoted in Campbell, *Talking at the Gates*, p. 38.

20. Mary McCarthy, "Baldwin," in *James Baldwin: The Legacy*, ed. Quincy Troupe (New York, Simon & Schuster, 1989), as quoted in Campbell, *Talking at the Gates*, p. 40.

21. Ruth R. Wisse, "The New York (Jewish) Intellectuals," *Commentary* 84, 5 (November 1987), pp. 28–38.

22. Lionel Trilling, in "Under Forty: A Symposium on American Literature and the Younger Generation of American Jews," *Contemporary Jewish Record* 7 (February 1944), p. 32.

23. Norman Podhoretz, *Making It* (New York: Harper and Row, 1967), p. 3.

24. Morris Dickstein, *Gates of Eden* (New York: Penguin, 1977), p. 155.

25. See pp. 235–38.

26. Susan Sontag, "What's Happening to America," *Partisan Review* 34, 1 (Winter 1967), p. 54.

27. Ironically, it was something akin to Hyman's view, or certainly a synthesis of Hyman's and Ellison's views, that informed Henry Lewis Gates's influential *The Signifying Monkey* (New York: Oxford University Press, 1988).

28. Richard Schlatter, contribution to "What's Happening To America," *Partisan Review* 34, 1 (Winter 1967), p. 50.

29. Dwight Macdonald, "Free & Equal," *Politics* 1, 1 (February 1944), p. 23.

30. *Politics* 1, 1 (February 1944), p. 31; Dwight Macdonald, "Resignation Letter," *Partisan Review* 10, 4 (July–August 1943), p. 382.

31. George Schuyler, "Free & Equal," *Politics* 1, 6 (July 1944), p. 182.

32. Walter A. Jackson, *Gunnar Myrdal and America's Conscience: Social Engineering and Racial Liberalism, 1938–1987* (Chapel Hill: University of North Carolina Press, 1990), p. 273.

33. See "The Cold War in Black America, 1945–1954," in Manning Marable, *Race, Reform, and Rebellion: The Second Reconstruction in Black America, 1945–1982* (Jackson: University Press of Mississippi, 1984), chap. 2.

34. Norman Podhoretz, "My Negro Problem—and Ours," *Commentary* (February 1963), p. 98.

35. For a fuller discussion of the key role played by neoconservatism in the retreat from race, see Michael Omi and Howard Winant, *Racial Formation in the United States* (New York: Routledge, 1994), esp. pp. 113–36.

36. Peter Steinfels aptly distinguishes between the broader, cultural coverage of *Commentary* and the relatively narrow range of public policy issues covered in *The Public Interest*, founded in 1965 by Daniel Bell and Irving Kristol and later an organ of neoconservatism. See *The Neoconservatives* (New York: Simon and Schuster, 1979), pp. 20–21. But relatively little of the fiction or criticism published in *Commentary* concerned race or African American experience. Culture remained more or less severed from race. Some of the articles on literature and the arts included Bernard Wolfe, "Uncle Remus and the Malevolent Rabbit" (July 1949); Irving Howe, "William Faulkner and the Negro" (October 1951); Steven Marcus, "The American Negro in Search of Identity—Three Novelists: Richard Wright, Ralph Ellison, James Baldwin" (November 1953); James Baldwin, "Life Straight in De Eye—Carmen Jones: Film Spectacular in Color" (January 1955) and "On Catfish Row: 'Porgy and Bess' in the Movies" (September 1959); Jervis Anderson, "Anger and Beyond: The Negro Writer in the United States" (March 1967); and John Thompson, "Baldwin: The Prophet As Artist" (June 1968).

37. Ralph Ellison, "The World and the Jug," in *Shadow and Act* (1964; repr. New York: Random House, 1972), p. 108.

38. Jackson, *Gunnar Myrdal and America's Conscience*, p. 291.

39. Several months after *Commentary*'s roundtable discussion, another debate at Town Hall between white liberals and black "militants" took place, the theme being "The Black Revolution and the White Backlash." This time the exchange was more openly contentious, replete with sarcastic, bitter attacks by some of the black participants, and bewildered, sometimes ingenuous defenses by the white liberals. The participants included James Wechsler, editor of the *New York Post*; Charles Silberman, an editor of *Fortune*; David Susskind, television commentator; Paule Marshall, novelist; Ozzie Davis, actor and playwright; Ruby Dee, actress; John Killens, novelist; and Leroi Jones, poet and playwright.

40. "Liberalism and the Negro: A Round-Table Discussion," *Commentary* 37, 3 (March 1964), p. 35.

41. Ibid., pp. 35–36. Hook misrepresents Baldwin here—what the novelist had actually said was that he could not imagine any black person not hating burdensome whites at least part of every day.

42. James Baldwin, *The Fire Next Time* (New York: Dell, 1988), p. 81.

43. Ralph Ellison, "The World and the Jug," in *Shadow and Act* (1964; New York: Random House, 1972), pp. 131–32. See also Irving Howe, "Black Boys and Native Sons," in *Selected Writings: 1950–1990* (New York: Harcourt Brace Jovanovich, 1990), pp. 119–39.

44. Katha Pollitt has recently called attention to a persistent aspect of this problem—the continued failure of progressive publications to diversify their editorial staffs. Pollitt observes,

> In the thirteen years I've been associated with *The Nation*, we've had exactly one nonwhite person (briefly) on our editorial staff of thirteen, despite considerable turnover. And we're not alone: *The Atlantic* has zero nonwhites out of an editorial staff of twenty-one; *Harper's*, zero out of fourteen; *The New York Review of Books*, zero out of nine; *The Utne Reader*, zero out of twelve. A few do a little better, although nothing to cheer about: *The Progressive*, one out of six; *Mother Jones*, one out of seven; *In These Times*, one out of nine; *The New Republic*, two out of twenty-two; *The New Yorker*, either three or six, depending on how you define "editorial," out of 100 plus. . . . *Ms.* comes off rather well, with three out of eleven, including the editor-in-chief, Marcia Ann Gillespie. (These figures do not include columnists, correspondents or contributing editors, who are overwhelmingly white, and at *The Nation* exclusively so.)

"Subject to Debate," *Nation* 260, 10 (March 13, 1995), p. 336.

45. See Omi and Winant, *Racial Formation in the United States*, pp. 113–36.

Chapter 10

1. Jack Kerouac, *On the Road* (1955; repr. New York: Penguin, 1985), p. 11. Among Sal's friends, whom he refers to as "intellectuals," are Carlo Marx (Allen Ginsberg) and Old Bull Lee (William Burroughs). Ginsberg, of course, was a student of Trilling's at Columbia in the mid-1940s, and Kerouac himself attended Columbia in the early 1940s (thanks to a football scholarship) before flunking out and joining the navy. Barry Miles, Ginsberg's biographer, makes the link to New York intellectual life, if not the New York intellectuals per se, in a chapter entitled "A Columbia Education: The Origins of the Beat Generation," in *Ginsberg: A Biography* (New York: Simon and Schuster, 1989).

2. "A severe and thundering strain of Jewish moralism took hold of these aging intellectuals as they contemplated the lurid spectacle of drugs, sex, rock, protest, and the rejection of authority. In the radicalism of the young they saw only an Oedipal psychodrama which pitted the sons against the fathers, though the story of Oedipus actually begins with the anxious father's attempt on the life of his infant son." Morris Dickstein, *Double Agent: The Critic and Society* (New York: Oxford University Press, 1992), p. 99.

3. Norman Podhoretz, "The Know-Nothing Bohemians," *Partisan Review* 25, 2 (Spring 1958), p. 305.

4. Ibid., p. 308.

5. Ibid., p. 311.

6. Ibid., p. 318.

7. Leslie Fiedler, "The New Mutants," in *Collected Essays*, vol. 2 (New York: Stein and Day, 1971), pp. 385–86.

8. Irving Howe made this point succinctly in "The New York Intellectuals," in *Selected Writings, 1950–1990* (New York: Harcourt Brace Jovanovich, 1990): "With uncharacteristic forbearance, Fiedler denies himself any sustained or explicit judgments of this 'futuristic revolt,' so that the rhetorical thrust of his essay is somewhere between acclaim and resignation. He cannot completely suppress his mind, perhaps because he has been using it too long" (p. 277).

9. Fiedler, "The New Mutants," p. 387.

10. David Bromwich, *Politics by Other Means* (New Haven: Yale University Press, 1992), p. 50; and Richard Rorty, *Contingency, Irony, and Solidarity* (New York: Cambridge University Press, 1989), p. xvi.

11. Alfred Kazin, *New York Jew* (New York: Vintage, 1978), p. 397.

12. Ibid.

13. Ibid.

14. Irving Howe, "The New York Intellectuals," *Commentary* 46, 4 (October 1968). Reprinted in Howe, *Selected Writings*, pp. 240–80.

15. See Trilling's preface to *The Liberal Imagination* (New York: Harcourt Brace Jovanovich, 1979).

16. Howe, "The New York Intellectuals," pp. 274–75.

17. Irving Howe, *A Margin of Hope* (New York: Harcourt Brace Jovanovich, 1982), p. 292.

18. Maxine Phillips, "Michael Harrington," in *Encyclopedia of the American Left*, ed. Mari Jo Buhle, Paul Buhle, and Dan Georgakas (Urbana: University of Illinois Press, 1992), p. 290.

19. Todd Gitlin, *The Sixties: Years of Hope, Days of Rage* (New York: Bantam, 1987), pp. 113–14.

20. James Miller, *Democracy Is in the Streets* (New York: Simon and Schuster, 1987), p. 115.

21. As quoted in Gitlin, *The Sixties*, pp. 117–18.

22. As quoted in ibid., p. 118.

23. Ibid.

24. Ibid., p. 117.

25. "Culture and the Present Moment: A Round-Table Discussion," *Commentary* 58 (December 1974), pp. 41, 44. As quoted in Mark Krupnick, *Lionel Trilling and the Fate of Cultural Criticism* (Evanston, Ill.: Northwestern University Press, 1986), p. 155.

26. It is worth noting that Polonius did not say, as Trilling claims, that one should be true to oneself *in order* to be truthful to others. Presumably, in Polonius's version one's relationship with others in the social world is normally personal enough to allow that others will tolerate and even respond favorably to the individual for whom the desire to be honest with oneself is paramount. A desire to be truthful to others is not needed because truthful expression toward them will *necessarily* follow: it is described by Polonius as a *result*, not a *reason*. If we grant that Trilling is correct in observing that gradually this ideal of sincerity deteriorates as the expectations of others increasingly take a toll on self-knowledge, to the point that the individual begins to fulfill a "role," this development only emphasizes how much an expanded, impersonal, and increasingly partisan public sphere shapes the self, circumscribing at the very moment it creates new desires and willful acts—including the intention to be, and seem, sincere. It would seem, then, that the "gratuitously chosen images of personal being" that are so important to Trilling are not quite gratuitous at all, although it can be said that they do exemplify an enviable degree of self-assertion within the social process of establishing an identity.

27. Krupnick, *Lionel Trilling and the Fate of Cultural Criticism*, p. 161. Krupnick also refers to Elaine Showalter's essay "R. D. Laing and the Sixties," *Raritan* 1 (Fall 1981), pp. 110, 117, in which she comments on the irony of Trilling attacking a devotee whose first book *The Divided Self* was deeply indebted to Trilling's famous essay on Keats.

28. Lionel Trilling, *Sincerity and Authenticity*, (1971, repr. New York: Harcourt Brace Jovanovich, 1980), p. 158.

29. "*Sincerity and Authenticity*: A Symposium," *Salmagundi* 41 (Spring 1978), p. 96.

30. See p. 12.

31. See Joyce Appleby, Lynn Hunt, and Margaret Jacob, *Telling the Truth about History* (New York: Norton, 1994); David Bromwich, *Politics by Other Means: Higher Education and Group Thinking* (New Haven: Yale University Press, 1992); Dickstein, *Double Agent:*

The Critic and Society; Frederick Crews, *The Critics Bear It Away: American Fiction and the Academy* (New York: Random House, 1992); John Patrick Diggins, *The Rise and Fall of the American Left* (New York: Norton, 1992); Russell Jacoby, *Dogmatic Wisdom: How the Culture Wars Divert Education and Distract America* (New York: Doubleday, 1994); Rorty, *Contingency, Irony, and Solidarity*; and Cornel West, *The American Evasion of Philosophy* (Madison: University of Wisconsin Press, 1989).

32. Cary Nelson, *Repression and Recovery: Modern American Poetry and the Politics of Cultural Memory, 1910–1945* (Madison: University of Wisconsin Press, 1989), p. 68.

33. Ibid., p. 124.

34. Ibid., p. 36.

35. Ibid., pp. 10, 12, 56.

36. Ibid., p. 103.

37. Ibid., p. 243.

38. Ibid., pp. 243–44.

39. Andrew Parker, "Ezra Pound and the 'Economy' of Anti-Semitism," in *Postmodernism and Politics*, ed. Jonathan Arac (Minneapolis: University of Minnesota Press, 1986), pp. 70–90.

40. Robert Casillo, *The Genealogy of Demons: Anti-Semitism, Fascism, and the Myths of Ezra Pound* (Evanston, Ill.: Northwestern University Press, 1988), p. 3.

41. Casillo, *The Genealogy of Demons*, p. 333.

42. Ezra Pound, *The Cantos* (New York: New Directions, 1970), p. 439.

43. Pound, *The Cantos*, p. 428.

44. Walter Kalaidjian, *American Culture between the Wars* (New York: Columbia University Press, 1993), p. 2.

45. Ibid.

46. Ibid., p. 10.

47. Ibid., p. 24.

48. Ibid., p. 10.

49. Max Hayward, *Writers in Russia, 1917–1978* (San Diego: Harcourt Brace Jovanovich, 1983), p. 261. Zhdanov's words are taken from his public report, printed in Soviet newspapers, of a Central Committee meeting that issued a decree expelling Akhmatova from the Union of Soviet Writers and banning further publication of her work in the Soviet press. Needless to say, further persecution followed.

50. Barbara Foley, *Radical Representations: Politics and Form in U.S. Proletarian Fiction, 1929–1941* (Durham, N. C.: Duke University Press, 1993), pp. 30–31.

51. Ibid., p. 63.

Chapter 11

1. Lionel Trilling, "On the Teaching of Modern Literature," in *Beyond Culture* (1965; repr. New York: Harcourt Brace Jovanovich, 1979), p. 12.

2. Lionel Trilling, preface to *The Liberal Imagination* (1950; repr. New York: Harcourt Brace Jovanovich, 1979), p. ii.

3. Trilling, "The Function of the Little Magazine," in *The Liberal Imagination*, p. 93.

4. Trilling, preface to *The Liberal Imagination*, p. vi.

5. Ibid.

6. Ibid., p. vii.

7. Lionel Trilling, Eulogy for Elliot Cohen, Riverside Chapel, New York City, May 31, 1959.

8. Lionel Trilling, *Contemporary Jewish Record* 7 (February 1944), p. 15.

9. *Partisan Review* 7, 1 (January–February 1940), pp. 24–40.

10. Lionel Trilling, "Marxism in Limbo," in *Speaking of Literature and Society* (New York: Harcourt Brace Jovanovich, 1980), pp. 101–2.

11. Lionel Trilling, "Hemingway and His Critics," in *Speaking of Literature and Society*, pp. 125–26.

12. Ibid., p. 129.

13. Lionel Trilling, "'Elements That Are Wanted,'" retitled "T. S. Eliot's Politics," in *Speaking of Literature and Society*, p. 166.

14. Trilling, "The Function of the Little Magazine," p. 94.

15. Ibid.

16. Lionel Trilling, preface to *Beyond Culture* (1965, repr. New York: Harcourt Brace Jovanovich, 1979), pp. iv–v.

17. Lionel Trilling, *Matthew Arnold* (1939; repr. New York: Harcourt Brace Jovanovich, 1977), p. 375.

18. Trilling, preface to *Beyond Culture*, p. iii.

19. Ibid., pp. iii–iv.

20. Ibid., p. iii.

21. Lionel Trilling, "The Leavis-Snow Controversy," in *Beyond Culture*, p. 138.

22. Lionel Trilling, *E. M. Forster* (Norfolk: New Directions, 1943), p. 23.

23. Ibid., pp. 11–12.

24. Ibid., p. 31.

25. Ibid., p. 14.

26. Ibid.

27. Ibid., p. 15.

28. Trilling, "The Function of the Little Magazine," p. 95.

29. Lionel Trilling, "Reality in America," in *The Liberal Imagination*, p. 10.

30. Lionel Trilling, "The Meaning of a Literary Idea," in *The Liberal Imagination*, p. 284.

31. Lionel Trilling, "Manners, Morals, and the Novel," in *The Liberal Imagination*, pp. 194–95.

32. Trilling, "The Meaning of a Literary Idea," p. 269.

33. Trilling, "Manners, Morals, and the Novel," p. 200.

34. Trilling, "Reality in America," p. 8.

35. Lionel Trilling, "Hawthorne in Our Time," in *Beyond Culture*, p. 174.

36. Lionel Trilling, "The Situation of the American Intellectual at the Present Time," in *A Gathering of Fugitives* (1956; repr. Oxford: Oxford University Press, 1980), p. 83.

37. Lionel Trilling, "Freud: Within and Beyond Culture," in *Beyond Culture*, p. 79.

38. Ibid., pp. 91, 93, 98.

39. Ibid., p. 98.

40. Lionel Trilling, "The Poet as Hero: Keats in His Letters," in *The Opposing Self* (1955; repr. New York: Harcourt Brace Jovanovich, 1979), p. 16.

41. Ibid., p. 25.

42. Lionel Trilling, "Wordsworth and the Rabbis," in *The Opposing Self*, pp. 115, 117.

43. Lionel Trilling, "George Orwell and the Politics of Truth," in *The Opposing Self*, p. 141.

44. Ibid., p. 144.

Index